An introduction to experimental design and statistics for biology

# An introduction to experimental design and statistics for biology

**David Heath**
*University of Essex*

UCL
PRESS

First published in 1995 by UCL Press
Second impression 1996

UCL Press Limited
University College London
Gower Street
London WC1E 6BT

The name of University College London (UCL) is a registered
trade mark used by UCL Press with the consent of the owner.


**British Library Cataloguing-in-Publication Data**
A CIP catalogue record for this book is available from the British Library.

**Library of Congress Cataloging-in-Publication Data**

Heath, David, 1946–
    An introduction to experimental design and statistics for biology
/ David Heath.
        p.    cm.
    Includes bibliographical references (p.         ) and Index.
    ISBN 1-85728-131-4 (HB). —ISBN 1-86728-132-2 (PB)
    1. Biometry.  2. Experimental design.  I. Title.
QH323.5.H39   1994
574'.01'5195—dc20                                            94-24461
                                                                CIP




ISBNs:    1-85728-131-4 HB
          1-85728-132-2 PB




Typeset in Times Roman and Univers
Printed and bound by
Butler & Tanner, Frome, England.

# Contents

# Acknowledgements

Many people have given me help, advice and encouragement; without them I could not have written this book. Graham Upton was always willing to explain patiently and clarify statistical points which I did not understand. He also read almost all of the text and identified many errors and misconceptions. At a very early stage, David Lees, Clive Ireland, Stuart Davies and William Booth read drafts of parts of the book and their positive comments encouraged me to continue. John Ratford read the whole of the book and made many useful suggestions which have been incorporated. Jane read the whole book pointing out ambiguities, faults in the flow of the argument and identifying errors of grammar and punctuation. Correcting these much improved the text. Any remaining faults are mine.

I should also like to thank all those people who helped with the actual production of the text. John Kent produced many of the diagrams and also Tables A1, A2 and A3; the other tables were typed by Vicky Rosenthal who also patiently incorporated the editorial revisions made during the production of the final version. Hilary Rasor battled with the large number of equations and their layout in the boxes; Rita Bartlett also helped with production of innumerable drafts. I am very grateful to Terry Ford who took the photographs for Plates 1, 2, 6 and 9 and to Carlos Rubbi for his help with Plate 3. Several people took the trouble to look out possible photographs for inclusion; I would particularly like to thank Bev Wilson (Plate 4), Terry Callaghan (Plate 5) and Tony Bradshaw (Plate 7). Finally I should like to thank all the students, who, over the years, have caused me to develop my ideas and explanations of statistics to a point where I felt they were worth publishing. In most cases the examples used in the book are based on the results of their practical and project work.

## Acknowledgements for tables

Table A4 is reproduced from Table A in Neave, H. R. and Worthington P. L. (1988) *Distribution-free tests* (Unwin Hyman) with permission of the publishers (Routledge).

Table A5 is an extract from the table of percentage points of the normal distribution in Neave, H. R. (1985) *Elementary statistics tables* (George Allen & Unwin) with permission of the publishers (Routledge).

Tables A6, A7, A11, A13 and A17 are extracted from the more extensive ones (Tables B1, B3, B16, B4 and B14) in Zar, J. H. (1984) *Biostatistical analysis*. Reprinted by permission of Prentice-Hall, Inc. Englewood Cliffs, NJ.

Table A10 is extracted from the more extensive table in Zar, H. J. (1972) Significance test-

vii

ing or the Spearman rank correlation coefficient, *Journal of the American Statistics Association* **67**, 578–80 with permission of the publishers.

Table A12 is extracted from Table 11.4 in Owen, D. B. (1962) *Handbook of statistical tables* (Addison-Wesley Publishers Co., Inc) with permission of the publishers apart from values for $\alpha = 0.1\%$ which were extracted from the table in Milton, R. C. (1964) An extended table of critical values for the Mann–Whitney (Wilcoxon) two sample statistic, *Journal of the American Statistics Association* **59**, 925–34 with permission of the publishers.

Table A14 is derived from Table 1 in Mackinnon, W. J. (1964) Tables for both the sign test and distribution free confidence intervals of the median for sample sizes to 1000, *Journal of the American Statistics Association* **59**, 935–56 with permission of the publisher.

Table A15 is derived from Table 1 in McComack, R. L. (1965) Extended tables of the Wilcoxon matched pair signed rank statistic. *Journal of the American Statistics Association* **60**, 864–71 with permission of the publishers.

Table A16 is reproduced from the table of critical values for the Kruskal–Wallis test (equal sample sizes) in Neave, H. R. (1985) *Elementary statistics tables* (George Allen & Unwin) with permission of the publishers (Routledge).

Table A18 is reproduced from the table in David, H. A. (1952) Upper 5 and 1% points of the maximum $F$-ratio, *Biometrika* **39**, 422–4 with the permission of the *Biometrika* trustees.

Table A19 is extracted from Table 3 in Harter, H. L. (1960) Tables of range and studentised range. *Annals of Mathematics and Statistics* **31**, 1122–47 with permission of the publishers.

# List of tables

# Introduction

As a biology student you may be put off statistics, an apparently difficult subject with abstract ideas, complex formulae and lengthy calculations. You also perhaps think of statistics as a peripheral subject and that putting effort into studying it leaves less time for what you are really interested in, namely biology! I think that these views, though understandable, are mistaken.

The basic ideas in statistics are relatively few and not that difficult to grasp, as I hope to show in the course of this book. Once you have mastered these ideas, you will find yourself using them over and over again. The formulae can be complex but you will see that they often have a common-sense explanation which helps us understand not only the statistics but also the underlying biology. Most of the calculations can be carried out easily on an ordinary calculator. Sampling and experimental design should not present too much of a problem. Much is common sense and once you are alerted to the issues, it is quite easy to design sound sampling programmes and experiments and, like solving puzzles, it can be satisfying.

I am a biologist by training and biology remains my primary interest. As an undergraduate, I went to several statistics courses but I often came out little the wiser. No-one explained why I needed statistics and the approach was very mathematical – a board covered in symbols and equations and scarcely a mention of an organism. Over the years I learned more because I had to use statistics, although I can't say I always understood what I was doing. Later, when I began to teach biology, I found myself trying to explain things like chi-squared to students in genetics practicals and correlation to students on a field course. They would listen and then say "But why?" or "But I still don't understand". In order to provide satisfactory answers I had to examine my own understanding of the subject and this process clarified the issues for me. About 5 years ago I started to teach a course on experimental design and statistics to first year undergraduates which meant that I brought these ideas together. This book is the result.

There are three different ways of approaching statistics. At one extreme, the recipe book method which goes something like this: here is the formula, put in the numbers and if the answer is bigger than some magical number then the difference is significant (whatever that means). The problem with this approach is that it does not promote an understanding of what is going on. At the other extreme we could go deeply into the mathematical background so that we would know, understood and could prove the derivation of every equation. For a biologist this would be quite unnecessary. Between these extremes there is a middle way which requires that we understand the rationale of the method and its limitations. This is the route we shall take in this book. We shall approach statistics in just the same way as any other technique or piece of apparatus – as a tool to enable us to see the biology more clearly.

Think of statistics as a spectrophotometer. If you are using a spectrophotometer you should not think of it merely as a black box with knobs to be twiddled according to a set of instructions. This approach is inflexible and does not help you when you come to interpret the results. On the other hand, to use a spectrophotometer successfully, you do not need to understand the underlying principles of physics and the equations of optics and electronics. What you do need is an understanding of how it works. For example, you need to know that a spectrophotometer works by measuring how much light is absorbed and that this is related to the concentration of the solution and the wavelength of the light. You also need to know that it can give a reading either of what light is absorbed or what is transmitted and that it is not absolutely accurate. Once you know this, it is easy to see why you have to use a wavelength of light specific to the substance whose concentration is to be measured, why the cuvette has to be the right way round and why you should not put fingerprints on the cuvette sides through which the light passes. It is also clear why the lid is lightproof, why transmission and absorption results go in opposite directions and why you need to check the machine every so often with a suitable blank. Its just the same with statistics: it is this sort of common-sense understanding of how and why the test works which I want to convey, using an absolute minimum of mathematical detail.

My aim is to write a readable book, with biology as the starting point. Broadly I shall try, wherever possible, to work from the biology to the issues of experimental design and statistics. In order that you can appreciate their wide application I shall take the examples from all areas of plant and animal biology, including biochemistry, cell biology, genetics, physiology and ecology. Whatever your interests in biology you will see the usefulness and relevance of experimental design and statistics. The biological examples will be relatively simple and of a sort that you might meet in first year practical classes, field courses and field course projects.

The ideas that underlie the various tests are first explained using diagrams, which lead on to the formula. These formulae can be off-putting and in these days of computer packages you could be forgiven for wondering if it is really necessary to deal with them. The answer is yes. The most important reason is that the formulae can usually be "unpacked" quite easily and the component parts are then found to have simple explanations which tell us what the test is doing and what biologically important information is going into it. Examining the components of the formula can throw valuable light on the best way to design an experiment.

A second reason for giving the formulae is that not everyone has access to a computer package all of the time, so it is useful to be able to carry out a test using the formula and a calculator. Indeed, if you think about how long it takes to get to a computer, switch on, load the program and so on, it is often more efficient to do simple tests on a calculator. I shall be giving hints and shortcuts for "hand" calculations. You will need a calculator with a standard deviation function so, if you are planning to buy one, make sure that it has this function (keys labelled $\Sigma x$ and $\Sigma x^2$ are also very useful).

I am very aware of how it is easy to underestimate the difficulties biologists have with statistical ideas and so I have proceeded slowly, sometimes saying the same thing twice in different ways or returning to an issue at a later point. As a result I have covered less ground than other books. However, the range of tests described should be sufficient for at least your first year as an undergraduate.

I have not included sections on the use of computer packages with details about the types of commands and outputs. This is because, in my experience, biology students find it just as difficult to get to grips with computing as they do with statistics. To try and develop both skills at the same time is, I think, asking for trouble. At Essex we use our own "menu driven" statistics package, so that all students have to do is choose options from a menu. Further details are available from the author.

Neither have I included any exercises for you to work through. One reason is because a central theme in the book is that the statistical analysis to be used on a set of results is an integral part of the design and execution of an experiment or sampling programme. The only way to appreciate this relationship and to solve the problems of experimental design is to have practical experience of doing experiments and collecting samples. To be provided with sets of results for statistical analysis divorces the data collection from the data analysis. This can be a dangerous step, because the output from a statistical test is only as good as the information that goes into it. The other reason is that it puts entirely the wrong emphasis on the purpose of statistics. It is the biology we are interested in and the purpose of statistical analysis is to enhance our understanding of it. Providing sets of data for analysis which do not relate to your course is unlikely to capture your interest. Furthermore, there is a much greater incentive to struggle with a statistical test when it is being applied to data from your own practical work.

Chapter 1 sets the scene by exploring the particular problems involved in biological work, explains why statistics and experimental design are so important and introduces a number of important terms. In Chapter 2, I use a simple example to expand on these points and to illustrate some of the central issues involved in statistical analysis. This is perhaps the most important chapter because, if you grasp the essential ideas explained there, you are well on the way to understanding the rest of the book. In Chapter 3 we look in more detail at variables, samples and populations, three important terms introduced in the first chapter. Chapter 4 is a standard account of the various measures and methods used to describe samples. Chapter 5 gives a biologically based exploration of three important models which can be used to account for variation and which form the basis for many statistical procedures. It also serves to consolidate some of the ideas dealt with in Chapter 2. Chapters 6 to 11 deal with the details of specific statistical procedures and tests. The material is organized into chapters according to the number of samples being analysed. Chapters 6 to 8 deal with procedures based on single samples. Tests for two independent samples are considered in Chapter 8, followed in Chapter 9 by tests for related or paired samples. Chapter 10 is devoted to tests involving three or more samples. The reason for this division is that it leads from the simple to the complex and it has a purely practical advantage: it will help you find the chapter you need because the one thing you should know is the number of samples you have.

In these later chapters the material progresses from what are called distribution-free or non-parametric tests to parametric tests. Again this reflects a trend from the simple to the complex. Distribution-free tests are often easier to understand, easier to calculate and have fewer and less restrictive assumptions. As a result, they are more widely applicable. Throughout the book, new ideas are introduced gradually, as and when they become relevant, and old ideas are revisited to help you see the connections.

What about sampling and experimental design? These are governed by a small number of general principles, rules and guidelines but the details of the ways in which they are applied differ according to the situation. Dealing in generalities often lacks concreteness and makes the issues less easy to grasp, so Chapter 2 uses a specific, simple example to highlight the problems and the types of solutions that have to be employed. A major part of Chapter 3 is given over to a more general treatment of this topic, along with details of basic procedures that underlie any well-designed sampling programme or experiment. These basic issues will be dealt with again in succeeding chapters when we deal with actual sets of data and when we cover other refinements which are appropriate to particular types of investigation. All these issues are brought together in Chapter 12 which is a summary of the main elements of experimental and sampling design in the form of two checklists. This chapter also has a key to statistical tests which will enable you to find out what test is appropriate to the experiment or sampling programme you are planning.

Although the material is organized in a logical progression, that does not mean you have to plough through the whole book. If you are new to the subject, then I suggest that you read at least the first four chapters. This should give you an understanding of most of the fundamental issues. By this stage you should be in a position to deal with the remaining chapters so you can move around these as necessary. If you are not sure which test to use, the key to statistical tests and procedures in Chapter 12 should enable you to locate the correct one. If you know which test to use, but not how to carry it out, the worked examples can be found in the boxes.

# 1
# Why biologists need sampling, experimental design and statistics

Our knowledge of the living world depends upon careful observation and experimentation, followed by analysis and interpretation of the results. Three indispensable tools used in this process are sampling, experimental design and statistical analysis. This chapter examines why they are so important and looks at some of the basic principles.

## What biologists do

Biology is the study of the living world, a subject which ranges from the molecules that make up cells to the structure and function of whole ecosystems. When the subject was young a biologist such as Darwin could successfully investigate topics as diverse as earthworms, barnacles, orchids and evolution, but as our knowledge has expanded it has become impossible for any individual biologist to deal with the whole of the subject.

Biology has become subdivided into a variety of separate specialised areas or disciplines, with individual biologists restricting themselves to the study of some small part of the subject. There are a number of different ways in which we can describe this subdivision. One is to use the different levels of organization of biological systems, starting at the level of molecules and progressing through cells, organisms and populations, to ecosystems. Such a subdivision is reflected in a range of different sorts of biologists, for example molecular biologists and population biologists. Another way of subdividing the subject is based on whether the organism being investigated is an animal, a plant or a micro-organism. There are biologists who would describe themselves as animal population biologists or plant molecular biologists, with even finer subdivisions and labels based on whether, for example, the animals of interest are vertebrate or invertebrate, insect or mollusc. Alternatively, we could use a scheme based on biological processes such as genetics, development or physiology. This too would produce a corresponding subdivision of biologists into types.

This is not to suggest that biology falls neatly into a number of labelled boxes. It does not! Rather, the divisions are artificial and the borderlines between the disciplines are fuzzy. What it does emphasize is the great diversity of the subject, something which might suggest that the different sorts of biologist have little in common. However, we find that this is not so if we look at the subject from a different angle. In fact all biologists are basically doing the same two things namely answering "What?" and "How?" questions.

"What?" questions involve the **description** of the characteristics of the particular part of the living world that is of interest, broadly "What is to be found?". So an animal ecologist might describe the number of individuals in a habitat and a cell biologist might describe the concentration of a specific protein in cells of a certain type at two different stages of the cell cycle. "How?" questions involve an **explanation** of what has been described, broadly "How can it be accounted for?". Our ecologist may want to know how a habitat supports the numbers observed. The cell biologist might want to know how the difference in time brings about the difference in protein concentration. As we shall see later on, these two types of question are quite closely related because explanation involves description under specified conditions.

In seeking answers to these two types of question all biologists encounter a similar set of problems, some of which are common to the other sciences, while others are unique to biological systems. It is these problems and the solutions to them which form the subject matter of this book and we start with the basic cause of the trouble – **variability**.

## Variability

Different types of biologists are interested in different sorts of things but they are all involved in making observations on the things of interest and describing them with respect to certain characteristics. Whether we are counting the number of bird species in different woods of a given size, counting the number of individual snails in quadrats on the seashore, measuring the dry weight of individual plants, measuring the concentration of a protein in 10 ml batches of solution or describing the ratio of red to white flowered plants in a genetic cross the observations will have one feature in common – variability.

What this means is that the things of interest, e.g. woods, quadrats, plants, and batches of solution, will not necessarily be identical with respect to the characteristic which is being described. For example not all woods of the same size will have the same number of species, not all individuals of a given plant species will have the same dry weight when grown under the same conditions and not all measurements of protein concentration in 10 ml batches of solution taken from a flask will give the same result. In fact, this feature of biological material is so universal that it gives rise to the general term which is used to refer to characteristics being described. They are called **variables**. The number of bird species in woods of a given size, the number of individual snails in a quadrat, the dry weight of individual plants, the protein concentration in a 10 ml batch of solution and the ratio of red to white flowered plants are all variables.

The basic problem with variability is that it gets in our way when we attempt to describe and explain. We'll see later why it gets in the way but to understand the problem we first of all need to know why things are variable.

### Machines and techniques – experimental variation

The first source of variability is common to all the sciences and is usually referred to as **experimental variation**. It is variability that is introduced by the experimenter or by the techniques and the equipment used. It arises because of **experimental error**. Whenever we count or measure anything we are unlikely to do it without error, in other words we will not

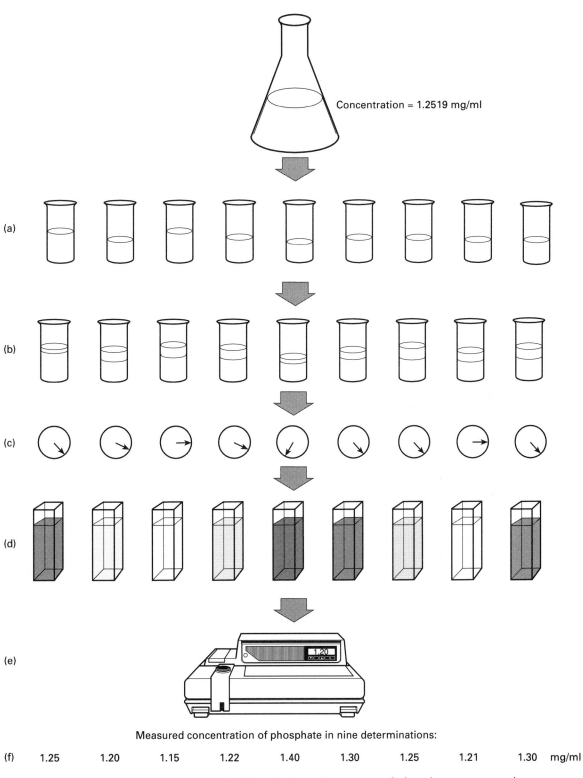

Concentration = 1.2519 mg/ml

(a)

(b)

(c)

(d)

(e)

Measured concentration of phosphate in nine determinations:

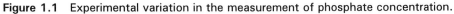

(f)    1.25        1.20        1.15        1.22        1.40        1.30        1.25        1.21        1.30    mg/ml

**Figure 1.1**   Experimental variation in the measurement of phosphate concentration.

always count or measure the true value. This may be because our measuring device, be it ruler, pipette or spectrophotometer has inherent limits as to its accuracy. In addition we may use the device in a slightly different way on different occasions, that is our technique may vary.

The simplest way to see how this can happen is to imagine making repeated measurements of exactly the same thing. Take a simple biochemical example which involves measuring the concentration of a chemical in solution (Fig. 1.1). To do this we have to pipette 10 ml of the solution from a flask, add a known quantity of a reagent to produce a coloured end-product at a concentration which is proportional to the concentration of the chemical of interest. The concentration of the colour and hence the concentration of the chemical is measured using a spectrophotometer. If you were to repeat this procedure several times using the same solution, would you get the same answer each time? The chances are that you would not, simply because of variation in the way in which you carry out the procedure on different occasions. Sometimes you will pipette slightly less than 10 ml, other times slightly more (Fig. 1.1a) and the quantity of the reagent added will not be exactly the same on different occasions for the same reason (Fig. 1.1b). You may leave the colour to develop for slightly different lengths of time (Fig. 1.1c), there may be slight differences in the optical properties of the different cuvettes (Fig. 1.1d), or the characteristics of the spectrophotometer may change as components age or the mains voltage fluctuates (Fig. 1.1e).

Each potential source of error may, on its own, produce a measurement which is either above or below the true value and, in any one determination of the concentration, errors at different stages of the procedure will be cumulative. Sometimes, by chance, the effects of errors in one direction will be more or less cancelled out by those of errors in the opposite direction, leading to a result fairly close to the true value. On other occasions, again by chance, the majority of errors will all be in one direction, leading to a result which is very different from the true value.

To show you the extent of this type of variability I have included at the bottom of the figure a set of results produced by one person (a first year student). They are concentrations of phosphate in solution measured by just the sort of procedure we have been discussing. The nine determinations were made on the same solution and the variability is obvious! Additional variability would have arisen if nine different students had taken part or if more than one spectrophotometer had been used.

The errors that lead to this variability are not like mistakes in spelling or calculations which, with care, are completely avoidable. Experimental variation is always with us. While it can be reduced by careful experimental technique it can never be completely eliminated. As a result the answer that we obtain will depend on which particular combination of errors our determination is subject to. Our description will be inaccurate.


## Variability in phenotypes and genotypes

A second source of variability arises from the differences in biochemistry, physiology, morphology and behaviour that exist between individual organisms of the same species. This is something that is very obvious in our own species and it is not hard to find examples in other organisms (Plate 1). Sometimes this type of variability is not apparent on a cursory inspection, however a close study of the individuals concerned will almost always reveal differences between them.

**Plate 1** Phenotypic variability in the leaves of the Southern Marsh Orchid *Dactylorhiza praetermissa.*

The almost universal presence of this type of variability is not surprising, considering what we know about the way in which these characteristics are produced. The characteristics of the individual comprise its phenotype, which is the result of the interaction between the individual's genotype and the environment in which it develops and lives. In sexually reproducing organisms the processes of meiosis and fertilization lead to the continual production of new combinations of genotypes, such that no two individuals are likely to be genetically identical. (There are some obvious exceptions to this such as identical twins, where two individuals have arisen from the same fertilized egg, and plants produced by means of vegetative reproduction. We will meet these and some other exceptions later on because they turn out to be quite useful when we are answering "How?" questions.) The presence of these genetic differences between individuals will lead to phenotypic differences between them, even if they lived in identical environments. This in itself is very unlikely because environments vary in both space and time. Because no two individuals are likely to experience the same set of environmental conditions any differences in the environment experienced by individuals will further increase their phenotypic differences.

The extent to which this will happen depends on the particular characteristic. Some phenotypic differences, such as blood group in our own species, appear not to be affected by the environment. If you have the genes for blood group O, then your phenotype is O irrespective of your environment. Other phenotypic differences between individuals in characteristics such as height, weight, behaviour and susceptibility to disease depend both on the genotype and the environment. As a result the same genotype can produce different phenotypes depending on the environment and, conversely, different genotypes can produce the same phenotype. An individual with genes for large body size will be large if there is plenty of food available, but will be smaller if food is limited; under these conditions it may be the same size as an individual with genes for small size which has access to plenty of food.

These interactions between the genotype and the environment are not restricted to the period of embryonic development, nor do they stop once the individual is an adult. Physiological characteristics such as hormone levels, reproductive state and nutritional state are aspects of an individual's phenotype which may change from hour to hour, day to day and month to month, because of changes in the environment. For example, human body temperature is said to be 36.8°C, but it differs slightly between males and females and between individuals of the same sex. Also, for a given individual, it varies according to where (in or on the body) it is measured, the time of day and, for females, the stage in the menstrual cycle. It also varies in relation to factors such as the level of physical exertion.

If individuals do differ from one another then, when you describe the characteristic, the answer which you get will depend on which individuals you have looked at. Again we have a problem in describing the variable.

## Variability in space and time

Characteristics such as the numbers of individuals per unit area of habitat and the number of visits per unit time by pollinators to a flower are also variables. Plate 2 shows a typical pattern of spatial variation in numbers per unit area. In order to understand how this third type of variability arises we need to think about how the organisms came to be where they were.

Organisms, their seeds, spores and larvae are often passively dispersed by air and water currents and we could imagine them simply "falling out" of the air or water onto the substrate. Contrary to what you might think, this will not lead to the individuals being regularly arranged on the surface because, by chance, some areas will receive more individuals than others. We will look into this in more detail in Chapter 5, but you can demonstrate this effect by simply dropping a handful of something like gravel onto the ground.

For real organisms these chance irregularities will be modified by a variety of factors. For example, eddy currents in the air or water may lead to much higher numbers being deposited in certain areas – think of the way in which the wind can produce piles of leaves.

**Plate 2**   Spatial variation in numbers of lichen colonies on roof-tiles.

Seeds may tend to fall near the plant that produced them, leading to high numbers of plants clustered around the parent and the same may be true of the offspring of some animals. Finally, animals in their dispersive stage are often capable of exercising choice in where they settle. If the surface is a mosaic of suitable and unsuitable patches then organisms will congregate in the suitable patches, again leading to variation in numbers in different areas. Even if none of this happened and organisms were regularly dispersed at the settlement stage, this is unlikely to persist for long, simply because of the patchy environment. Individuals will survive (and reproduce) better in the suitable patches again producing variation in the numbers from one place to another. Note that one thing which may determine the suitability of a patch is the number of individuals of your own species that are already there.

Rather similar arguments apply to the distribution of events in time. The number of mutations occurring in a population of fixed size will vary by chance from one generation to the next even though the underlying mutation rate is constant. Similarly, if we record the number of pollinator visits to each of 10 identical flowers in a 5-min period, again purely by chance, the number of visits to each flower will not necessarily be the same.

## Why description needs statistics

As mentioned earlier, the presence of this variability makes it impossible to describe "things" exactly, that is to put a reliable numerical value to the characteristic of interest. Why is this?

### Estimating

First let's look again at the example of the measurement of phosphate concentration that we were using earlier to illustrate experimental variation. Remember that an error at one stage can reduce or increase the final value by a varying amount and that there are several sources of error in any one determination. We can imagine a huge number of potential determinations which could be made, each one characterized by its own combination of experimental errors. The effect overall will be to produce a whole range of possible values for the concentration. The answer that we get when we try to measure the concentration will depend on which of these potential determinations we happen to make.

If you have difficulty with the idea of potential determinations, the second example is more concrete and concerns the different types of cells found in vertebrate blood. These include red blood cells and several different forms of white blood cells (e.g. basophils, neutrophils, acidophils and macrophages) which all have different functions, differ in size and staining properties and can be recognized under the microscope.

Suppose we wanted to describe the proportions in which these cell types occur in healthy frogs, that is, to find out what the percentage of each type of blood cell is. We could take a small quantity of blood, stain it and examine it under the microscope (Plate 3). Then we could simply count how many cells were of each type in a field of view. But would this give us the answer? We can see that the different types of cells are not evenly distributed across the slide, and that by chance, the proportions of the different types will be different in different fields of view. It follows that the answer we get will vary, depending on which field

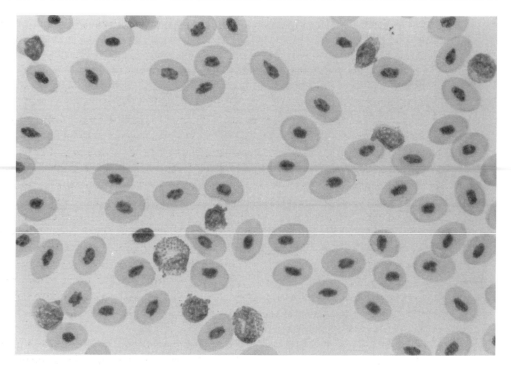

**Plate 3**  Variability in frog blood cells.

of view we choose. We could conceivably count every field of view on this slide, but then this slide is only one of many slides which we could have made using a small quantity of blood from a healthy frog and there are many healthy frogs which we could have used. In other words there are a huge number of possible fields of view which we could examine, that is a huge number of observations which we could make.

These two examples illustrate the fundamental problem of describing "things" which are variable. Later we will have to define carefully what a "thing" is. Here the "things" are determinations and fields of view. The problem arises because, usually, in any one investigation there are a large number of the "things" which could be described. In the absence of variability each "thing" described would yield the same result and so a single observation would be all that was needed to describe the characteristic of all the "things". In contrast, when there is variability then a single observation will be of very limited use as a description of the "things" because it could by chance be atypical, giving us a value which is unrepresentative of the "things" in general. The same will be true, though to a lesser extent, if we increase the number of observations and calculate something like an average. We could still end up with atypical values. Logically the only way to get round this problem would be to go to the other extreme and to measure or count all of the "things", that is to make every possible observation. Then we would be able to describe them exactly, that is we could say what the true value was.

This is rarely possible in practice (for reasons that will be considered in more detail in Chapter 3) and so we have to adopt the compromise strategy of describing the variable in a subset of the "things". This subset is called a **sample**. The complete set of "things" is called the **population**, a word that we shall define rather carefully when we define what "things" are. We then use the description from the sample to make **inferences** about the variable in the population.

While a sample of several observations will give a better picture than a single observation it is still not likely to provide us with the true value of the variable because a sample will still consist of a chance mixture of typical and less typical observations. We can clearly see this happening in the first example. The values of 1.25, 1.20, 1.22, 1.25, 1.30, 1.15, 1.40, 1.30 and 1.21 mg/ml which were obtained are a sample of nine observations from a very large population of (potential) measurements. Experimental error accounts for the variation between potential measurements and the act of sampling (i.e. of making just a few of the possible observations) produces a chance selection of these values. Some of these are likely to be quite close to the true value, others may be quite distant from it, but since we do not usually know the true value we cannot tell which are which. We might, by chance, have hit on the true value with one of our determinations, but again we cannot tell whether this is so.

This chance variation in the make-up of a sample is called **sampling variation** or sometimes **sampling error**, although the latter suggests an avoidable mistake. Because this variation is an absolutely inevitable consequence of making observations on a subset of things which show variability I prefer the term sampling variation. The result of sampling variation is that we can never know the true value of the phosphate concentration in our example. All we can do is to obtain an **estimate** of the true value. Similarly, the proportions of the different types of blood cells can only be estimated.

There is a further problem associated with using a sample which we can illustrate with the blood cell example. How do we choose the fields of view in which to count the cells? The difficulty here is that a personal, subjective element may creep into our decision. For example, we might be tempted to choose fields of view in which there was predominantly only one type of cell because these are easier to deal with. Alternatively, we might tend to choose those fields which have a great diversity of cell types because they are more interesting. Because we want to use the sample as a description of the population, the subset of observations in the sample should be representative of all the possible observations, but whenever we choose items there is always going to be the possibility of **bias** in the way that they are chosen. If there is bias then certain sorts of observations will be over- or under-represented in the sample, that is they will occur more (or less) often in the sample than they do in the population. This will result in samples which consistently overestimate or underestimate the true value of the variable in the population.

The twin problems of choosing a sample which is representative and the chance variation in the make-up of the sample are what get in the way of our attempts to describe. Fortunately, we can get round the problems. To remove any possibility of a subjective element in the choice of sample we need to design a **sampling programme or strategy** which is totally objective. The second problem is not so easy and we have to accept that we can never know the true value of what we are interested in. What we can do is to get a measure of the closeness of an estimate to the true value, that is, a measure of the **reliability** of the estimate. This requires **statistical analysis**.

## Detecting differences

Our previous examples of description have been rather simple in that they have involved describing things under only one set of conditions, e.g. phosphate concentration in one solu-

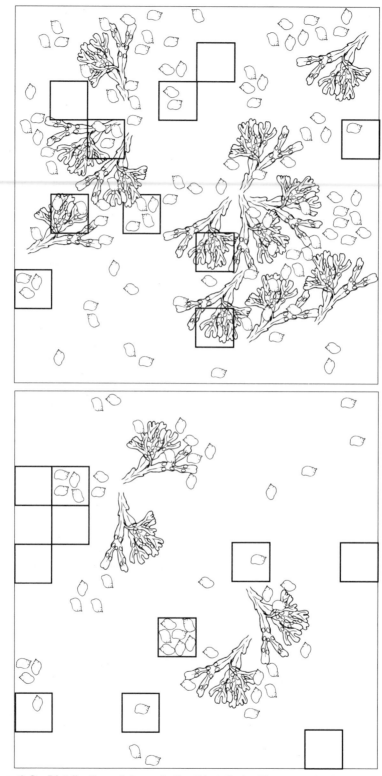

**Figure 1.2** Distribution of dog whelks (*Nucella lapillus*) on a rocky shore. (a) Site 1; (b) site 2.

tion and proportions of cell types in healthy frogs. In these circumstances our sample has been taken from a single population. It may be that this is all that we want to know, but an equally interesting question may be "If the conditions are different does the variable have a different value?" For example, "Do diseased frogs and healthy frogs have different proportions of the blood cell types?" and "Is phosphate concentration different in two solutions?". The problems encountered here arise for the same reasons as before and can be illustrated with the phosphate concentration example.

The solution was made up very carefully using accurate equipment so that the concentration was 1.2519 mg/ml (note that we still do not know the concentration exactly because there will have been unavoidable errors in weighing, etc.). Suppose you had made only two determinations and they happened to yield concentrations of 1.15 and 1.20 mg/ml, this would lead to an underestimate of the true concentration. Now imagine that you had another solution which, in fact, had exactly the same concentration, and again you took two measurements, using the same method. The range of possible values would be the same as before but, by chance, the errors in your determination could lead to values of 1.30 and 1.40 mg/ml. This would lead to an overestimate of the true value. Because of experimental variation and the chance element involved in sampling only two of the possible observations, you could be led to think that the solutions were different when they are actually the same.

Again we can make this more concrete by using an ecological example. The dog whelk (*Nucella lapillus*) is a common snail in the intertidal zone of rocky shores. Suppose we want to know whether the number of snails per unit area is different in two sites. At site 1 (Fig. 1.2a) there are 100 snails in an area of 25 m², but there is variability in numbers between different areas within the site. In some areas snails are very common, in other areas they are rare or absent. Finding and counting snails is a laborious task and our ecologist does not have time to cover the whole site and so decides to count the snails in a representative sample of areas, each of which is 0.5 m × 0.5 m. This means that the site contains 100 such areas which could be used and the value for the site as a whole (= the true value) is one snail per 0.5 m × 0.5 m area. The sample in this case consists of 10 areas. These areas, shown in the figure, are chosen in an appropriate way to avoid bias (which will be described in more detail in Chapter 3) and are marked out using a quadrat.

This sample of 10 areas contains 15 snails giving an estimate of 1.5 snails per 0.5 m × 0.5 m quadrat a value which just by chance is higher than the true value for the site as a whole. If we had chosen a different 10 areas we might have obtained a value which was lower than the true value. Some combinations of 10 areas will, by chance, give us the true value, but in a real-life situation this is not much use because we would not know what the true value was. Now imagine that we repeat this at a different site in order to investigate whether the number of snails (per unit area) is different. In Figure 1.2b there are only 50 snails altogether but again there is considerable variation, and by chance our sample of 10 areas contains 14 snails giving an estimate of 1.4 per quadrat for the whole area.

On the basis of the samples (which are all that we would have in practice) snail numbers (per unit area) in the two sites appear to be very similar, although they are in fact quite different. Sampling variation is getting in the way of our attempts to detect a difference. This scenario is obviously just one of a whole range of possibilities. For example, the numbers in the two sites could in reality be very similar but we could by chance take a sample from site 1 made up of atypically high values and a sample from site 2 made up of atypically low values. On the basis of the samples, the numbers in the two habitats would look different.

Situations like these, in which we are trying to detect differences, are very common in biological investigations. Now we have two populations and we have to take a sample from each. Again we need a method of choosing the "things" that make up each sample so that each sample is representative of the population. For example, there is more seaweed at the first site and it looks as though the snails are more numerous in areas where there is weed. Looking for snails in seaweed is much more difficult than looking for them on bare rock and so there is the danger that our hard-pressed ecologist could tend to avoid the weedy areas in the first site. If this happened there would be a subjective element in the choice of areas and the effect of this would be to bias the result. It would make the numbers in the two areas appear more similar than they actually were. We need a sampling strategy which removes this possibility.

Once we have done the practical work we are left with two possible reasons for what we have found. The difference between the two samples could have arisen because of a difference between the two populations. Alternatively, the two populations could be the same and the difference between the samples might merely due to chance variation in their make-up. To distinguish between these alternatives we will again need statistical analysis.

# Explanation

Explanation involves identifying the **cause(s)** of what has been described, and this involves finding out what **factors** influence the variable. We face an immediate problem which is that biological systems are complex so that there are many possible factors which could be implicated. Our explanation cannot just be a "hand waving" exercise in which we say what we think the cause is, we have to demonstrate unambiguously whether or not the factor does have the postulated effect. We start with a tentative or provisional explanation, a **hypothesis**, which then has to be tested. If it passes the test then the explanation can be accepted as correct, always provided that the test was fair!

Sometimes the factor to be investigated is identified by the very nature of the investigation. For example, if we want to know whether the drug Hypobar reduces blood pressure then Hypobar is the factor to be investigated. If, on the other hand, we want to know what causes some people to have lower blood pressure than others, then there is a whole range of possible genetic and environmental factors which could be investigated. In this sort of situation it would make sense to begin by trying to identify which were the most likely candidates. One way of doing this is to make our initial descriptions more focused or more comprehensive.

## Identifying possible causes

This approach is particularly common in ecology and we can illustrate it with our example of differences in snail numbers in the two areas of the shore. Suppose that by using a suitable statistical test, we have been able to demonstrate that there is a clear difference in the numbers of snails between the two areas so that an explanation is obviously called for. How should we proceed?

We could first use our knowledge of the biology of the snails to make a list of the possible causes, for example food availability, the number of predators and the nature of the substratum. We could then revisit the two sites and describe them with respect to these factors in order to see whether there were any differences between the sites. This would of course require sampling and statistical analysis. (In practice we would almost certainly have thought of these issues right at the start and gathered this additional information at the same time as we counted the snails.) If we found that the site at which the snails were more common had a greater availability of food then we might be tempted to conclude that a difference in food availability was the cause of the difference in snail numbers.

The approach outlined above would be particularly appropriate if we had no prior information about what factors might be important, but suppose we have already singled out a factor for investigation. In this case we could use a different strategy. We could deliberately choose two sites that obviously differed with respect to the factor, for instance one site where food was abundant, the other where it was not. We could then count snails in a sample of quadrats from each site. Again if we found a difference in snail numbers between these two areas, we might conclude that differences in food availability caused differences in density.

Investigations of this type, which make use of naturally occurring variation in the factors of interest are known as **observational studies**. They are widely used but, unfortunately, they have a major drawback. They cannot prove that the factor which is singled out is responsible for the difference in the variable of interest. The fact that snails are more common where their food source is more abundant does not prove that differences in the amounts of food cause differences in the number of snails. There are alternative explanations. It could be that the degree of exposure to wave action varies between the sites and this is what causes the differences in numbers of snails. If degree of exposure also affects the abundance of the food plant then food abundance would appear to be directly linked to snail numbers, even though there is no causal connection.

Observational studies are not confined to ecology. For instance we might want to know whether differences in eye colour in *Drosophila* cause differences in behaviour. It would be relatively easy to measure some aspect of behaviour in red-eyed and white-eyed flies, but any difference detected could not be attributed to eye colour. The gene for eye colour may affect other variables such as body size or activity. The genetic difference between individuals (which causes the difference in eye colour) could therefore cause differences in other variables and it is these which affect the behaviour. In general the only way to demonstrate unambiguously what the causal factor is, is to adopt a different approach.

## Proving causal connections

The only way to verify the hypothesis that factor $X$ causes change in variable $Y$ is to use an **experiment**.

Factor $X$ may have been identified from a knowledge of previous work on the system currently under investigation or from studies on a related problem. For example our ecologist may know of work done in other snail species which shows that food supply in the form of vegetation affects the numbers of snails, or may have noticed (or even measured) a difference in the amount of vegetation between the two sites. Alternatively, the factor identified for investigation might just be a hunch! Wherever the idea comes from, the essence of the experimental approach is to try and vary only the factor of interest while keeping other possible causative factors the same. Because the experimenter deliberately modifies the factor of interest, rather than making use of naturally occurring variation in it, this type of approach is called a **manipulative experiment**. In the simplest experiments we would vary just one factor, so that it was present at just two different levels; for example, high and low levels of food, drug present and absent, low pH and high pH. Our experiment would have two different **treatments**. If one of the treatments corresponds to some sort of "natural" level of a factor it may be labelled as a **control**.

We set the experiment up so that the only difference between the two treatments is in the factor of interest. It follows that if we subsequently observe a difference (in the variable) between the two treatments then it must have been an effect of the treatment, that is, it must have been caused by the factor of interest. Note two things about this. First, the success of the experiment depends absolutely on the treatments differing only in the factor of interest. If there are other differences between the two treatments that we may or may not know about then we cannot be sure what causes any observed effect. Achieving this in practice is not as easy as it sounds and it involves careful thought and planning. This (and other issues) is what is meant by **experimental design** and we shall see that it has some parallels with the design of sampling strategies which we considered earlier. The second point is that in deciding whether the treatment has an effect on the variable, we are involved again with description and all its associated problems. Our experimental results will need statistical analysis.

## Types of experiment

The classic "laboratory" experiment involves the experimenter manipulating the factor of interest to obtain the different treatments, although the experiment is not necessarily carried out in a laboratory. Whether or not it is depends on the type of biological material and the variable under investigation.

Physicochemical factors such as temperature, light intensity, pH and drug dose can be manipulated directly by the experimenter to produce two or more different levels and their effects on the functioning of enzymes, micro-organisms and plant and animal tissues might well be investigated in the laboratory. If the focus was on the effects of these types of factor on whole animals or plants then a greenhouse or animal house would be used or, in the case of humans, a hospital ward. Levels of biotic factors such as disease organisms, competitors or predators can also be altered by the experimenter in the "laboratory" to give different treatments and their effects on individuals examined. In suitable organisms, such as microbes and *Drosophila*, we can also investigate the effects of different treatments on variables such as the growth rate of populations.

Manipulative experiments of this type do not have to be carried out in the laboratory or the greenhouse. In some cases we may want to know what factors are affecting a variable under more natural conditions and in these cases we can carry out **field experiments**. These are widely used in agriculture where we may want to investigate the effect of fertilizer on plant yield or of pesticide application on levels of infestation. It is, for example, relatively easy to expose sets of plants in a field to contrasting fertilizer levels and then to measure the weight of plant material produced in the different treatments.

This sort of approach can sometimes be used in completely natural environments by artificially modifying the factor of interest. To investigate whether food supply has an effect on the numbers of snails in a habitat, we may be able to alter food availability (by removing or adding food) so as to produce areas with contrasting levels of food. A number of classic studies on the effects of predation and competition on rocky shore communities have in-

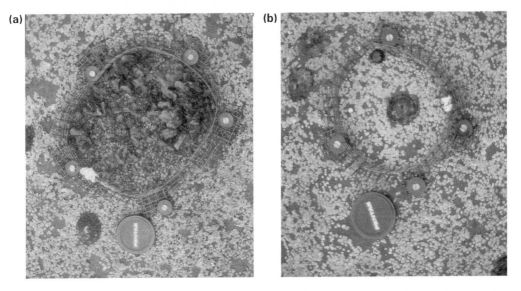

**Plate 4** Field experiment on the effects of limpet grazing on the abundance of seaweeds. (a) Limpets excluded by cage (b) limpet enclosed by cage.

**Plate 5** Field experiment on the effects of elevated temperatures on the growth of arctic vegetation.

volved removing/adding predators or competitors from small areas which are then "fenced" to stop animals moving in or out (Plate 4). The polythene tents in Plate 5 act as mini greenhouses and are being used to investigate the effects of increased temperature on the growth of arctic vegetation. There are some untented areas marked out with pegs which you can just about see and some of these are control areas. Clearly the feasibility of this type of field experiment depends on the nature of the investigation, it would be very difficult to use this approach on a tropical rainforest!

I have used simple examples to illustrate the principles of the experimental approach, but these can be extended to more complex experiments in which several factors might be varied with several levels of each. This gives rise to a larger number of treatments. For example, the experiment illustrated in Plate 5 was also being used to investigate the effects of increased precipitation and nutrient availability. Extra water was applied to some of the tented areas and to some of the untented areas, others received extra nutrients and some received extra water and extra nutrients. Some areas were left completely undisturbed and acted as a control. Altogether there were nine treatments.

## More on observational studies and manipulative experiments

Although manipulative experiments represent the ideal, they are not always possible. Often in ecological work it is impracticable or too costly to do a manipulative experiment. Suppose we want to investigate whether the number of species found in ponds is affected by the size of the pond. To do a manipulative experiment we would need to create a number of ponds of two different sizes, sit back for several years and then count the number of species in the two types of pond. There would be serious practical problems relating to time and cost! An observational study, on the other hand, would use existing large and small ponds which were otherwise similar.

There may be ethical reasons why a manipulative experiment cannot be done and the classic example of this is our attempts to understand the cause of lung cancer. Using an observational study one can show that lung cancer is more prevalent in smokers but this does not mean that smoking causes lung cancer. Perhaps people have a disposition to smoke because of some physiological or psychological characteristic which also disposes them to develop lung cancer. To show that smoking causes cancer would require a manipulative experiment which would involve selecting two equivalent groups of people and exposing one group to cigarette smoke. Because this cannot be done we have to fall back on other sorts of evidence. These include the fact that some chemicals present in cigarette smoke can be shown to cause tumours in other animals, that the incidence of lung cancer differs for different types of smoker (cigar/pipe/cigarette) and that smokers who give up smoking have a lower incidence of lung cancer than those who do not give up.

In other cases, a manipulative experiment can be done but great ingenuity and care is needed. The males of some species of swordtail fish have coloured, elongated tails, which are used to attract females. In order to explain the evolution of this characteristic we need to show that females prefer males with long tails, in other words that long tails cause males to be chosen. The factor is tail size, but to investigate its effect we need to have fish with different sized tails.

If we find a genetic mutant without a tail then the gene may also cause differences (from tailed individuals) in other aspects of its biology which may render it unattractive to females. If we remove the tails from some tailed fish then the shock of the surgery may affect their behaviour, or the loss of their tails may alter their swimming pattern. In either case there could be additional differences between tailed and tailless fishes which could affect their attractiveness. As a result, we could not attribute any female preference to tail size. The ingenious solution adopted in one such study was to take a closely related species in which the males are tailless and to give them artificial tails, using minor surgery. Pairs of males of similar size and colour were chosen and one was given a coloured tail and the other given a transparent tail of the same size and shape, which had been shown to be invisible to females. The males in this latter group were therefore tailless as far as females were concerned but this would be the only difference between the two groups. There would be no genetic differences, both tailed and "tailless" males had undergone surgery and any effect of the artificial tails on swimming would be the same in both groups. Just in case there were any subtle differences between the tailed and "tailless" members of each pair, their artificial tails were swapped and the experiment was repeated. As a result any differences in female preference must be caused by the difference in tail size.

Manipulative experiments do have shortcomings. First, they tend to simplify things because they focus on a single factor (or at most three or four factors). In reality, even the simplest biological systems are likely to be affected by many factors, the effects of which may interact in subtle and unpredictable ways. These interactions will not be detected in a simple experiment. Second, the need to manipulate the factor means that these experiments are always somewhat artificial. Enzymes and tissues have to be extracted from the organism and treated in test tubes, organisms have to be kept under unnatural conditions and the cages used to exclude grazers on the sea shore may alter other aspects of the habitat. As a result we may have to be careful about extrapolating from the results of our experiment to what happens in natural conditions.

## Sampling programmes, experimental design and statistics

I hope that you now have an overview of why these three issues are important, but there is a further crucial point. The design of the sampling programme or of the experiment and the details of the subsequent statistical analysis have to be considered together. They are not separate issues.

As we will see there are many different statistical tests each one capable of answering a different question. There are different tests for different sorts of variable and each test involves assumptions about the properties of the variables to be analysed and the way in which the results have been gathered. Tests may also differ in the number of observations required. These specifications will add an extra dimension to the design of the experiment or the sampling programme. This may involve additional thought about the details of how it is set up, the numbers of observations that need to be made and the way in which the variable is actually described.

As biologists we want to get down to the practical biology as quickly as possible. The issue of how exactly we are do it and, even more distantly, how we might analyze the results seem to be distractions. However, it cannot be overemphasized that both the details of the design and of the statistical analysis need to be carefully considered *before* the practical work is carried out. The essential starting point in all this is a very clear idea about the question to be answered. Without this initial investment of time and thought, it is easy to carry out an experiment which cannot answer the question either because the practical work itself has a flaw or because the results cannot be statistically analysed. This is perhaps the hardest lesson to learn.

In the next chapter we will put some of these ideas into practice.

# 2

# Some basic ideas in experimental design and statistical analysis

The first chapter outlined the general problem. In this chapter we shall examine in more detail what is meant by experimental design and why we need statistics and also look at some of the basic ideas common to all statistical procedures. The issues to be covered are absolutely central to the whole subject and by the end of the chapter you should be well on your way to understanding the major ideas involved. In order that we can concentrate on these issues I have deliberately chosen a rather simple piece of biology – habitat choice in woodlice.

## Deciding on the research hypothesis

Experiments are usually carried out in an attempt to explain some previous finding. In this case our starting point is the pattern of dispersion of woodlice which are terrestrial crustaceans commonly found in grassland, leaf litter, under stones and under the bark of dead trees. What is noticeable about their distribution is that it is very uneven; for example some pieces of bark will have rather few woodlice underneath, whilst others will be absolutely crawling with animals. How can this be explained?

Well, one possibility is that there is nothing to explain! Perhaps we are just looking at an uneven pattern of dispersion of the sort that we already know can occur by chance. Normally we would check whether or not this was the case by carrying out a statistical test, but we do not yet know enough to be able to do this. What we will have to do, is to assume that the difference in numbers between places is too large to be due to chance variation and that there must be some other reason for the uneven pattern of dispersion. One possible explanation (and you will recognize this from the previous chapter) is that the woodlice are initially distributed more evenly and that they then die at different rates in different places. Another possibility is that they choose to congregate in those places where the environment is particularly suitable which could be because of differences in light, humidity or food levels. Alternatively the presence of a few animals in one place could automatically make that place a preferred environment for other woodlice.

In biology it is not unusual to come up with a list of possible explanations for a given phenomenon and it is a good idea to always try and do this. It would be very difficult to test all these explanations simultaneously, so in practice we have to make a decision about which is the most likely and test that one first. A key step at this stage is to define very care-

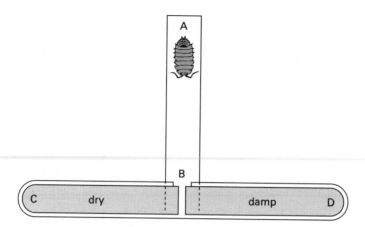

**Figure 2.1**  A simple choice chamber.

fully what the objective of the experiment is. Our list of other possible explanations can be used to sharpen our focus on the objective by drawing attention to what we are *not* investigating. It will also be invaluable when we come to think about the actual design of the experiment.

Let's assume that our objective is to test whether or not humidity is the factor. Our tentative hypothesis is that the difference in woodlouse numbers between places is caused by woodlice having a humidity preference. This is called the **research hypothesis**. On the basis of this, we decide on an experiment to test whether woodlice have a preference for either damp or dry areas. A suitable experimental set-up is shown in Figure 2.1, which consists of a simple choice chamber. A woodlouse is placed in the chamber at A (facing towards B) so that at the junction it can choose to turn left or right. The floor at C and D is covered with two separate pieces of filter paper, one in each arm. One of these pieces of filter paper is dampened, so that when the lid is placed on the chamber a damp micro-environment is created in that alley. To carry out the experiment we put a woodlouse at A and record which way it turns at B. This constitutes a single trial, which we would repeat several times with different woodlice.

In this experiment, as in all experiments, there are two separate issues to consider. First, before we carry out the experiment we must ensure that it has been designed properly. Secondly, assuming that this has been done, we need to take into account the fact that our results will be subject to sampling variation. To show whether or not the results obtained support the idea that woodlice have a preference for one of the humidity regimes we will have to carry out a statistical analysis.

## The design of the experiment

In this simple experiment the factor that we are interested in is humidity and the variable is the direction in which a woodlouse turns. There are two treatments, high and low humidity, and we want to show whether the difference between the two treatments affects the direction of turn. As we saw in the last chapter any variable is likely to be affected by many factors and the same is true of the variable in this case (i.e. direction turned). Humidity is not the only factor that could affect it. Perhaps woodlice tend to move towards the north or away from strong light, or to where the walls are rough rather than smooth. Alternatively, they could be genetically programmed in some way so that they tend to turn right. Perhaps, it could be that in handling the animals we tend to damage the legs on one side, which results in a tendency for them to veer one way or the other. You may think that some of these ideas are more plausible than others, while some sound decidedly far-fetched, but they are all possibilities and none can be ignored at this stage. Perhaps you can think of some others as well but in fact we do not need an exhaustive list, because, as we will see, these factors fall into just a few categories. What we do need to do is to take them into account when designing the experiment because we have to be sure that none of them can affect the result.

It goes without saying that the experiment needs to be designed correctly before we actually carry it out, but at the design stage it is often helpful to imagine that we have actually done the experiment in a particular way and obtained some results. We then ask ourselves the question "Can these results only be explained by the humidity difference between the alleys?". If the answer is no, then the design is inadequate. So let's suppose we have used the design described above and that, out of 20 woodlice, 19 turned to the damp and one to the dry. At first sight it looks as though they prefer the damp alley but, with our simple design described above, this would be only one of several possible explanations which are outlined below.

(a) Woodlice prefer high humidity.
(b) Woodlice have no preference for humidity variation but there is some other difference between the two alleys for which they do have a preference.
(c) Woodlice have no preference for any differences that there might be between the alleys but they have an intrinsic tendency to go in one direction rather than another.
(d) In some way our observations were biased so that we recorded individuals as turning to the damp when in fact they did not.
(e) Humidity does not affect choice so that in the long run equal numbers of animals will turn in both directions. The fact that, in our experiment, more animals turn in one direction is due to sampling variation.

Explanation (e) is a statistical explanation that will be considered in the next section, but the fact that there are four other explanations for our imaginary result means that we would not be able to say which, if any, is the correct one. The design of our experiment is flawed and since it cannot answer our original question, the experiment in this form would not be worth doing. So how should it be modified?

## Removing the possible effects of other factors

In order to rule out explanation (b) we simply need to ensure that the only difference between the two alleys is a difference in humidity. One way to do this would be to take a great deal of care in making the chamber, to try and ensure that both alleys were identical and that it was placed on a completely level bench, equally lit from all directions, shielded from the earth's magnetic field, etc. As you can imagine this approach is going to be quite costly, but it suffers from a more fundamental flaw. How do we know that we have thought of every factor that might be relevant? After all, we do not know what is important to woodlice.

A second approach overcomes both these objections by recognizing that there are always likely to be other differences (besides humidity) between the two alleyways. However, possible effects of these differences can be eliminated by doing the experiment in such a way that any such effects will cancel out. This can be achieved simply by having the damp filter paper in the left alley for one half of the trials and in the right alley for the other half of the trials. On balance the humid treatment will not have rougher walls or a sharper corner to be negotiated. This procedure will also deal with explanation (c), because a damp turn is as likely to be a right (or a north) turn as it is to be a left (or a south) turn. The factor right/left (north/south) has been evened out across the two treatments of interest.

This specification on its own is still not adequate because it does not take account of possible changes in things like light intensity over the course of the experiment. You could imagine a situation in which the left alley is less illuminated at the start of the session and the right alley less illuminated at the end, because of the changing position of the sun. If the trials with the damp filter paper in the left arm are all carried out at the beginning of the session and the woodlice all turned to the left we would not be able to tell whether this was in response to the low light or the increased humidity. One solution to this would be to alternate the damp treatment between the left and the right alleys at every trial in a regular way, but this would not be ideal. It is better to make the sequence random (we shall see how to do this and why it is important in Chapter 3). There is one other factor that this design will not adequately deal with and which will require a minor modification. It will be more convenient to deal with it later (p. 32).

Arranging the experiment in such a way that variation in other factors which might conceivably affect the variable is averaged out over the two treatments is one aspect of what is meant by experimental design. In this case we ensure that the only consistent difference between the damp and the dry treatments is a difference in humidity. If we do not do this then any experimental results cannot be used to answer our original question. We would not be able to say for sure whether it is the difference in humidity or the difference in some other factor that is causing the woodlice to turn into the damp alley of the maze. There is a third approach which we could use which you may have thought of and that is to use a control. We could proceed as already described, but in addition carry out a number of trials in which both alleys were damp (or dry). This would show the extent to which other factors were affecting the direction turned and we could then compare the two sets of results. This would be quite acceptable, but it would need a more complicated statistical analysis than we can cope with at this stage.

## Dealing with bias

Even if we did the experiment as set out above we could be biased in the way we make our observations and this is particularly likely to happen if we do not have a clear idea about what constitutes a result. I deliberately described the outcome of each trial in a vague way, using words like "turned" and "ended up", without defining what they meant. This failure to clearly define the outcomes leads to the possibility of bias, because we may not record the outcome of each trial correctly.

In some trials the woodlouse will turn the corner and proceed to the far end of the alley without stopping. In other trials a woodlouse will get so far round a corner and then stop. It may remain stationary, move off along its original track or reverse and eventually turn in the opposite direction. With some of these possibilities there is an ambiguity about how we should record the outcome. If a woodlouse stops before reaching B then that particular trial has to be abandoned and recorded as a "no turn". The same would be true in the situation where an animal gets to B and then stops without showing any sign of turning. But what about a situation where it just angles its body slightly to one side or puts its head round the corner before stopping? These are not clear-cut outcomes and, if we make a decision about each one individually as it occurs, there is a danger that our personal preferences may intrude.

Suppose for some reason we have the idea that woodlice do prefer a damp environment then, if a woodlouse angles slightly to the damp before stopping, we might record that as a damp turn. When one angles the same amount to the dry before stopping we might be tempted to record it as a "no turn", seeking to justify our decision on the grounds that it was not angled quite as much as in the previous case! If we were consistently biased in this way we would artificially increase the number of damp turns that we recorded. The simplest way to overcome this problem is to have a stated, easily applicable, criterion of what constitutes a turn. A suitable criterion in our case would be that we record only the first turn made and that the whole of the body has to be past a line, real or imaginary, drawn across the end of each alley at B.

## Other design issues

To complete our design we need to consider two other issues which are common to any investigation. However, at this point I shall only mention them briefly because to deal with them fully will require more background. The details will be dealt with in later chapters.

The first relates to the number of woodlice to be used; we will use 20. The second concerns how we choose the 20 woodlice. The experiment has to be restricted to a few specific woodlice, but remember that the purpose is to try and explain what woodlice in general do. In other words we want to be able to make inferences from the limited results of our experiment. It follows, almost as a matter of common sense, that the animals we use in our experiment ought to be representative of woodlice in general. For example, if the woodlice under the bark in the natural environment are a mixture of different sizes then our experiment should use animals which are representative with respect to size. We will see how to do this in Chapter 3, but for the moment let's assume we have sorted it out satisfactorily so that we can proceed to do the experiment.

## Statistical analysis

The results of the experiment were as follows. Out of the twenty trials 14 resulted in turns into the damp alley ("damp turns") and three were "dry turns". In addition there were three "no turns", as defined earlier, which we can discard. Given that we have done the experiment correctly, do the results show that the humidity difference has an effect on the direction taken at the T junction? Our initial answer would almost certainly be to say yes on the basis of the fact that the numbers of turns in the two different directions are rather unequal. However, a little thought will show that the answer is not quite so simple because of the problem of sampling variation.

## How can the results be explained – sampling variation or the effect of humidity?

The point at issue can be illustrated by assuming that the humidity difference has *no* effect and imagining what sort of result we might then have obtained. This assumption allows us to formulate a hypothesis called the **null hypothesis** which is often denoted by $H_0$. Although this assumption might seem odd, making it turns out to be an essential step in any statistical analysis. We shall see why shortly. It stands in contrast to the **alternative hypothesis**, denoted by $H_1$, which is formulated on the assumption that the factor does have an effect.

So let's imagine that the humidity difference has no effect on the direction turned, which means that the null hypothesis is true. If this is so then an individual woodlouse is just as likely to turn into the damp alley as it is to turn into the dry one, that is, the chance (or probability) of it turning in either direction is equal. Conventionally we express a probability ($p$) on a numerical scale of 0–1, where 0 means that there is no chance of the event happening while 1 means that we are absolutely certain that it will happen. When two things are equally likely to occur, then they both have a probability of 0.5. We can therefore write the null hypothesis more formally as

Null hypothesis: probability of a damp turn = 0.5

Likewise we can write down the alternative hypothesis as

Alternative hypothesis: probability of a damp turn $\neq 0.5$

The reason for starting from a null hypothesis is that we have absolutely no difficulty in making a sensible statement about exactly what the value of $p$ would be under these conditions. It must be exactly 0.5. Being able to specify a value for $p$ turns out to be an essential starting point for the subsequent analysis. Note that we cannot make any such statement about the value of $p$ under the alternative hypothesis, because there is an infinite number of different values which $p$ could have and we have no sensible reason for choosing one of them. The alternative hypothesis as you can see is really a translation into a more exact form of our original research hypothesis that humidity does affect behaviour.

## Possibilities and their expected relative frequencies

If an individual does have an equal chance of turning in either direction how many of the two types of turn will we get if we carry out several trials? To answer this it will be easier if we begin by considering an experiment consisting of only a small number of trials. Once we have established the principle, we can return to our actual experiment.

Figure 2.2 depicts what can happen with an experiment of just four trials and shows us that there is more than one answer to this question. The first woodlouse has an equal chance of turning to the damp or the dry, let's suppose it turns to the damp. For the second woodlouse the choice is exactly the same, it too could turn to the damp or the dry, let's suppose it also turns to the damp, while the third woodlouse turns to the dry and the fourth to the damp. This sequence of events producing three damp and one dry turn is shown in bold on the right-hand diagram. The diagram also shows that there are a number of other possible results. For instance the first woodlouse could just as well have turned to the dry

**(a)**

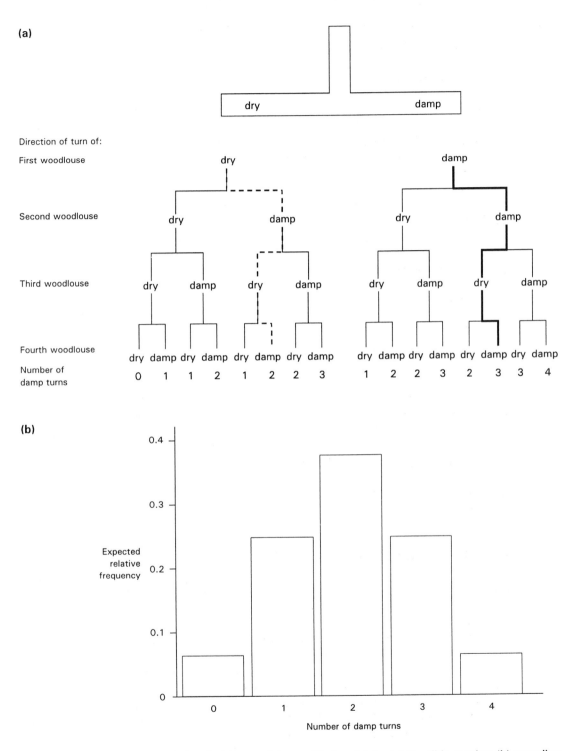

**(b)**

**Figure 2.2** Possible results of a choice experiment with four trials. (a) Possible results; (b) sampling distribution under the null hypothesis.

and, if it had done so (while the other woodlice behaved as before), the result would have been as shown by the dotted sequence in the left-hand diagram. This is just as likely to have occurred and would have given two damp turns and two dry turns. A similar argument applies to all the other 16 possible sequences of turns. They are all equally likely to occur.

If we now focus on the line which gives the total number of damp turns for each sequence, it shows us clearly that, even though the probabilities of the two outcomes (i.e. types of turn) are equal, we shall not necessarily get equal numbers of the two outcomes in any one experiment. We can in fact get any number of damp turns from 0 to 4! It also shows us that a second set of four trials will not necessarily give us the same result as the first set, while a third set could give a different result again. What we are seeing here is sampling variation. Our imaginary experiment actually involves taking a sample of four turns from a huge population of possible turns in which both types of turn are equally probable. The make-up of this sample is determined by chance.

The same principle will hold for our experiment of 17 trials, though there will obviously be many more possible sequences. At first sight this leads to a rather worrying conclusion because it tells us that if the null hypothesis is true, then all results are possible including 14:3! In other words a result of 14:3 could arise even if the humidity difference has no effect. This is diametrically opposed to our initial common-sense conclusion that the result showed that humidity did have an effect.

The key to resolving this discrepancy lies in recognizing a further feature of Figure 2.2 which is that, although all the different results are possible, they will not be equally frequent. The reason for this is that there is only one sequence out of the 16 which produces no damp turns, but there are four different sequences giving one damp and three dry turns, and six sequences which give two damp turns.

What this means is that if we imagine repeating the experiment many times, then in the long run one out of every 16 experiments will produce a result of no damp and four dry turns. By the same argument, four out of every 16 experiments will yield a result of one damp turn. This can be expressed slightly differently. In the long run we would expect that the proportion of results with no damp turns would be 0.0625 (= 1/16), while the proportion with one damp turn would be 0.25 (= 4/16). We will call these proportions the **expected relative frequencies**. These are presented in the form of a graph in Figure 2.2b, in which the height of each column is directly related to the relative frequency of each type of result. Remember that this has been arrived at on the basis of the null hypothesis being true so it gives us all the possible results involving samples of four turns and their expected relative frequencies if the woodlice had no preference for damp or dry conditions. For this reason it is called the **sampling distribution under the null hypothesis**.

Although we have produced it by imagining that we have taken a large number of samples of four turns, in fact we would need to repeat our experiment an infinite number of times to get exactly this distribution. It is therefore a theoretical distribution which can be used to tell us what the relative frequencies of each result are expected to be in the long run. You can see that it is symmetrical and centred on equal numbers of turns in the two directions. It has a hump in the middle and a tail at each side which simply means that the result 2:2 will be the most frequent and that more uneven, extreme results will be less frequent.

## From expected relative frequencies to probabilities

To see where this gets us we need to look at the sampling distribution from a slightly different perspective. If, in theory, the expected relative frequency of samples (of four trials) with no damp turns is 0.0625, then this result will not be very common. In the long run such a result will only occur about six times in every 100 experiments of four trials. It follows that if we do only one experiment the chance of our getting a result like this is not very high. We can express this chance or probability on a scale of 0 (no chance) to 1 (complete certainty) and it bears a very simple relation to the expected relative frequency. The chance or probability of getting such a result in one experiment is 0.0625, which is numerically exactly the same as its expected frequency derived from the sampling distribution.

This transition, from the expected (or theoretical) relative frequencies with which results would occur in the long run, to the probability of getting a particular result in a single experiment, is one that we will be using a great deal. You will need to be totally at ease with this idea. If you have problems with it just think of how we use the idea in everyday life. If the relative frequency (or proportion) of raffle tickets with winning numbers is 0.01 (i.e. 1/100) then in the long run out of every 100 tickets one is a winner. If you buy just one ticket you have a probability of 0.01 of getting one with a winning number.

Although we have used a small number of trials, this analysis shows us a general principle. There are some extreme results which have a low probability of occurrence if the null hypothesis is true, which simply means that in any one experiment we would not be likely to get such a result. This information on the probabilities of different results begins to explain our initial reaction to the result of 14:3, which was to conclude that woodlice did have a preference. It was a common-sense reaction to what is clearly a rather extreme result. A result like this would be rather unlikely to occur if individuals do not have a preference and rather more likely to occur if individuals do have a preference. Our decision, on balance, was in favour of the latter interpretation. Putting that into our new terminology of the null and alternative hypotheses, what we were saying was, that a result of 14:3, although possible, was not very likely to occur if the null hypothesis were true (i.e. if woodlice have no preference). Conversely, it was a more likely result if the alternative hypothesis were true. As a consequence our initial reaction was, in effect, to reject the null hypothesis as an explanation of the result. In doing so we automatically accept the alternative and conclude that woodlice do have a preference.

The only problems with this sort of common-sense approach is that it is all rather subjective. We have a feeling that the result is unlikely, but we do not know exactly how unlikely. Nor do we have any standard procedure for deciding how unlikely something has to be before we conclude that the null hypothesis is an unacceptable hypothesis.

We deal with these two problems by using a statistical test which takes the null hypothesis as its starting point and follows the same logical sequence as above. The only difference is that it uses a mathematical model to calculate the expected relative frequencies involved, that is to produce the sampling distribution under the null hypothesis, and a set of agreed rules to define what constitutes likely and unlikely results. Using a statistical test we can make an objective decision about whether the result we have obtained is a likely or an unlikely occurrence if the null hypothesis is true. If the result is judged to be a likely occurrence, then we can accept the null hypothesis as a perfectly reasonable explanation for the result. If the result we obtained is an unlikely occurrence (again assuming that the null hypothesis is true) then the null hypothesis is not an acceptable explanation for the result. In this case we would reject the null hypothesis and in doing so automatically accept the alternative hypothesis which, to remind you, is the hypothesis that humidity does affect direction of turn.

## Models and the sampling distribution

To employ a statistical test we need the sampling distribution under the null hypothesis, so that we can see what all the possible results are and find their expected relative frequencies or probabilities. We can then see which are the likely and unlikely results. We could get this by drawing a diagram like the one in Figure 2.2 every time, but that would obviously be very time-consuming, especially when the number of trials is large.

Fortunately we do not have to do this because, as you may already have recognized, this distribution is one that is very well known to mathematicians. It is, in fact, an example of a **binomial distribution** and it can be described or specified by a formula which gives us the expected relative frequencies of all the different results. At this stage we do not need to tangle with the mathematical details of the formula, but nevertheless there is an important point to make about the mathematics. If we are going to use a mathematical model to produce a sampling distribution, the model needs to be an accurate description of the biological system being investigated. This means that the model has to incorporate all the relevant biological features. If it does not, then the sampling distribution it generates will not be "true to life" and will not form a reliable basis for making decisions about the null hypothesis.

So why are the possible results of an experiment like ours expected to follow a binomial distribution, or, to put it another way, why is the binomial distribution an adequate mathematical model for the sampling distribution of the number of damp (or dry) turns? Put simply, the conditions which will give rise to a binomial distribution are that each trial has two possible outcomes, which are mutually exclusive and that the outcomes of successive trials are independent. As we shall see these conditions adequately describe the biology of our woodlouse experiment.

First, each trial has only two possible outcomes, namely a turn to the damp and a turn to the dry. Secondly, these two outcomes are mutually exclusive which means that the outcome is either a damp or a dry turn it cannot be both. Thirdly, for the outcomes of successive trials to be independent, the outcome of one trial must not affect the outcome of any other trial. In our case this means that an individual should not be influenced by how a previous woodlouse has turned. If, for example, woodlice followed tracks left behind by those that had gone before, then later woodlice would tend to turn in the same direction as earlier ones and the outcomes of successive trials would not be independent. Ensuring that this cannot happen is the missing element in our experimental design and an appropriate modification can be incorporated very easily. All we need to do is to clean out the chamber between trials. So on this basis it is reasonable to assume that under the null hypothesis the sampling distribution for the results of our experiment should be a binomial distribution.

To specify the actual binomial distribution, that is, to calculate the expected relative frequencies of all the different results, we need two pieces of information. These are the number of trials which is the same as the sample size ($n$) and the probability ($p$) of one of the outcomes. In our example one of the outcomes is a damp turn, so $p = 0.5$. The values of $n$ and $p$ can be either plugged into the formula for the binomial or, more conveniently in our case, used in sets of tables which give us directly the appropriate probabilities for the different results.

## Using binomial tables

Table A2 in the Appendix is a set of binomial tables arranged according to the number of trials, which is the same thing as the sample size. Part of it is reproduced in Table 2.1. Just to get our bearings, let's look at the table for a sample size of 4 ($n = 4$).

Along the very top and bottom of the table is a row of highlighted numbers labelled $p$ and these are a selection of values for the probability of one of the outcomes occurring. In our case one of the outcomes is a damp turn, which has a probability (under the null hypothesis) of 0.5. This puts us in the column at the extreme right of the table. Down each side of the table is a column of highlighted figures labelled Y which is the number of times the outcome could occur in an experiment of n trials.

We can see that in an experiment of four trials there are five possible values of $Y$ ranging from no damp turns (out of four) to four damp turns (out of four). The highlighted figures in the left-hand column go in the opposite direction to those in the right-hand column, but in this particular case (when $p = 0.5$) we can use either column. In what is called the body of the table (in the column under the appropriate value of $p$) and in the row opposite to the desired value of $Y$ are the expected relative frequencies of each of the five possible results (that is the expected relative frequency of 4, 3, 2, 1 and 0 damp turns). The values are exactly the same as the ones we obtained from Figure 2.2.

**Table 2.1** Expected relative frequencies for binomial distributions with $n = 4$ and $n = 17$.

| $p =$ | .01 | .02 | .03 | .04 | .05 | ... | .333 | .35 | .40 | .45 | .50 | |
|---|---|---|---|---|---|---|---|---|---|---|---|---|
| $n = 4$ | | | | | | | | | | | | |
| $Y = 0$ | .9606 | .9224 | .8853 | .8493 | .8145 | ... | .1975 | .1785 | .1296 | .0915 | .0625 | 4 |
| 1 | .0388 | .0753 | .1095 | .1416 | .1715 | ... | .3951 | .3845 | .3456 | .2995 | .2500 | 3 |
| 2 | .0006 | .0023 | .0051 | .0088 | .0135 | ... | .2963 | .3105 | .3456 | .3675 | .3750 | 2 |
| 3 | .0000 | .0000 | .0001 | .0002 | .0005 | ... | .0988 | .1115 | .1536 | .2005 | .2500 | 1 |
| 4 | .0000 | .0000 | .0000 | .0000 | .0000 | ... | .0123 | .0150 | .0256 | .0410 | .0625 | 0 = Y |
| $n = 17$ | | | | | | | | | | | | |
| 0 | .8429 | .7093 | .5958 | .4996 | .4181 | ... | .0010 | .0007 | .0002 | .0000 | .0000 | 17 |
| 1 | .1447 | .2461 | .3133 | .3539 | .3741 | ... | .0086 | .0060 | .0019 | .0005 | .0001 | 16 |
| 2 | .0117 | .0402 | .0775 | .1180 | .1575 | ... | .0345 | .0260 | .0102 | .0035 | .0010 | 15 |
| 3 | .0006 | .0041 | .0120 | .0246 | .0415 | ... | .0863 | .0701 | .0341 | .0144 | .0052 | 14 |
| 4 | .0000 | .0003 | .0013 | .0036 | .0076 | ... | .1510 | .1320 | .0796 | .0411 | .0182 | 13 |
| 5 | .0000 | .0000 | .0001 | .0004 | .0010 | ... | .1963 | .1849 | .1379 | .0875 | .0472 | 12 |
| 6 | .0000 | .0000 | .0000 | .0000 | .0001 | ... | .1963 | .1991 | .1839 | .1432 | .0944 | 11 |
| 7 | .0000 | .0000 | .0000 | .0000 | .0000 | ... | .1542 | .1685 | .1927 | .1841 | .1484 | 10 |
| 8 | .0000 | .0000 | .0000 | .0000 | .0000 | ... | .0964 | .1134 | .1606 | .1883 | .1855 | 9 |
| 9 | .0000 | .0000 | .0000 | .0000 | .0000 | ... | .0482 | .0611 | .1070 | .1540 | .1855 | 8 |
| 10 | .0000 | .0000 | .0000 | .0000 | .0000 | ... | .0193 | .0263 | .0571 | .1008 | .1484 | 7 |
| 11 | .0000 | .0000 | .0000 | .0000 | .0000 | ... | .0061 | .0090 | .0242 | .0525 | .0944 | 6 |
| 12 | .0000 | .0000 | .0000 | .0000 | .0000 | ... | .0015 | .0024 | .0081 | .0215 | .0472 | 5 |
| 13 | .0000 | .0000 | .0000 | .0000 | .0000 | ... | .0003 | .0005 | .0021 | .0068 | .0182 | 4 |
| 14 | .0000 | .0000 | .0000 | .0000 | .0000 | ... | .0000 | .0001 | .0004 | .0016 | .0052 | 3 |
| 15 | .0000 | .0000 | .0000 | .0000 | .0000 | ... | .0000 | .0000 | .0001 | .0003 | .0010 | 2 |
| 16 | .0000 | .0000 | .0000 | .0000 | .0000 | ... | .0000 | .0000 | .0000 | .0000 | .0001 | 1 |
| 17 | .0000 | .0000 | .0000 | .0000 | .0000 | ... | .0000 | .0000 | .0000 | .0000 | .0000 | 0 = Y |
| | .99 | .98 | .97 | .96 | .95 | ... | .667 | .65 | .60 | .55 | .50 | $= p$ |

Let's now use these tables to obtain the sampling distribution under the null hypothesis for the actual experiment where the number of trials was 17 (remember that three trials had to be discarded as "no turns"). The appropriate table is also reproduced in Table 2.1. We see that $Y$ could have any value between 0 and 17, in other words out of 17 turns we could get any result ranging from 17 damp turns ($Y = 17$) to no damp turns ($Y = 0$). There are 18 possible results, and by looking in the body of the table in the column under $p = 0.5$ we can find their expected relative frequencies. Values are given to four decimal places, a zero simply means that the probability is less than 0.00005. The first figure is the expected relative frequency of samples of 17 turns, all of which are to the damp and, as we might expect for such an extreme result, the value is very low, so low in fact that it is given as zero (the actual value is 0.0000076). For 16 damp turns (in a sample of 17) the value is 0.0001, for 15 damp turns it is 0.0010, etc. The results with the highest expected frequencies are 9 and 8 which are equally likely to occur. Starting at the other end of the column we get exactly the same values, which is not really very surprising. After all, when the number of damp turns is 2 ($Y = 2$), the number of dry turns is 15. Since the chance of a woodlouse turning into the dry alley is the same as the chance of it turning to the damp alley, the expected relative frequency of a result of 15:2 should be the same as that of a result of 2:15.

The complete sampling distribution has been drawn in Figure 2.3. It has a hump in the middle and falls away symmetrically on either side to produce two "tails" containing the more extreme results. Figure 2.3 is the sampling distribution under the null hypothesis; that is, it is the array of all possible results that we could get if the null hypothesis were true and the relative frequencies with which they are expected to occur. These frequencies are of course equivalent to the probabilities of getting a particular result in a single experiment. We can use them to find out how likely we would be to get the result (number of damp turns $Y = 14$) that we actually obtained in our experiment. However, before we do this it will be useful to introduce another term.

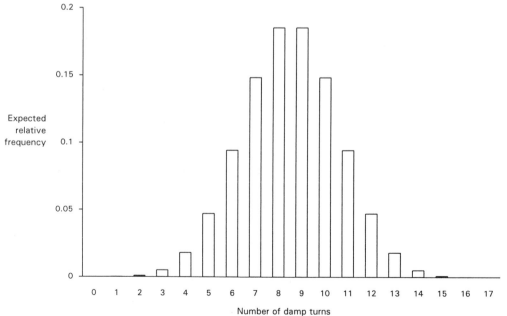

**Figure 2.3** Sampling distribution under null hypothesis for the woodlouse choice experiment ($N = 17, p = 0.5$).

## Test statistics

The result from our experiment, that is $Y$ = number of damp turns = 14, is a simple example of a **test statistic**. A test statistic is a single number obtained from the experimental results which has a known sampling distribution under the null hypothesis; Figure 2.3 is the sampling distribution for our test statistic. It is a "test" statistic because we can use it to test whether the null hypothesis is a reasonable basis for the results we obtained. The test simply involves checking the value of the test statistic from the experiment against the sampling distribution of all possible values to see whether it can be classed as a likely or an unlikely event. If the value of the test statistic that we have obtained is judged to be unlikely, then our decision will be that the null hypothesis is not an acceptable explanation for it.

Our test statistic here is quite simple because we can obtain it immediately from the results of the experiment. As we will see later, most test statistics are rather more complicated because they involve using a formula to calculate something from the results. Because of this it is convenient to label the value obtained as the **calculated value** (of the test statistic). This serves to distinguish it from another value which we will meet shortly.

We are now in a position to judge whether the calculated value of the test statistic is likely or unlikely to occur if the null hypothesis is true, and thus to decide whether to accept or reject the null hypothesis as an explanation of the result obtained. This raises the issue of how we make this distinction between likely and unlikely, which underpins our final decision. The danger here is that we might be tempted to make the decision in a subjective way (that is, in a way which is influenced by our prejudices). To avoid this and to reach an objective decision, we need to adopt a standard procedure, which is agreed by convention and set out before we do the experiment. If we then stick to this procedure, there will be no possibility that our conclusion will be affected by personal preference and other workers will know on what grounds we have made our decision.

## A preliminary rule for deciding whether to accept or reject the null hypothesis

At this point we will use a simple rule which has the virtue of being easy to follow and apply. Later on, when we have learned a bit more, we will examine the pitfalls of this approach and develop a more sophisticated rule. The procedure is quite straightforward. We divide up the sampling distribution of the test statistic into two regions even before we carry out the experiment. The first of these regions comprises all the values of the test statistic which we consider to be likely if the null hypothesis is true (that is if humidity does not affect behaviour). This is called the **acceptance region** and is shown in Figure 2.4. If the result that we do get falls anywhere in this region then our decision is to accept the null hypothesis. The other region comprises all those values of the test statistic which we think we would be unlikely to get if the null hypothesis were true, i.e. the more extreme results that should occur only rarely. This is the **rejection region** or the **critical region**. If our actual result falls in this region, our decision is to reject the null hypothesis. In our example the sorts of results which would be unlikely if the null hypothesis were true are those where the number of the two types of turn are very unequal, e.g. 17:0, 16:1, 15:2 and 2:15, 1:16, 0:17. The rejection region will therefore comprise both the left-hand and the right-hand tails of the sampling distribution, as shown in Figure 2.4.

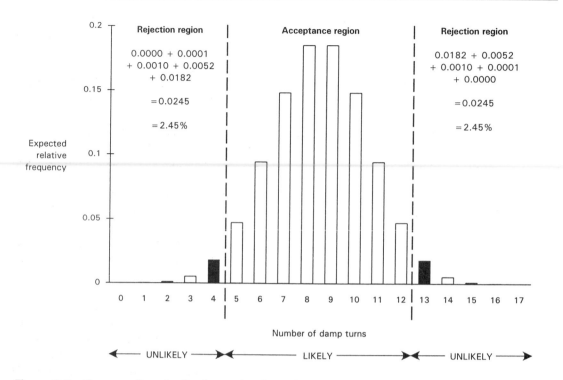

**Figure 2.4** The sampling distribution under the null hypothesis showing acceptance and rejection regions and critical values (solid bars).

The only issue which remains is why the dividing lines between these regions are put where they are. This obviously depends on how big we choose to make the rejection region relative to the acceptance region and this is something which is set by convention. The convention which is most commonly used in biological work is that the rejection region is set at 5%; that is, the rejection region comprises 5% of the total sampling distribution. The size of the rejection region is called the **significance level**, commonly abbreviated to $\alpha$. This abbreviation is usually qualified by adding subscripts and superscripts to distinguish between different sorts of rejection region (we will deal with this in more detail later). In this example it would be written $\alpha_2$, the subscript 2 signifying that the rejection region takes in both tails of the sampling distribution.

If we do make $\alpha_2$ equal to 5%, which values of the test statistic would fall into the rejection region? To find this out, we first need to remember that relative frequencies can be converted to percentages by multiplying by 100. We can then work in from both sides of the sampling distribution in Figure 2.4, adding up the individual relative frequencies for the different values of $Y$ until we obtain a total of 0.05, which is the same as 5%. The actual relative frequencies involved have been taken from Table 2.1 and are written on Figure 2.4. By the time we have got as far as 13:4 on the right-hand side of the distribution, the sum of these relative frequencies is $0.0245 = 2.45\%$. At this point we have included 2.45% of the total distribution. With the equivalent tail at the left-hand side of the distribution this makes a total of 4.9%, which is as close to 5% as we can get in this particular case. We cannot make the rejection region exactly 5% of the total because of the nature of the distribution. If we were to move the dividing lines to the next position, to the left of 12:5 and to the right of 5:12 the size of the rejection region would be 14.34%. In a case like this, we err on the side of caution and set the significance level at less than 5% (we shall see why this is cautious in Chapter 6). In a case like this the significance level which we are aiming for (i.e. 5%) is called the **nominal significance level**; the actual significance level is 4.90%. With this significance level the rejection region includes results of four or fewer damp turns and results of 13 or more damp turns.

Our actual result was 14 damp turns. It lies in the rejection region and is therefore judged to be an unlikely occurrence if humidity does not affect behaviour. Our decision is to reject the null hypothesis and to accept the alternative hypothesis. We can put this back into a more common-sense context. If humidity has no effect then the number of damp turns is too extreme to be reasonably attributable to sampling variation and the result is better explained by humidity having an effect. We conclude that humidity variation does affect the direction in which the woodlice turn; they do have a preference for the damp arm of the maze. Just to repeat the logic of this, because it is an important point, a result in the rejection region is an unlikely result if humidity has no effect (i.e. if the null hypothesis is true), but it is a likely result if the alternative hypothesis is true. So our decision in this case is in favour of the alternative hypothesis.

Note, however, that we cannot be absolutely certain of our decision, because a result of 14:3 is possible (though improbable) if the null hypothesis is true. In our experiment the null hypothesis could have been true but we could have been "unlucky" and by chance ended up with one of those extreme results that are bound to occur from time to time. If this were to have been the case we would not, of course, have been aware of it and we would have rejected the null hypothesis even though it was true. You can probably see that the opposite could also happen. Humidity could affect choice so that the probabilities of the two types of turn were not equal but, by chance, we could happen to have a sample of turns which has similar numbers of the two types. As a result we would accept the null hypothesis even though it was untrue. The possibility of coming to the wrong conclusion in this way is inherent to all statistical tests so they never give us a decision about which we can be absolutely certain. All they enable us to do is to decide in an objective way whether or not the odds favour the treatment having an effect. We will come back to all these issues in Chapter 6 when we have some more background.

## Using critical values

You could be forgiven for thinking that this whole procedure for working out what is included in the rejection regions is very long-winded. In practice the use of statistical tables can be made much easier by using **critical values**. We can think of these as the value(s) of the test statistic (two in this case) which actually "mark" the borderline between the acceptance and the rejection regions. In our example they are 4 and 13 and they are marked on Figure 2.4. If you like, they are the least extreme results which would lead us to reject the null hypothesis and they enable us to formulate an easy to use **decision rule**. The exact form of this rule will vary from test to test but in our example it would be as follows, "If the calculated value of the test statistic is equal to or more extreme than either the upper or the lower critical value then reject the null hypothesis". Our critical values are 4 damp and 13 damp turns and the calculated value of the test statistic is 14. This is more extreme than 13, so we should reject the null hypothesis.

For all the statistical tests we shall be using, critical values of the test statistics have been calculated for us and it is the critical values which are given in the various tables in the Appendix. All we will have to do is to compare our calculated value of the test statistic with the appropriate critical value and use the appropriate decision rule.

Note that once we have set the significance level at, say, 5% this defines what the critical values will be, but this does not mean that the probability (under the null hypothesis) of getting the critical values is 0.05. The 5% refers to the proportion of the distribution relating to the critical value *and* all the more extreme values. This means that the probability of 0.05 is a cumulative probability; it is the probability of occurrence of the critical values *plus* the probabilities of all more extreme values. We shall need to refer to probabilities of this type frequently and it will be too cumbersome to keep saying "the probability under the null hypothesis of getting a value of the test statistic which is as extreme as, or more extreme than, the one obtained is 0.05". A cumulative probability related to a critical value in this way is referred to simply as an **associated probability**. We shall denote this by the symbol **P**.

## Starting from the alternative hypothesis

You may by this stage be wondering why we start from the null hypothesis. After all, it seems that we don't really want it to be true and that our aim is to reject it eventually. Why not start with the alternative hypothesis? We could then ask whether our result was a likely or unlikely occurrence under the alternative hypothesis and consequently make a decision about whether it should be accepted or rejected.

There are two reasons for not doing it this way round. First, there are situations where we may want the null hypothesis to be true (we shall meet some of these later). You might think that even in these situations we could still start from the alternative hypothesis although we would be hoping to reject it rather than accept it. However, there is a second, much more fundamental reason, why we can never start from the alternative hypothesis. To do this we would need to produce the sampling distribution under the alternative hypothesis, that is the frequency distribution of all possible results (assuming the alternative hypothesis were true). When we started from the null hypothesis this was easy because we knew exactly what the probability of a damp turn would be (i.e. 0.5). We then knew that the sampling distribution would be a binomial with $n = 17$ and $p = 0.5$. The problem with starting from the assumption that the alternative hypothesis is true is that we do not know exactly what the probability of a damp turn will be. We know that it will not be 0.5, but further than that we cannot say. This in turn means that we cannot produce the sampling distribution. It would still be a binomial distribution with $n = 17$, but without knowing the value of $p$ we would not know where to look it up in our tables.

Now at this stage you may say to yourself that you do know what the probability of a damp turn is under the alternative hypothesis because you can get it from the experiment. We observed 14 damp turns out of 17 so the proportion of damp turns is 0.823, which means that the probability of a damp turn is 0.823. There are two problems with this line of reasoning. The first is the one that we met in Chapter 1, namely that because of the chance element which accompanies sampling, a sample only gives an estimate of the true value of a variable. Our experiment, which utilized a sample of 17 turns, can only tell us roughly what the probability of a damp turn is and there is no way in which we can ever find the value exactly. The best we could do is to calculate upper and lower limits for our estimate which enable us to quantify its reliability, that is, its closeness to the true but unknown value. This also involves thinking about the probability distribution of all possible results and is something we shall explore in Chapter 7. The second problem is that even if we could find $p$ from our experiment, to use it would involve a circular argument, because we would be testing the result of the experiment against itself.

## The main points summarized

So far so good, I hope. There are three basic ideas to grasp. One is the use of a mathematical model to produce the sampling distribution of all possible values of the test statistic assuming that the null hypothesis is true. This is a distribution of the expected relative frequencies with which all possible values would occur in the long run. The second idea is the way in which we use this to find the probability associated with a particular value occurring in a particular experiment. The third is the idea of using this probability to make a decision about whether a particular result is likely or unlikely. If you have got the gist of this then you are well on the way to understanding what follows. You will be able to polish up on the details because all these ideas crop up again several times later on! Incidentally, all the formulae in the rest of the book are merely mathematical devices for condensing the results obtained from observational studies and experiments into calculated values of test statistics, which can be checked off against sampling distributions.

The flow diagram shown in Figure 2.5 summarizes the relationships between the main points (with some simplifications). Note the close links between the design of the experiment and the choice of a test statistic, both of which are sorted out before the practical work is done! Note, too, the dotted arrows which indicate that ideas about possible answers feed into the experimental design and into the statistics. Also, in practice, thinking about the design and the statistics often highlights deficiencies in the original research hypothesis, which results in modifications being made to it. Sometimes you go round in circles (for a while)!

## Back to the biology

For the biologist, experimental design and statistical analysis are tools to be used to enhance an understanding of a biological issue so we should not lose sight of our original piece of biology. Our simple experiment and statistical test have shown us that we could reasonably reject the null hypothesis and, therefore, conclude that the treatment did affect the variable. That, if you like, is the statistical meaning of the result, but what is its biological meaning?

The experiment has shown us that these particular woodlice, under these particular conditions, have a preference for that particular high humidity, but has that explained why woodlice congregate in certain places? The answer is no. Our result has now raised a whole lot of new questions. Does humidity vary from one place to another and are places where woodlice are common more humid than places where they are rare? Does the behaviour observed in our highly artificial experiment occur under more natural conditions or was

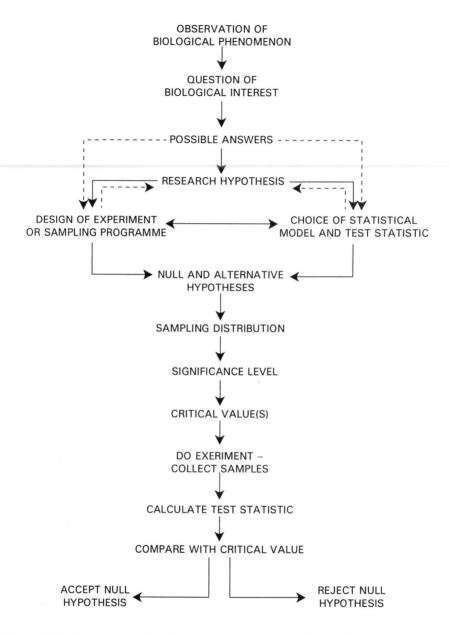

**Figure 2.5** Flow diagram for designing experiments and sampling programmes and statistical testing.

what we observed merely an artefact produced under laboratory conditions? How do wood-lice detect humidity differences? These new questions require further investigation if we are to understand fully our initial observation.

# 3

# Variables, populations
# and samples

The three words in the title have been used frequently in the first two chapters. We know that there is something called a population, although it appears to be a population of "things" rather than a population of organisms. We also know that we take a sample of these "things" and we perhaps have an inkling of why this is done and some of the dangers associated with sampling. Some characteristic of each "thing" in the sample is then described. This characteristic is called a variable for the simple reason that the "things" described are not necessarily the same with respect to the characteristic. Before we can proceed further we need a much clearer understanding of the meaning of these words and of an additional related term.

## Sampling units

To begin with we need to make a distinction between the sample and the units which make up the sample. To do this we will return to the examples of phosphate concentration, types of blood cell, numbers of snails and woodlouse turns. These involve very different sorts of biology, but we can always identify a basic unit which is under investigation – a thing whose properties we wish to describe. This unit is built into the question we are trying to answer. It is the unit on which each observation is to be made and on which the description is to be based. In the first example in Chapter 1 the phosphate concentration was measured in a 10 ml quantity of solution so the unit which is described is a 10 ml quantity of solution. In the blood cell example the unit was a single blood cell because for each cell we had to decide what type it was. In the example of the snails on the shore we counted the number in a quadrat, so the unit is a quadrat. We shall call this unit a **sampling unit**. If we were interested in the weight of individual plants then the sampling unit would be an individual plant; if it was the number of visits made by pollinators in a time interval then the time interval would be the sampling unit. In the woodlouse example, the unit was a turn and even though this was an experiment, we shall still call it a sampling unit. The reason for this, alluded to earlier, is that experiments involve sampling, a point we shall return to later in this chapter.

In practice we are likely to describe several such units which together will constitute the sample, that is we repeat our observations. For this reason the sampling units are sometimes called **replicates**. The **sample size** is equal to the number of sampling units or replicates in the sample and is usually denoted by $n$. For example, in the woodlouse experiment the sampling unit was a turn and the sample was made up of 17 of them, so the sample size was

17 (i.e. $n = 17$). In the phosphate example, the concentration was measured in nine separate volumes of 10 ml, so $n = 9$.

We need to have this term because it enables us to make an important distinction between two aspects of samples which are often confused: the size of the sample $n$; and the number of samples, denoted by $k$. In the snail example, where we were comparing numbers in two different parts of the shore, we took a sample of 10 quadrats from each site. The number of samples was therefore 2, so $k = 2$, and the size of each sample was 10. In experimental work the number of samples ($k$) will be the same as the number of different conditions (i.e. treatments) being investigated. For example, the field experiment shown in Plate 5 had nine treatments, so the number of samples was 9. The size of each sample, that is the number of areas in each treatment, was 6. We shall be dealing with this distinction again later in the chapter.

# Variables

The variable is the property of the sampling unit which we describe. Using our previous examples the variables are phosphate concentration, type of blood cell, number of snails per quadrat and direction of turn.

Although there are obviously a huge number of variables, you need to realize that they can be classified into just five different types which have different basic properties. This classification is important because, amongst other things, the properties of a variable determine the sorts of statistical procedures which can be applied to it. The reason is that (as we saw in Chapter 2) the statistical test we use is always based on a mathematical model of the biological situation, that is, it involves assumptions about the nature of the variable. Applying a statistical test when the variable does not meet the assumptions can lead us to the wrong conclusion, so we have to ensure that the test is appropriate for the type of variable. The five types of variable are **nominal**, **ordinal**, **discontinuous**, **continuous** and **derived**.

## Nominal variables

This sort of variable is perhaps the simplest and quickest way of describing something and it is illustrated by the woodlouse turning example discussed in Chapter 2. The variable was the direction of turn which we could describe as being either towards the damp or towards the dry. The variable existed in a small number of clearly defined and separate alternative states (in this case only two) and the difference could best be expressed by putting the sample units (turns) into named categories (damp and dry). Hence variables of this type are called nominal (or category) variables and they are very commonly encountered in biology. Other examples occur in genetics where we classify individual organisms as male or female, or as normal winged or vestigial winged. Likewise in ecological work we might describe individuals as parasitized or non parasitized, or as alive or dead.

This type of variable is not restricted to just two categories – we can have any number. In genetics we might be able to distinguish four phenotypes, for instance, blood group A, B, AB and O, in the progeny of a cross or in a population. Amongst the frog blood cells shown in Plate 3 we can identify several different types. In a behavioural study we might be able to describe what individuals were doing by using a number of different categories of behaviour (e.g. sleeping, feeding, preening, or mating). In an ecological survey in which we are identifying species, each sampling unit (= individual organism) is assigned to a taxonomic category, so there will be as many categories as there are species.

You can see why this type of variable is easy to use. We do not have to measure or count any property of the sample units, we just classify them on the basis of their attributes, so that the difference between them is qualitative rather than quantitative. However, you can see that it also has limitations. The examples used involve properties that fall naturally into clear-cut categories with no intermediates, but this will not always be the case. Mendel was able to put his pea plants into tall and dwarf categories, but heights of individuals often show a complete range of values. It is also a rather crude form of description. For example, it allows no distinction to be made between a lightly parasitized and a heavily parasitized individual, which both go in the "parasitized" category. Note that we may choose for convenience to label the categories with numbers rather than names, but the values used have no numerical meaning. If we label male and female categories as 1 and 2, respectively, it would not mean that females had twice as much of some characteristic as males.

When, however, there are quantitative differences between the sampling units we can describe them by assigning a number to each one which conveys information about the relative magnitude of the variable. Often this can be done by counting or measuring some property, but in some cases this is impractical and we find it useful to adopt a different approach.

## Ordinal variables

This sort of variable is used in situations where there are obvious quantitative differences between sampling units but the variable cannot be counted or measured. For example we cannot measure or count how unpleasant some chemical compound smells. What we can do is to give a panel of judges a range of compounds and ask them to put them in order of unpleasantness. The position in the order of the particular compound of interest can then be described by a number which we call a **rank**. These sorts of variable are common in behavioural and psychological studies. We might want to know whether biology students and arts students had different attitudes to an issue such as genetic engineering or the use of animals in research. There is no way we can measure the importance that a student attaches to, say, the moral dimension of the issue, all we could ask them to do is to put a number of points in order of importance.

Other variables such as the level of anxiety or aggression of individuals could perhaps be measured in some way, but it might be easier to rank the individuals concerned. Even a variable such as germination time is not very easy to measure, because it would be required that we could identify a point in a continuous process that enabled us to say that a seed had germinated. We would then need to watch the seeds continuously which might be inconvenient. It would be much easier to note the order in which the seeds germinated, and this could even be done retrospectively if we had missed a couple of days over the weekend. A slower germinating variety will tend to come later in the order and have larger rank values. Skin pigmentation in our own species is another case where measuring is technically quite difficult, but where it would be easy to order the individuals from fair-skinned to dark-skinned. The same is true of the degree of purple pigmentation shown by the leaves in Plate 1, which is a complex variable involving the number of spots, their size and their level of pigmentation.

In essence, ranking the sampling units to get an ordinal variable, rather than making measurements or counts, is equivalent to using the finishing order in a race, rather than the actual finishing times. This type of variable has advantages and disadvantages. In some instances it is the only way of describing something, in others it is quicker than counting or measuring and does not require sophisticated equipment. It also allows the use of statistical tests that are quite easy to understand and use and which often have rather few underlying assumptions! But it does have drawbacks. First it is subjective, so that there is a danger that our preconceptions may influence the results. Secondly, although we put a numerical value to each sampling unit, we need to be careful about what the numbers mean and how we use them. If we had actually measured the germination time in days, then we could say that a value of 5 (days) means that germination is twice as fast as a value of 10 (days). We cannot say this when the numbers relate to the order of germination; the seed ranked fifth could have germinated on day 7, the one ranked tenth could have germinated on day 39. Neither can we relate the results of this experiment to those of some separate experiment because the ordering is internal to a particular experiment and is not related to any absolute scale.

Finally, because the method is relatively crude, we may have difficulty in distinguishing between sampling units, i.e. in deciding whether one should come before or after another in the order. If two or more sampling units are indistinguishable in this way, then they are said to be **tied** and are given the same rank. Where this problem is particularly acute it may be easier to assign each observation to one of a small number of ordered categories. You may have used this sort of variable if you have assessed the abundance of a plant species, such as a grass, in a quadrat. Although it is possible to count individual plants there are often practical problems concerned with (a) deciding what constitutes an individual and (b) the laborious nature of counting large numbers of stems or shoots. So, rather than counting or measuring abundance in each sampling unit (quadrat), you can assign each quadrat to a category which reflects how common the species is. One such scheme is the ACFOR system, which employs five categories (abundant, common, frequent, occasional and rare); other schemes employ more categories. The quantitative relationship between the categories is clear and could be represented numerically by labelling the categories 5, 4, 3, 2 and 1. Note that this only tells us that a quadrat ranked 4 has more plants than one ranked 2. It does not mean that it has twice as many.

## Discontinuous variables

Numerical values can also be assigned to sampling units when we can count some property of each unit. We can count the number of species in a pond, the number of snails in a quadrat, the number of seeds in a pod, the number of colonies of micro-organisms on a sterile plate, or the number of yeast cells in a known volume of culture medium. Each sampling unit can be described by a whole number. This type of variable is called a discontinuous, or discrete, variable because it can only take integer values (a quadrat cannot have 1.5 snails). With this sort of variable the numbers used convey more than just the order; they also tell us the size of the difference. A quadrat with four snails has twice as many as a quadrat with two snails.

## Continuous variables

As the name suggests these variables can take any value (within certain limits). Typical examples of this type of variable would be measurements of environmental variables such as temperature and pH, dimensions such as body length and beak width and physiological/ biochemical characteristics such as oxygen consumption, blood sugar concentration and protein content. Each sampling unit is again described by a number but, unlike a discontinuous variable, it is not restricted to whole-number (integer) values. It can take any value between certain limits; phosphate concentration can be recorded as 1.21 mg/ml, body temperature can be recorded as 36.8°C and the length of a dog-whelk shell can be measured as 2.87 cm.

Because variables like this can have any value, we need to realize that when we record a temperature as 36.8°C we are not recording the temperature exactly. Although it might be 36.8000000000 . . . °C, it could also be 36.81690 . . . °C or 36.800024 . . . °C. Any measuring device has limits to its **precision**, where precision is defined as the limits between which we know that an actual measurement lies. The smallest division on the scale of a clinical thermometer corresponds to 0.1°C, so with this we can measure temperature to the nearest 0.1°C. This is more precise than an ordinary thermometer used for recording room temperature in which the smallest scale division is 1°C. An electronic thermometer might read to the nearest 0.01°C.

The degree of precision associated with a measurement is conveyed in a standard way. In recording body temperature as 36.8°C the position of the last digit tells us the precision. The last digit (8) is in the first position after the decimal point so it refers to "tenths of a degree". This indicates that the measurement is precise to within 0.1°C and implies that the true value lies somewhere within an interval of 0.1°C, which extends equal amounts on either side of the recorded value. So recording the temperature as 36.8°C implies that the true value lies somewhere between 36.750000 . . . °C and 36.84999 . . . °C. These are the **implied limits**. If we had measured it with an electronic thermometer and recorded the temperature as 36.85°C, this implies that the true value lies somewhere between limits of 36.8450 . . . °C and 36.8549 . . . °C. Our measurement is now more precise.

In theory, with a continuous variable, if we measured with sufficient precision, every observation would be different. In practice, we find that some observations have the same value because of the limitations of our measuring device. This raises the question about how precise we ought to be when measuring something, and a useful rule of thumb is given by Sokal & Rohlf (1981). They suggest that the number of unit steps between the smallest and the largest measurement should be between 30 and 300. So in the case of the dog-whelk shells, if the smallest shell was 1.90 cm and the largest 3.30 cm, then recording length to the nearest 0.1 cm is not satisfactory because it would give fewer than 30 unit steps. On the other hand, recording to the nearest 0.01 cm is acceptable because it gives 140 unit steps. There is a simple reason for employing this rule. If our measurements are sufficiently imprecise then all sampling units will be recorded as having the same value, even if they are different and we would not be able to detect any differences between samples. On the other hand, excessive precision costs time and money. The 30–300 rule is a reasonable compromise.

Precision is one factor that affects the **accuracy** of a measurement, defined as its closeness to the true value. The other factor is **bias**. If the thermometer has been calibrated incorrectly, so that for example the whole of the scale is too far down the stem, then all the temperatures will be recorded as too high, even though they are quite precise. Similarly, you have to calibrate a spectrophotometer by making up a standard solution with a known concentration. If you do this incorrectly then even the most precise machine will give inaccurate results. Because of imprecision and bias, measurements of a continuous variable are always inaccurate to a certain extent.

### Derived variables

Sometimes the original variables are not in the most convenient form and need to be converted into derived variables which are more easily understood and compared. Examples are **ratios**, **proportions**, **percentages** and **indices**.

A ratio expresses the relationship between two variables. The ratio of damp turns to dry turns is 14:3 and we could describe the two phenotypes in a genetic cross as being in the ratio of 45:12. It is usually more convenient to describe a ratio as a proportion or a percentage in which one of the numbers is related to the total. The total number of turns is 17, so the proportion of damp turns is $14/17 = 0.823$. We could express tail length as a proportion of body length and gain in weight as a percentage of the initial weight. A percentage is the number out of 100, obtained by multiplying the proportion by 100.

An index is a single number that combines information on several variables. For example, physical fitness can be measured by an index of physical efficiency which combines information on the duration of a bout of standardized physical activity and the pulse rate at defined time intervals after it. A diversity index incorporates information on the relative abundance of several species.

### Some general points about variables

These different types of variables are interrelated and often interconvertible. For example, we could measure body height, rank individuals with respect to body height or put them into categories such as short, medium and tall. While describing something using a nominal or an ordinal variable is quick and cheap it conveys less information than counting or measuring. As a general rule, the more information we have on each sampling unit, the more detailed our description will be and the easier it will be to detect differences. So, wherever possible, we should collect our data by counting or measuring. The observations can always be converted later to ranks or categories, but the reverse process is not possible.

The act of describing each sampling unit in terms of one of the types of variable yields what is commonly called an **observation**. It is also called a **datum**. Typically, a scientific investigation will involve one or more samples, each of several sampling units, so we will have several observations or **data** (in the plural) at our disposal. A more technical word for each observation is a **variate**, a term which allows us to make a distinction between **univariate** and **bivariate** data. If the sampling units are individual plants and we describe them in terms of one variable (e.g. height) then the data are univariate; if we also count the number of flowers on each plant then the data are bivariate.

## Populations

We have seen that biologists categorize, rank, count or measure some property of each sampling unit. Sampling implies the use of a small part of something as representative of the whole and in this case the whole is the population, a term which needs further explanation. This is particularly necessary for us as biologists because we already have a clear definition of the word population. For this reason there is scope for considerable confusion.

### Ecological and statistical populations

To biologists a population is an ecological entity, a group of individual organisms of a given species, living in a particular area or habitat at a particular time. This definition, although simple enough in theory, is usually difficult to apply in practice because it can be hard to tell where the boundaries between populations lie.

In the statistical sense the term has a different meaning, which may relate to the ecological definition but more often does not. This is because a statistical population is to be thought of either as a population of sampling units or, even better, as a population of variates. If the sampling units happen to be the individual organisms of a biological population then the statistical population could be thought of as equivalent to the ecological population in some cases (see below). In other situations a single ecological population can give rise to several statistical populations, while several ecological populations can produce a single statistical population. A statistical population need not be anything to do with an ecological one. A statistical population may actually exist, whereas in other cases it may be imaginary or hypothetical! The easiest way to illustrate these points is to use some examples, which will also throw some more light on the meaning of sampling units.

The snake's-head fritillary *Fritillaria meleagris* was once a widespread plant in damp meadows in England but, because of changes in agricultural practices and the destruction of its habitat, it is now confined to only a few localities. As a result, the ecological populations are very clearly defined and separated from one another so in this case we have no problem in defining the ecological population. However, the statistical population involved depends on the question being asked (Box 3.1). If our question is "What proportion of the plants in the ecological population are flowering?", then our sampling unit is an individual plant and the statistical population is all the plants in the ecological population. If we want to describe the number of seeds produced by individual plants, then the sampling unit is an individual plant and the statistical population comprises all the plants which could produce seed or,

---

**BOX 3.1 POPULATIONS AND SAMPLING UNITS**

| Question | Sampling unit | Statistical population |
|---|---|---|
| What proportion of the plants are flowering? | An individual plant | All the plants in the ecological population |
| How many seeds per flower? | An individual plant in flower | All the plants in flower |
| How many seeds per white-flowered plant? | An individual white-flowered plant | All the white-flowered plants in flower |
| How many plants/m² in the field? | An area of 1 m² | All the areas of 1 m² in in the field |
| How long are the stamens? | A stamen | All the stamens |
| How much time do bees spend on a visit to a flower? | A visit by a bee to a flower | All the visits made by bees to flowers |
| How many bees visit in a 5-minute observation period? | A 5-minute observation period | All the 5-minute observation periods which could be made |

more strictly, the collection of values which we would get if we counted the seeds on all the plants. Some of the plants have white flowers and others have purple flowers. If we want to know whether flower colour affects numbers of seeds there are now two populations: all the white flowered plants and all the purple flowered plants! These and some other examples are set out in Box 3.1. If we are describing a community by counting the numbers of several different invertebrate species in quadrats then there is one population, a population of quadrats, though there are several biological populations involved. Similarly, if we want to know how many species there are in ponds of a certain size, the sampling unit is a pond and the population is a population of ponds.

## Statistical populations in the laboratory

Things get more complicated in a laboratory situation because populations under these circumstances are hypothetical, that is, they are populations of observations that could occur. In our woodlouse turning experiment, the sample of 17 sampling units is a small subset of all the possible turns that could have been made by woodlice under these conditions. The population is all the possible turns we could have looked at. If we examine the germination time of seeds, then the population is all the times which we could get, if we looked at all seeds of this variety treated in this way. Similarly when you do a genetic cross and determine the phenotype of each offspring, the sampling unit is an individual but the population is all the possible individuals that could have come from a cross of that type, which is clearly a very large number. As a final example, when we measure a concentration our sample might consist of three volumes of 10 ml taken from the unknown solution. The sampling unit is a volume of 10 ml and the population comprises all the values of the concentration that could be obtained. Given that the actual concentration in every 10 ml will be the same if the solution is properly mixed this imaginary population comprises all the different values that could arise because of the sorts of errors we talked about in Chapter 1. This example highlights the fact that a statistical population is really a population of variates.

Populations have been dealt with at some length because of the particular confusion which can arise between these two meanings of the word. We shall be using the word population a great deal in the following chapters and, unless stated otherwise, it will always refer to a statistical population. Statements such as "the samples have been taken from the same population" and "the populations are different" are to be interpreted in a statistical rather than an ecological context.

# Samples

A population comprises all the sampling units which are actually or potentially available for description. A sample is simply the subset of sampling units which we do describe. It is a collection of sampling units which are obtained and described in a specified way or, alternatively, the collection of data which is produced. In this section we look at the essential features of sampling.

## Why use samples?

It should now be obvious why we have to use samples even though we are interested in populations. Clearly in some cases the population is exceedingly large, indeed it may be infinite, and under these conditions we could not make every observation even if we wanted to. Where the population is of a more manageable size there are still good practical reasons for taking samples. We probably could count the flowers on all of the plants in the field of fritillaries but it would take a lot of time and therefore money in terms of ecologists wages. We will see later that the cost of this increase in sampling effort will not be repaid because a law of diminishing returns operates. After a certain point, taking larger and larger samples only yields small increases in our knowledge of the population so that it is not cost effective.

A further reason for sampling is that, at least in an ecological context, sampling is potentially damaging to the system under investigation. To examine every sampling unit in our example, the ecologist would have to walk over the whole area. This would lead to trampling of the vegetation, possible damage to the plants under investigation and alteration to their environment. This would clearly be undesirable in terms of conservation, but even if this is not an issue, the unnatural disturbance that could accompany intensive sampling could lead to changes in the variable being studied. This type of effect will also be a possibility if material has to be removed from the site for laboratory analysis.

## Replication

This term simply means that a sample should normally consist of more than one sampling unit, that is, we should have more than one replicate. The reason for this is that there are four different questions which we might want to ask. To answer them requires a knowledge of the amount of variability within the sample and replication provides this information.

(a) One question which is interesting in its own right is "How variable is the variable?". We can only assess this by looking at the variation within a sample (i.e. the variation between sampling units). Clearly the only way to do this is to take more than one sampling unit.

(b) Alternatively, our question might be "What is a typical or average value of the variable in the population?". Now a single observation might provide an answer that is reasonably close to the population average but of course it could, by chance, be atypical. Increasing the number of observations will have the effect of cancelling out the effect of sampling variation and will lead to a sample that is more representative. The average value of a sample of several observations is likely to be nearer to the average value of the population. Thus replication is needed here as well. However, replication is only needed if the population is variable; in the absence of variability one sampling unit will suffice. Your blood is so well mixed that there is no appreciable variation in, say, the number of white blood cells per unit volume at any one time. This is why the doctor can get away with a sample of a single volume.

(c) Another question might be "How reliable is an estimate, such as the sample average?". The reliability tells us how close the estimate is likely to be to the population average. We shall see in Chapter 7 that we can only answer this question if we know something about the variation within a sample (that is the variation between sampling units).

49

(d) Finally, we might want to know whether there is a difference between two (or more) populations and to do this we need to carry out a statistical test on our samples. As we shall see in Chapters 9 to 11 a knowledge of variation within samples turns out to be an essential piece of information for this purpose. Our ability to tell whether or not there is a difference between two populations depends not just on what we know about the differences between the two samples, but also on the variation within each sample. Replication is the only way in which we can get this information.

We shall see all these issues illustrated in later chapters; the point to take on board at this stage is that it is essential to have several replicates. This has practical implications. If sampling actually involves physically collecting material, you must keep the material from different sampling units separate. Thus, suppose you are sampling invertebrates from a stream and you take four replicates, then you must not put them all together in the same bucket. If you do, then you lose all the information on variability between replicates. (Incidentally it cannot be recreated by dividing the contents of the bucket into four, because this will not mirror the natural variability found in the stream.) The same is true for the data obtained from each sampling unit which must be recorded separately as well.

Replication raises a further question: "How many replicates should be taken?", or "How big should the sample be?" Unfortunately there is no easy answer. The bigger the better might be one maxim, but we have already seen how constraints such as time, money and disturbance may make this impossible. The answer also depends on how variable the populations are, how reliable we want our estimate to be, how big the true difference is between the two populations and how sure we want to be of detecting the difference! Another issue concerns the statistical test(s) which we may want to use. Although they can all deal with large sample sizes, some of them cannot be used with samples below a minimum size. Furthermore, tests involving larger sample sizes (of, say, 50 or more) may be based on fewer assumptions and thus have wider applicability.

Many of these points will be illustrated in later chapters – the important point to note here is that deciding on the sample size is an important element in the design of an experiment or an observational study. Note that this is concerned with the number of sampling units, not the number of samples. The number of samples taken is a reflection of the number of treatments under investigation. Having decided on the number of replicates we now have to choose which ones to include in our sample.

## Choosing which sample units to take

Practical considerations mean that we have to take samples, however our real interest is in the population from which the samples are taken. It follows from this that the sample needs to be representative of the population or, if you like, the sampling units chosen should be a typical "cross-section" of all of the units available. This is true both when sampling in an ecological context and when allocating material to experimental treatments. So how do we decide which sampling units to choose?

We could simply try to make them a reasonably representative collection. The danger of this approach was illustrated in Chapter 1 in the brief discussion on how to sample areas on the shore. However hard we try to be fair, all sorts of subjective factors may affect our choice. For example, areas may differ in their accessibility, in the ease with which they can be worked and in their attractiveness and, as a result, we may have a tendency to over-sample some areas and undersample others. We may be aware of some of these factors and try to compensate for them, but it is impossible to do this satisfactorily. Anyway, in many cases we will not be aware of the subtle ways in which our behaviour is being influenced. We cannot hope to get a representative sample in this way.

One possible way round these sorts of problems is to take sampling units at regular intervals. We could imagine laying a grid of squares over the area to be sampled and counting the snails in every fifth (or ninth, or whatever) square. Since the position of each sampling unit is identified in advance without our seeing it, this would not be subjective. It would also provide a good even coverage of the area.

I mentioned this as a possible way of organizing the sequence of trials in the woodlouse experiment, that is, by regularly alternating the damp arm of the maze from the left to the right. However, there is a general problem with this method. Some aspect of the natural or experimental environment or the experimental material or the experimenter could also change in a regular way, either in space or in time. If it does, then there is the possibility that our regularly spaced sampling units will coincide with particular environmental conditions. If the illumination in the laboratory fluctuates regularly because of regular changes in voltage, you should be able to produce a scenario in which alternating the damp alley of the maze from left to right leads to the damp alley being on average darker. Not very likely, but conceivable and therefore best avoided.

The only way to overcome these problems and to obtain a representative sample is to use **random sampling** which simply means that every sampling unit has an equal chance of being included in the sample. You can think of this as a raffle, in which each sampling unit is individually identified by a numbered ticket. All the tickets are then put into a hat and thoroughly mixed up. Then, a few of the tickets are withdrawn without any knowledge of what number they carry. You can in fact use the hat-and-ticket routine if the total number of sampling units is not too large, but there is another way, using **random numbers**, which is usually more convenient.

A set of random numbers is a set of digits produced in such a way that each digit is equally likely to be 0, 1, 2, . . . ., 9. These can be presented in tabular form as in Table A1, but they can also be produced by most calculators. In either form any given digit (e.g. 0) is just as likely to follow (or be preceded by) any digit (including itself). What this means is that any sequence of digits taken from this table by reading in any direction (up, down, left, right, diagonally) is a random sequence. The same should be true of a sequence from your calculator. If you want to check this, you can count up how many times each digit occurs and also how many times a given digit is followed by the same or any of the other 9 digits.

These numbers can be used to choose sampling units randomly from a population or to assign them randomly to an experiment. The easiest way to see the principle of using these numbers is to consider some examples.

## Choosing sampling units at random

Suppose you want to describe the number of plants of a particular species per unit area ($0.25\,m^2$) along the centre of a 50 m stretch of footpath. The sampling unit is a $0.5\,m \times 0.5\,m$ quadrat and we can imagine the population as a linear array of 100 sampling units each $0.5\,m \times 0.5\,m$. Having decided that a sample size of 10 ($n = 10$) such units will suffice,

**BOX 3.2  USING RANDOM NUMBER TABLES TO CHOOSE SAMPLING UNITS**

**(a)  Part of Table A1**

| 1 | 75523 | 70787 | 04873 | 98093 | 57054 | 26622 | 88599 | 26925 | 19605 | 11720 |
| 2 | 23399 | 42116 | 44736 | 21504 | 92732 | 96640 | 56980 | 59465 | 54342 | 02903 |
| 3 | 34142 | 65196 | 53174 | 75802 | 01310 | 56580 | 16431 | 79507 | 44423 | 05190 |
| 4 | 42872 | 62003 | 72242 | 30428 | 57935 | 06491 | 96314 | 29103 | 59548 | 86793 |
| 5 | 44614 | 58545 | 26468 | 33215 | 25384 | 16767 | 18290 | 07341 | 11075 | 52477 |

**(b)  Choosing 10 quadrats at random along a footpath**

*Line 1*    75   52   37   07   87   04   ~~87~~   39   80   93   57

*Random positions*

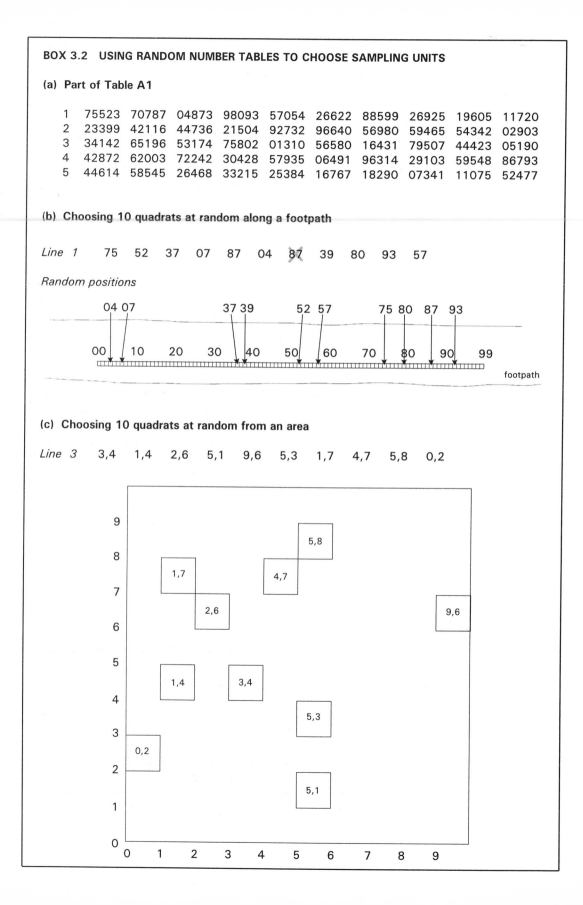

**(c)  Choosing 10 quadrats at random from an area**

*Line 3*    3,4    1,4    2,6    5,1    9,6    5,3    1,7    4,7    5,8    0,2

how can these be chosen at random? We first think of these 100 sampling units being labelled in sequence from one end to the other with numbers from 00, 01, 02, . . ., 99. If we then read off pairs of random digits along the top line of the table (as illustrated in Box 3.2a,b) the first 10 pairs will identify 10 units at random. Their position "on the ground" can be found using a tape measure. If a pair of digits which has already been used occurs again it is ignored, as has happened for the pair 87.

You may be wondering why the first label used is 00 and not 01. The answer is simple. Each label has to have an equal chance of being chosen, because this is the essence of random sampling and this will only be true if each label has the same number of digits. If we start from 01, then the last sampling unit has a three-digit label (100). Because there are more possible three-digit labels than there are two-digit labels, the chance of picking this (last) sampling unit will not be the same as that for the others.

If we wanted to choose another 10 sampling units in the same study, perhaps arranged along the edge of the footpath, then we should not re-use the same random numbers. We should carry on along the same row from where we left off. At the end of a row, we can either go to the start of the next row, or read backwards along it from the right-hand end. The same procedure can be used for other sizes of population and sample. Suppose that we wanted to select a sample of 20 from a population of 72 you would simply read off the first 20 pairs of digits which were between 00 and 71.

If we are sampling in two dimensions, as in the mollusc example in Chapter 1, then we can identify each potential sampling unit using a numerical "grid reference". The sample units can then be selected by taking a random sample of grid references. In this example the population consisted of 100 sampling units $0.5\,m \times 0.5\,m$, arranged in a $5\,m \times 5\,m$ square (Box 3.2c). With the bottom left-hand corner of the square labelled as (0, 0), each quadrat can be identified by a two-figure reference for its bottom left-hand corner. Reading off pairs of digits horizontally along the third line in the table, the first digit gives the co-ordinate on the $x$ axis and the second that for the $y$ axis. These are the random quadrats marked on the example in Figure 3.1a. The next 20 digits in line 3 give the positions in Figure 3.1b.

The two diagrams in Box 3.2 illustrate one of the features of random sampling, namely the uneven coverage of the population. By chance, on the footpath there are no quadrats between 10 and 30 and there are three between 70 and 90. On the shore there are rather few quadrats on the right-hand side of the site.

## Allocating units at random in experiments

Suppose that we had 87 woodlice and we wanted to choose 20 at random for our experiment. To do this we would have to give each animal a number between 00 and 86. The first 20 pairs of digits from the table would identify the experimental animals. To randomize the order of the woodlouse experiment we note that there are two alternatives: the damp alley can be to the left or to the right. We can simply denote the first by an even random number and the second by an odd random number (or vice versa). Line 2 of Table A1 gives the sequence of random numbers shown in Box 3.3b, meaning that in the first trial the damp turn should be to the left, in the second to the right, in the third to the right, etc. We do not need to find a number for the last turn; it must be to the left.

We can use a similar idea to assign experimental subjects to two treatments. Suppose we want to look at the effect of the amount of liquid drunk on the concentration of urea in the

**BOX 3.3   USING RANDOM NUMBERS TO ALLOCATE EXPERIMENTAL MATERIAL TO TREATMENTS**

**(a)  Part of Table A1**

| | | | | | | | | | |
|---|---|---|---|---|---|---|---|---|---|
| 1 75523 | 70787 | 04873 | 98093 | 57054 | 26622 | 88599 | 26925 | 19605 | 11720 |
| 2 23399 | 42116 | 44736 | 21504 | 92732 | 96640 | 56980 | 59465 | 54342 | 02903 |
| 3 34142 | 65196 | 53174 | 75802 | 01310 | 56580 | 16431 | 79507 | 44423 | 05190 |
| 4 42872 | 62003 | 72242 | 30428 | 57935 | 06491 | 96314 | 29103 | 59548 | 86793 |
| 5 44614 | 58545 | 26468 | 33215 | 25384 | 16767 | 18290 | 07341 | 11075 | 52477 |

**(b)  Allocating two treatments in a random sequence (using row 2)**

| | 2 | 3 | 3 | 9 | 9 | 4 | 2 | 1 | 1 | 6 | 4 | 4 | 7 | 3 | 6 | 2 | 1 | 5 | 0 | – |
|---|---|---|---|---|---|---|---|---|---|---|---|---|---|---|---|---|---|---|---|---|
| Damp turn on | L | R | R | R | R | L | L | R | R | L | L | L | R | R | L | L | R | R | L | L |

**(c)  Allocating 24 students at random to two different treatments (using row 3)**
**(E, experiment; C, control)**

| Michael Allen | 3 | E | Lucy Heath | 9 | E | Sylvia Parnell | 5 | E |
|---|---|---|---|---|---|---|---|---|
| Nicola Bevan | 4 | C | Sarah Irvine | 6 | C | Suzanne Perkins | 8 | C |
| Ruth Bollons | 1 | E | Paul Jones | 5 | E | Wendy Price | 0 | C |
| Andrew Brown | 4 | C | Susan Little | 3 | E | Clive Reed | 2 | C |
| Sam Cooper | 2 | C | Elizabeth Moore | 1 | E | Graham Simms | 0 | C |
| Richard Dixon | 6 | C | Alistair Norton | 7 | E | Mandy Smith | 1 | E |
| Anne Eaton | 5 | E | Brian Ogilvie | 4 | C | Paul Taylor | 3 | C |
| David Green | 1 | E | William Parker | 7 | E | Peter Williams | 4 | C |

**(d)  Allocating four treatments (A, B, C, D), each of five replicates, in a random sequence**
**(using rows 4 and 5)**

*Row 4*

| Digits | 42 | 87 | 26 | 20 | 03 | 72 | 24 | 23 | 04 | 28 | 57 | 93 | 50 | 64 | 91 | 96 | 31 | 42 | 91 | 03 |
|---|---|---|---|---|---|---|---|---|---|---|---|---|---|---|---|---|---|---|---|---|
| Replicates | A | A | A | A | A | B | B | B | B | B | C | C | C | C | | | | | | |
| Labels | 2 | 7 | 6 | 0 | 3 | 12 | 4 | 3 | 4 | 8 | 17 | 13 | 10 | 4 | 11 | 16 | 11 | 2 | 11 | 3 |

*Row 5*

| Digits | 59 | 54 | 88 | 67 | 93 | 44 | 61 | 45 | 85 | 45 | 26 | 46 | ... | 38 | ... | 29 |
|---|---|---|---|---|---|---|---|---|---|---|---|---|---|---|---|---|
| Replicates | C | D | | | | D | D | | | | | | | D | | D |
| Labels | 19 | 14 | 8 | 7 | 18 | 4 | 1 | 5 | 5 | | | | | 18 | | 9 |

Sequence in which measurements are made

| 0 | 1 | 2 | 3 | 4 | 5 | 6 | 7 | 8 | 9 | 10 | 11 | 12 | 13 | 14 | 15 | 16 | 17 | 18 | 19 |
|---|---|---|---|---|---|---|---|---|---|---|---|---|---|---|---|---|---|---|---|
| A | D | A | A | B | D | A | A | B | D | C | C | B | B | C | D | C | B | D | C |

urine of students and we have 24 students available. How can we allocate them at random to two treatments (an experiment and a control) so that there are equal numbers in each? All we need to do is to list the names (in any order, alphabetical will do) and then assign to each one in turn a random number taken from the tables. Because there are only two treatments we could simply note whether the digit is odd or even, having decided in advance which treatment the odd- and even-numbered sets would be allocated to. Reading along the third row of the table would give an allocation as shown in Box 3.3c. Once we have allocated the 12 subjects to one treatment, the remaining ones are automatically allocated to the other treatment. In this case the last two subjects are allocated to the control.

If there were three treatments then we would proceed as follows. A subject is allocated to the first treatment if it has a random number between 0 and 2, to the second treatment if its number is between 3 and 5 and to the third treatment if its number is between 6 and 8. The digit 9 is redundant in this case.

Randomization, particularly in experimental work, can involve more than just the allocation of material to treatments; in fact it should be used at any subsequent stage at which factors other than those under investigation could cause unwanted variation. In a biochemical experiment in which we are measuring enzyme activity at four different pH values (A, B, C, and D), the measurements are subject to variability due to ageing of solutions, changes in technique and changes in the machine. If we have five replicates in each treatment, it would be a mistake to do the measurements in the order AAAAABBBBBCCCCCDDDDD, because it introduces another factor, time. The five measurements on treatment D are all made much later than say those on treatment A. If we find a difference in activity between A and D we cannot tell whether it is due to the difference in pH which is what we are interested in. It could be caused by the ageing of one of the reagents. We should therefore randomize the order in which observations are made. This could be done by assigning the numbers 00 to 19 at random to the 20 replicates and then doing the measurements in the order specified by the numbers.

There is a slight practical problem with using random numbers in a case like this. We have to use two-digit labels, but we only need the twenty between 00 and 19. A lot of the two-digit labels turned up by the tables will be greater than 19 and will have to be discarded, making the whole process rather inefficient. There is a simple way round this. Any two-digit numbers between 20 and 39 which turn up can be applied to labels between 00 and 19 by merely subtracting 20; for digits between 40 and 59, subtract 40; between 60 and 79, subtract 60, and between 80 and 99 subtract 80. In this way you can use all the pairs. The third example in Box 3.3d illustrates this in detail, for example random number 42 becomes 02. Again any label which is repeated is discarded.

The final example shows how to allocate replicates at random to positions in, say, a greenhouse. In an experiment on the possible effects of a period of low temperature on germination, two samples of seeds could be allocated at random to the two treatments. One sample of six replicates is kept at 5°C for 2 weeks while the other is kept at room temperature. Then both sets are placed on damp filter paper in petri dishes in the greenhouse. Because light and temperature almost certainly affect germination and these factors vary in the greenhouse, it would be unwise to put all the replicates of one treatment in a similar position, so we randomize their positions.

We first number the positions from 00 to 11 and then work through the 12 dishes in a systematic way giving each one a random number between 00 and 11 which specifies what position it should occupy. Again to make the most efficient use of the table of numbers, we use a procedure similar to that described in the last example. There are 12 positions to fill, so we can afford to identify each one by eight two-digit labels. This will mean we can draw on 96 (= 8 × 12) of the 100 two-digit labels available. Digits 00–07 identify position 1, 08–15 identify position 2, and so on. Position 12 is coded for by digits 88–95.

## The need for randomization

This type of random sampling is called **simple random sampling** and carrying out experiments using this sort of random allocation is called a **completely randomized design**. The experiments illustrated in Plates 5 and 6 are of this type. It represents the simplest form of randomization, but it is not necessarily the most efficient. We shall discuss this briefly later in the chapter.

Adequate randomization is absolutely essential for any sampling programme or experimental design. There are two reasons for this. First, randomization ensures that the sample or the allocation is not biased. As a result, the description based on the sample will be an unbiased description of the population. Secondly, it is only with randomization that we can get a fair measure of the variability between replicates. Remember that this is something which is required in subsequent statistical procedures.

It is vital to realize that there is one assumption which is common to all statistical procedures, namely that samples are taken and allocations to treatments are made at random.

**Plate 6** A completely randomized experiment on the growth of bean plant, which differ in the amount of chlorophyll in their leaves.

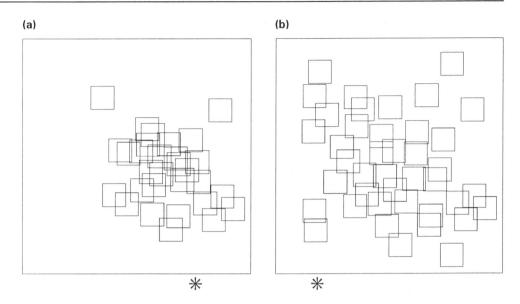

**Figure 3.1** Non-random positions in which quadrats fell when thrown from point *:
(a) thrown over the shoulder unseen; and, (b) thrown looking towards the area.

This is because all the procedures are based on the idea of a sampling distribution which, as we saw in the woodlouse example, is a distribution of all possible random samples. If we want to use this to test a sample, then for a fair test the sample should be one that is taken at random. Without this safeguard the results of any statistical analysis will be misleading.

In some instances when the data do not meet the assumptions of a statistical test, you can get round it by manipulating the data after it has been obtained. One thing you cannot do is to correct for an initial lack of randomization. It has to be built in to the actual practical work from the beginning. So, if in doubt, randomize with respect to any additional factors which might affect the variable of interest. As the examples have shown, these factors may relate to the initial properties of the material under investigation, the way in which it is maintained during the course of the experiment, or the way in which the measurements are made.

OK, you might say, I accept that things have to be randomized, but do I have to go through this apparently time-consuming procedure every time? Yes! Supposedly random procedures, such as throwing quadrats over the shoulder, are not random (Fig. 3.1a). Nor do you get a random pattern when you can see where to throw a quadrat (Fig. 3.1b). Although the latter looks reasonable, the area in the centre has been oversampled while the peripheral areas are undersampled. This is a predictable result, because you avoid throwing the quadrat too near the edge in case it falls outside the area. Similar studies on the allocation of experimental material show that only the use of random-number tables gives real randomization.

## More on sampling

Unfortunately, simple random sampling is not always the best or the most practicable method of random sampling and a completely randomized experimental design is not necessarily the most efficient. Consider, for example, the problem of obtaining a random sample of snake's-head fritillary plants. Using simple random sampling, we would need to give a number to every plant and then use random numbers to choose a sample. Since there are several thousand plants this would be impracticable. To get round this we use **multistage random sampling**. In principle you would have to do the following. Select at random, say, 20 areas $10\,\text{m} \times 10\,\text{m}$. Within each of these select at random one area $1\,\text{m} \times 1\,\text{m}$, and then within these select at random one area of $0.2\,\text{m} \times 0.2\,\text{m}$. Number all the individual plants in this area and chose one of these numbers at random to identify the sampling unit. Repeat this for each $10\,\text{m} \times 10\,\text{m}$ area.

A second problem with simple random sampling is that, by chance, the coverage may not be very even and some areas may be missed altogether. We noted this in the two examples given in Box 3.2. To achieve a more even coverage we can divide up the population of sampling units into what are called **strata**. A stratum is simply a subset of the sampling units which are in some way similar to one another. We then take a simple random sample from each stratum. This is called **stratified random sampling**.

For example, we could divide the site on the shore into four smaller squares, so we would have four strata, and sample at random in each. Or we could divide it into three horizontal bands corresponding to upper, middle and lower shore and again sample at random in each of these strata. Another possibility would be to divide the area up into two strata, comprising weed-covered and bare parts of the shore.

The advantages of stratified random sampling are two-fold. First, it enables us to get information separately for each stratum and thus to see whether they are similar. Secondly, it makes the estimate from the whole sample more reliable. The reason for this can be seen by considering an extreme case. Imagine that all the quadrats within a stratum have exactly the same number of snails, but that there are differences in snail numbers between the strata. In this situation a single sample unit from each stratum would describe the population exactly. If we take the same number of sample units by simple random sampling then, by chance, one (or more) strata might not be sampled. Under these circumstances, our sample will not describe exactly the population.

Similarly, in a completely randomized design, one treatment may be concentrated on one part of the greenhouse bench, or the measurements on it may all fall early in the sequence. You can see this happening in the examples we have used. The use of a **randomized block** experimental design is the equivalent to stratified random sampling. The block is equivalent to the stratum.

The term "block" comes from field experiments on the growth of crop plants. Suppose we want to compare the yield of four varieties of wheat. To get replicates we could sow seed of each variety on several small plots of ground. However, if the soil in one part of the field is more fertile, then with simple random allocation, the plots of one variety could end up being concentrated in the more fertile area of the field. Clearly this would not be desirable. To overcome this, we could divide the field into smaller areas or blocks, within each of which soil fertility will be more uniform. Each block can then divided up into four plots each of which is sown with one of the four varieties. Allocation of varieties to plots in the block is done at random. Plate 7 shows two blocks of an experiment each with 16 plots.

We can apply this principle to any situation. Take, for instance, the experiment on the effects of drinking on urea concentration. Suppose we felt that there was likely to be a great deal of variation between sexes in urea concentration then we could divide our students up into two blocks (male and female). We would then allocate the males at random to the two

**Plate 7** A randomized block experiment on the effects of different fertiliser treatments on plant growth on nutrient-poor mine waste. Two blocks are shown each of sixteen plots, one for each of the sixteen treatments.

treatments and then the females. If, on the other hand, we thought that body weight or diet was likely to be a source of variation then blocks could be formed of individuals who were alike in weight or in diet.

Likewise the greenhouse bench could be divided up into three blocks at different distances from the window, each with space for four dishes. Two experimental and two control dishes could then be allocated at random to these four spaces. A block could be a time interval. In the enzyme activity experiment we could divide the time into five blocks (e.g. before coffee, after coffee, after lunch, etc.) and allocate one determination at each pH in a random order in each time interval. Anything which we think might be a source of unwanted variation in an experiment could be treated as a block in this way.

These more sophisticated methods are generally more efficient than simple random samples/completely randomized designs, but they do require more sophisticated statistical analysis. These will not be dealt with except for one special case, the so-called paired design, which we will be looking at in Chapter 9.

## More on replication: independence and interference

Having seen why we take random samples we need to go back to replication. Although this seems to be a simple procedure, it is often done incorrectly.

The best way to illustrate this is with the example of the germinating seeds which we have already used when talking about randomization. In that case we put one seed per petri dish, which you might think was rather wasteful of dishes and space. Why not put all six control seeds in one dish and all six experimental seeds in another, and then randomly assign the two dishes to locations on the bench. Surely this would be equivalent to the other design because we would still have information from the same number of seeds in each treatment and it would have the advantage of being easier to carry out.

The problem here is that the six sampling units (seeds) in each sample are not true replicates because they are not **independent** of one another. Remember that one of the purposes of replication is to get a measure of the variation in germination time within each population. This variation arises because of inherent differences between the seeds and differences in the conditions under which they are germinated. Six seeds kept in one dish are all under the same conditions and provide no information about the variation in germination time due to factors such as differences in light level and temperature across the bench. The germination times will tend to be much more similar to one another than they would be if the seeds were in separate dishes in different places, so they are not independent observations. We cannot use them as six separate observations; they are in fact only equivalent to a single true replicate. If we put three seeds in each of two dishes we have the same problem – there are only two true replicates.

This is described as **pseudoreplication** and it is easy to do it unless you think carefully about the purpose of replication. For instance counting seeds in 10 random flowers on one random plant is not the same as counting seeds in one random flower on 10 random plants. The first one is adequate if the population of interest is all the flowers on one plant, but it does not sample the variation in seed number due to genetic and environmental differences between different plants. Likewise, making 10 determinations of protein content on tissue extracted from one organism is not the same as doing one determination on each of 10 extracts from the same organism. Neither of these is the same as one determination on one extract from each of 10 different organisms. The first one only allows you to make inferences about variability inherent to the machine being used, while the second one gives information on variability due to the extraction technique. In each case they would be proper replicates for these purposes. However, neither tells us anything about variability between organisms and so they could not be used for assessing differences between two groups of organisms. If used for this purpose they would be pseudoreplicates.

We also need to consider the possibility of **interference** between units. The ways in which this could happen are as varied as the sorts of variable that we might use, so I will give only a few examples. A germinating seed may produce a substance that diffuses through water or air and inhibits or enhances the germination of neighbouring seeds. A rapidly growing plant variety may shade a slower growing variety in a separate but nearby pot, thus reducing its growth rate. The sight or sound of an active animal may cause increased activity in animals in neighbouring cages, and so on. Obviously if the purpose of the experiment is to detect the presence of these sorts of effects then there is no problem, but in other situations they can enhance or reduce the effect of the treatment of interest. The design of the experiment needs to take the possibility of this sort of effect into account.

## Making inferences from samples

I hope by now that you have a much clearer understanding of the terms sampling unit, variable, population and sample. What I want to do now is clarify another relationship between populations and samples. Although we only have a sample at our disposal, what we are really interested in is the population. We have to use the information in our sample to make inferences about the variable in the population. The point I want to make here is a matter of common sense, but it is frequently overlooked. Statistical inferences only apply to the population which we have sampled.

Suppose we are interested in describing human body temperature and so we take the oral temperature of a random sample of 10 first year biology undergraduates in a practical class at the University of Somewhere. What inferences can we make? The population we are sampling is the oral temperatures which we could have observed in all the students in the class, at that particular time of day and month. So this is all we can make statistical inferences about. We cannot use this sample to say anything about oral temperatures of these students at other times of the day or the year, nor will it tell us anything about core body temperature. Neither can we use it to make inferences about temperature in arts students at the University of Somewhere nor in biology students at the University of Somewhere Else.

If we want to make inferences about body temperature in humans in general, then we have to sample an appropriate population. What this means is that the question of interest defines the population and that both of these need careful thought. Should we want to generalize from our sample to some different population, then we may be able to do this by carefully employing our biological knowledge, but any such "inferences" will not have a statistical basis.

Just to round off this discussion, it is probably worth pointing out that in a few situations a sample may include the whole population. There are only three ecological populations of the military orchid (*Orchis militaria*) in Britain, all of which are very small. It would be possible to count the number of flower spikes on each plant or the number of plants in every quadrat. Because this is a count and not a measurement, our sample will contain every possible observation. Similarly, we can get the number of males and females in all captive giant pandas. As you can see, your chances of coming across such instances are rather remote but they do highlight an important relationship between samples and populations.

Consider the following two statements: (a) the number of military orchid plants per quadrat is greater at site A than site B; and (b) the number of snake's-head fritillary plants per quadrat is greater at site C than site D. At first sight these look equivalent but they are not. In the first case the numbers in every quadrat will have been described at both sites so the sample includes the whole population. We can describe both populations of quadrats exactly and there is absolutely no doubt that the two populations are different. As a result no statistical test is required. In the second case not every quadrat will have been described. The statement is an inference from two samples about two populations of quadrats. Now there is some doubt about whether the two populations are different. They could be the same and the difference between the two samples could be due to sampling variation, so we need a statistical test.

# 4

# Describing samples

The basic output of any scientific investigation is a collection of observations or data, derived from one or more samples. These (note the plural) data are collected in the laboratory or in the field and recorded in some form. This usually involves writing them down. In this chapter we look first at ways of recording data. We then move on to ways of summarizing and presenting the information from a sample.

## Recording data

Careful recording of data is an essential part of any scientific investigation and is as important as careful practical work. There are two ways of recording data as it is produced.

### Lists and tally charts

The most obvious way is simply to write down for each sampling unit the value of the variable, in the form of a list (Box 4.1). This is particularly appropriate when the number of observations is small and when we are not sure what values the variates might take. As you can imagine this is not efficient when there are many observations which have the same value, because you find yourself writing down the same number over and over again. Under these circumstances it is worth planning ahead and recording your data in the form of a tally chart (Box 4.1). This will also save time when you come to analyze your results.

To produce a tally chart, you first draw up a list, in a row or a column, of all the possible observations that you might come across. This is particularly easy for nominal variables, such as the progeny types of a genetic cross or the taxonomic groups encountered when sorting a sample of freshwater invertebrates. It is also easy with discontinuous variables, such as the number of micro-organism colonies on sterile plates where you can quickly find out what the range of values is. You can also use it with a continuous variable, although you will need to have found out roughly what the biggest and smallest observation is and used the 30/300 rule (p. 45) to decide how precise your measurements are to be. Having set up the table, you then enter a tally mark for each sampling unit in the appropriate row/column. These can be grouped together in fives by using four vertical marks and one horizontal mark (this saves time later).

---

**BOX 4.1 RECORDING DATA**

**1. List**

(a) Direction of turns made by woodlice:
damp, damp, dry, damp, damp, damp, damp, damp, dry, . . . , etc.

(b) Number of dog whelks (*Nucella lapillus*) in fifteen 0.5 m × 0.5 m quadrats:
0, 15, 0, 1, 1, 2, 0, 7, 0, 6, 2, 0, 5, 1, 1

(c) Activity of acid phosphatase (μM/min) at pH 3:
11.1, 10.0, 13.3, 10.5, 11.3

**2. Tally chart**

(a) Direction of turns made by woodlice:

damp ⊥⊦⊺ ⊦⊦⊺ ||||

dry |||

(b) Number of micro-organism colonies on sterile plates:

Number of
colonies

0 ⊦⊦⊺ ⊦⊦⊺ ⊦⊦⊺ ||||

1 ⊦⊦⊺ ⊦⊦⊺ ⊦⊦⊺ ⊦⊦⊺ ⊦⊦⊺ ⊦⊦⊺ ||||

2 ⊦⊦⊺ ⊦⊦⊺ ⊦⊦⊺ |||

3 ⊦⊦⊺ ⊦⊦⊺ ||||

4 |||

5 |

6 |

7

Although a tally chart is a quick way of recording data, it does have one disadvantage in comparison to a list. A list preserves the data in the order in which they were obtained, something which is lost in a tally chart. The order can give useful clues about both the biology and the experimental technique. For example, we would expect the three dry turns in the woodlouse experiment to be scattered amongst the 14 damp turns; if they were the last three observations we might be suspicious. It could suggest that the behavioural pattern was changing over the course of the experiment or that we were becoming progressively more biased in our recording. Similarly, in Figure 1.1, replicates with high and low readings of phosphate concentration should occur in a random sequence. If the low values are concentrated towards the end of the sample, we might suspect that the spectrophotometer was going out of adjustment or that some systematic change was occurring in one of the reagents. This sort of knowledge could help in the interpretation of the results and suggest improvements to later experiments.

However you record the data, it is most important to do it tidily and unambiguously and with clear information stating where, when and under what conditions the data were obtained. It is no good spending time and effort designing and executing an investigation if it is not matched by care in recording the results. Most of us go through a stage in which we scribble results down on a handy bit of paper and then a few days later find that we cannot make head or tail of them. I speak from bitter experience!

Remember to make a clear distinction between a situation where an observation is made and the result is zero and one where the observation is missing for some reason. A result of zero must be used in the subsequent analysis, unlike an observation which is not made.

Statistical analysis cannot correct for defects in the data which are put into the calculations, so try to develop the habit of drawing up a suitably titled blank table in advance of doing the practical work. This will often clarify your ideas on the aim of the exercise as well as focusing your attention on aspects of the design, such as the type of variable, the sampling unit and the number of replicates. Ideally, drawing up such a table should be part of the design process. It can also clarify what exactly you have to do in the practical session and so save wasting time during the practical work. For more complex experiments and project work in which the same sorts of data will be gathered many times, photocopied blank data recording sheets are essential.

Data in the form in which they are recorded at the bench or in the field are called **raw data**. In this form, and especially if the sample is of a reasonable size, it is usually difficult to see what the major features of the data are. It is also difficult to compare two or more sets of data. To bring out these features and to facilitate comparisons we need to process the data, which involves condensing or summarizing it.

## Summarizing and presenting data in frequency distributions

A **frequency** is simply the number of times something occurs. A **frequency distribution** simply reveals how many times each different value of the variable occurs in the sample. It can be presented in either a **tabular** or **graphical** form, but before we deal with these, it will be helpful to have a shorthand way of denoting a variable. A variable will be identified by a capital letter and in most cases we shall use $Y$. Thus $Y$ could stand for the variable "number of dog whelks in a quadrat" or the variable "activity of acid phosphatase".

Some textbooks and calculators use the symbol $X$, but we shall follow Sokal & Rohlf (1981) and reserve this symbol for so-called independent variables. $Y$ will then stand for a dependent variable. The relationship of these terms is explained in Chapters 8 and 11. Individual observations, i.e. variates, will be identified as $Y_1$, $Y_2$, $Y_3$, etc. Thus $Y_1$ could stand for the number of dog whelks in the first quadrat.

### Tabular frequency distributions

Finding the frequency with which each value occurs is very straightforward, especially if you have recorded them on a tally chart as the observations were made. If not, you will have to go back through the list of data. In a tabular frequency distribution, we list the different values of the variable in one row (or column). We then record the frequencies in a second row (or column) opposite the appropriate values of the variable. There are two examples in Box 4.2.

Constructing a frequency distribution simplifies the data by effectively reducing the number of observations that we have to deal with, for example the data from 90 plates have been reduced to eight figures. The number of times each value occurs can be expressed either as a frequency (denoted by $f$) or as a **relative frequency**, that is, as a fraction of the total number of observations. A relative frequency can be expressed either as **proportion**, that is, a number between zero and one (a decimal fraction) or as a number between zero and one hundred, that is, as a **percentage**. Proportions are converted to percentages by multiplying by 100. A percentage (so many out of 100) is perhaps easier to visualize and percentages are in everyday usage. As we saw in Chapter 2, proportions have the advantage of being on the same scale as probabilities, which is an advantage in the context of statistics.

Relative frequencies are useful when we want to compare two samples of different size as shown in Box 4.2b. Using frequencies, it is difficult to see the differences in faunal composition between the two habitats, but these emerge clearly when we use relative frequencies.

Note that converting to relative frequencies leads to a loss of information, because the size of the sample has disappeared. Without this information, the figures can be misleading

---

**BOX 4.2 TABULAR FREQUENCY DISTRIBUTIONS**

(a) **Frequency distribution of the number of micro-organism colonies ($Y$) on plates.**

| Number of colonies ($Y$) | 0 | 1 | 2 | 3 | 4 | 5 | 6 | 7 |
|---|---|---|---|---|---|---|---|---|
| Frequency ($f$) | 19 | 34 | 18 | 14 | 3 | 1 | 1 | 0 |

(b) **Composition of invertebrate fauna in two stream habitats: riffles and pools.**

| | Frequency | | Percentage | |
|---|---|---|---|---|
| | Riffles | Pools | Riffles $n = 496$ | Pools $n = 706$ |
| Oligochaeta | 38 | 141 | 7.7 | 20.0 |
| Hydrobiidae | 17 | 28 | 3.4 | 4.0 |
| Gammaridae | 11 | 31 | 2.2 | 4.4 |
| Chloroperlidae | 37 | 27 | 7.5 | 3.8 |
| Nemouridae | 3 | 24 | 0.6 | 3.4 |
| Baetidae | 63 | 15 | 12.7 | 2.1 |
| Ephemerellidae | 56 | 63 | 11.3 | 8.9 |
| Ephemeridae | 11 | 55 | 2.2 | 7.8 |
| Ecdyonuridae | 104 | 100 | 21.0 | 14.2 |
| Caenidae | 3 | 18 | 0.6 | 2.5 |
| Leptophlebidae | 32 | 76 | 6.5 | 10.8 |
| Leptoceridae | 33 | 72 | 6.7 | 10.2 |
| Rhyacophilidae | 63 | 52 | 12.7 | 7.4 |
| Elminthidae | 25 | 54 | 5.0 | 0.6 |
| Total | 496 | 706 | (100.1) | (100.1) |

because we have no idea about the extent of sampling variation. This is something which we will be dealing with in more detail in Chapter 7, here we shall be guided by common sense. In this particular case, both of the samples of invertebrates are reasonably large and so the fact that oligochaetes make up 7.7% of the sample from the riffles and 20.0% from the pools probably reflects a real difference between the two habitats. However, the same percentage figures would result from sample sizes of 13 and 20 and numbers of oligochaetes of 1 and 4. With such small samples we would be foolish to conclude that the percentages reflected a real difference between the habitats, but without the information on sample size we cannot make this distinction. When presenting data as relative frequencies, always include the sample size.

---

**BOX 4.3  A FREQUENCY DISTRIBUTION FOR A CONTINUOUS VARIABLE. BODY LENGTH ($Y$ MM) IN A SAMPLE OF 400 FEMALES OF THE ISOPOD *SPHAEROMA RUGICAUDA***

| Original measurements | | Grouping into 15 classes | | Grouping into 10 classes | |
|---|---|---|---|---|---|
| $Y$ (mm) | Frequency ($f$) | Class mark (mm) | Frequency ($f$) | Class mark (mm) | Frequency ($f$) |
| 3.6 | 2 | 3.65 | 5 | 3.7 | 9 |
| 3.7 | 3 | | | | |
| 3.8 | 4 | 3.85 | 9 | | |
| 3.9 | 5 | | | | |
| 4.0 | 12 | 4.05 | 20 | 4.0 | 25 |
| 4.1 | 8 | | | | |
| 4.2 | 14 | 4.25 | 30 | 4.3 | 42 |
| 4.3 | 16 | | | | |
| 4.4 | 12 | 4.45 | 29 | | |
| 4.5 | 17 | | | | |
| 4.6 | 21 | 4.65 | 44 | 4.6 | 61 |
| 4.7 | 23 | | | | |
| 4.8 | 22 | 4.85 | 44 | | |
| 4.9 | 22 | | | 4.9 | 77 |
| 5.0 | 33 | 5.05 | 69 | | |
| 5.1 | 36 | | | | |
| 5.2 | 26 | 5.25 | 50 | 5.2 | 86 |
| 5.3 | 24 | | | | |
| 5.4 | 28 | 5.45 | 50 | | |
| 5.5 | 22 | | | 5.5 | 67 |
| 5.6 | 17 | 5.65 | 29 | | |
| 5.7 | 12 | | | | |
| 5.8 | 10 | 5.85 | 14 | 5.8 | 26 |
| 5.9 | 4 | | | | |
| 6.0 | 2 | 6.05 | 4 | | |
| 6.1 | 2 | | | 6.1 | 5 |
| 6.2 | 1 | 6.25 | 1 | | |
| 6.3 | 0 | | | | |
| 6.4 | 1 | 6.45 | 2 | 6.4 | 2 |
| 6.5 | 1 | | | | |
| Total | 400 | | 400 | | 400 |

Tables with large numbers of columns are cumbersome and difficult to work with so you may need to make a further modification. This is especially so with continuous variables. We saw in Chapter 3 how we ought to use between 30 and 300 unit steps to ensure sufficient precision in our measurements, but this will give an unwieldy frequency table. A second useful rule of thumb is to aim for between 12 and 20 classes in a frequency table. This is achieved by grouping together two or more adjacent classes of observation. The procedure is illustrated in Box 4.3.

There were 30 unit steps originally, so a convenient grouping will be to have two unit steps in each class, thus reducing the number of classes to 15. Remember that, with a continuous variable, the recorded value of a measurement does not tell us precisely what the value is. Rather it specifies a range or a class of values within which the measure lies. The width of this range is the **class interval**. The recorded measure is the **class mark**, which marks the centre of the interval.

In this example, the first size class includes any length between 3.5500 . . . mm and 3.6499 . . . mm. These are the **implied limits** of the class, which for convenience are written as 3.55 and 3.65 mm. The class interval for the original measurements is 0.1 mm. If we pool together adjacent classes in pairs, the class interval becomes 0.2 mm. The first class now includes any length between 3.5500 . . . mm and 3.7499 . . . mm. The class mark of this new class is halfway between the class marks of the two original classes, that is, 3.65 mm. Once we have the first new class mark, the other ones can be obtained by repeatedly adding the new class interval: $3.65 + 0.2 = 3.85$ mm, $3.85 + 0.2 = 4.05$ mm, etc.

The effect of this procedure is two-fold. First it is much easier for the eye to scan the column of 15 frequencies and to get an impression of the pattern. Second the pattern itself is clearer. If there are too many classes, then the number of observations in each class will be small and chance variation in numbers between classes will make the major features difficult to see. On the other hand, with too few classes, major features tend to disappear altogether.

The 12–20 rule is only a rough guide and need not be adhered to slavishly. Smaller samples may benefit from the use of fewer classes. In the third column of Box 4.3, the original classes have been grouped together in threes, to give 10 classes. Although it is not immediately apparent from the tabular distributions, there is one problem with pooling classes; it can alter the appearance of the distribution (Fig. 4.2).

## Rounding off

This is a convenient point to deal with the mechanics of rounding off. According to my calculator, the proportion of animals in the riffle sample which were oligochaetes is 0.0766129 (7.66129%). There are two reasons why it would not be sensible to present it as such.

First, from a purely practical standpoint it takes up a lot of space. Secondly, it gives a false impression. Remember that we are using the information in the sample to tell us something about the variable in the population; that is we want to know something about the proportion of oligochaetes in the population. The sample proportion is only an estimate of the population proportion, which is something we can never know. In Chapter 7 we shall see that it is possible to get an idea about how close the estimate is to the population value.

This closeness is called the reliability of the estimate and it is rather similar to the idea of the precision of an individual observation. To present the estimate as 7.66129% suggests that the estimate is very reliable. It might lead the reader to conclude that the population proportion lay between values of 7.661285% and 7.661295%. We shall see later that with a sample of this size, the reliability will be nowhere near this good, so we need to round the value off to a more reasonable number of significant figures. The rules for rounding off are given in Box 4.4. Since reliability will increase with sample size, the larger the sample, the more significant figures we can retain in our estimate. Again a rough and ready rule is given in Box 4.4.

Only apply this rule when reporting the result. If the result is to be used in some subsequent calculation, then do not round off until the final answer has been produced. For simplicity in presentation, values for the intermediate stages of calculations which are reported in the examples in this book have been rounded off. However the final value will have been calculated on a pocket calculator using non-rounded intermediate values. Having to write down intermediate values which have not been rounded is very tedious, but you can avoid this by using the memory on your calculator. You should make sure that you know how to do this.

---

**BOX 4.4   RULES FOR ROUNDING OFF (ONLY TO BE USED ON THE FINAL ANSWER TO A CALCULATION, NOT AT INTERMEDIATE STAGES).**

**1.   The mechanics of rounding off; digits to be retained are underlined.**

(a)   Digits to be retained followed by a digit less than five:
   - **digits retained unchanged:**
     e.g. proportion of Ephemeridae in riffles   =   11/496   =   $\underline{0.022}177419$
     =   0.022

(b)   Digits to be retained followed by digit greater than five or a five followed by non-zero digits:
   - **last digit retained increased by one:**
     e.g. proportion of Oligochaeta in riffles   =   38/496   =   $\underline{0.076}612903$
     =   0.077
     e.g. proportion of Leptoceridae in riffles   =   33/496   =   $\underline{0.066}532258$
     =   0.067

(c)   Digits to be retained followed by a five alone or a five followed by zeros:
   - **if last digit to be retained is even it remains unchanged**
     e.g.   $\underline{3.8}5$ = 3.8
   - **if last digit to be retained is odd it is increased by one**
     e.g.   $\underline{0.45}500$ = 0.46

**2.   Reporting results (e.g. proportions, means, etc.).**

(a)   Proportions:
   Sample size
   Less than 100:   report with two significant figures, e.g. 0.61
   100–1000:   report with three significant figures, e.g. 0.614

(b)   Means:
   Sample size
   Less than 100:   report with one more significant figure than original measurements
   100–999:   report with two more significant figures than original measurments

---

69

## Graphical frequency distributions

Frequency distributions can also be presented in a graphical form and the most common types are illustrated in Figure 4.1. For nominal and discontinuous variables, the appropriate format is a **bar chart** (Fig. 4.1a & b). The variable (taxonomic group, number of colonies) is on the horizontal axis, the frequency on the vertical axis. Frequencies can be the number of occurrences or relative frequencies. The height of the bar is proportional to the frequency

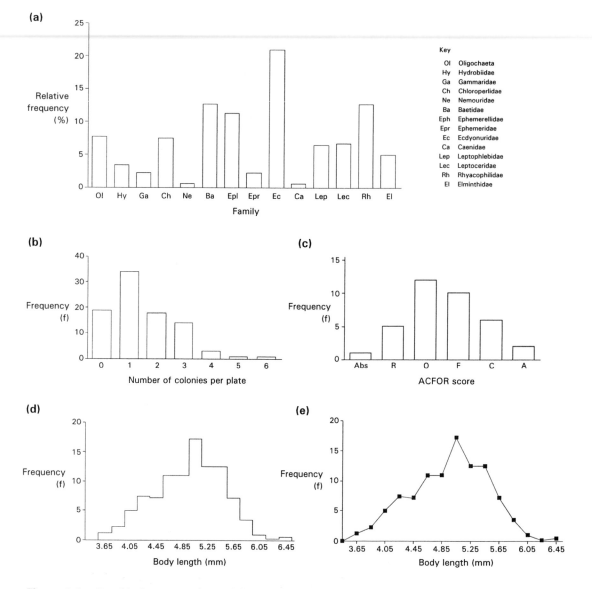

**Figure 4.1** Graphical presentations of frequency distributions. *Bar charts:* (a) percentage composition by family of sample of freshwater invertebrates from riffle habitat; (b) frequency of micro-organism colonies on 90 plates; (c) frequency of ACFOR scores recorded by 36 students in survey of abundance of *Achillea millefolium*. *Histogram* (d) and *frequency polygon* (e), frequency distribution of body length in female isopods *(Sphaeroma rugicauda)*.

and the bars do not touch those on either side, which indicates that the variable is discontinuous and that intermediate values cannot occur. A frequency distribution of an ordinal variable or an ordered category variable would be presented in the same way (Fig. 4.1c).

The appropriate graphical presentation for a continuous variable is the **histogram** (Fig. 4.1d and Fig. 4.2) in which the adjacent blocks do touch, which signifies that there is no real division between the classes. The classes on the horizontal axis are identified by their class marks. The **frequency polygon** (Fig. 4.1e) is an alternative format. Here the fre-

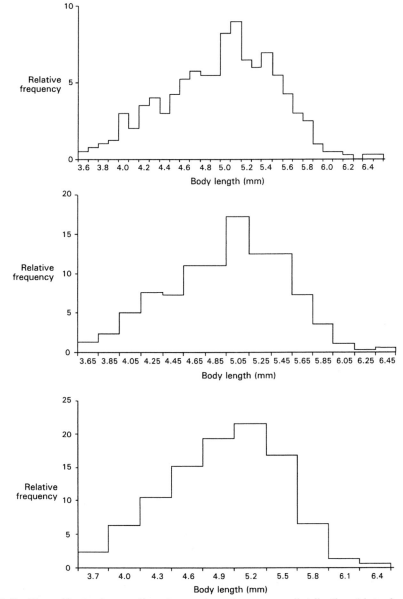

**Figure 4.2** The effect of grouping classes in a frequency distribution (data from Box 4.3).

71

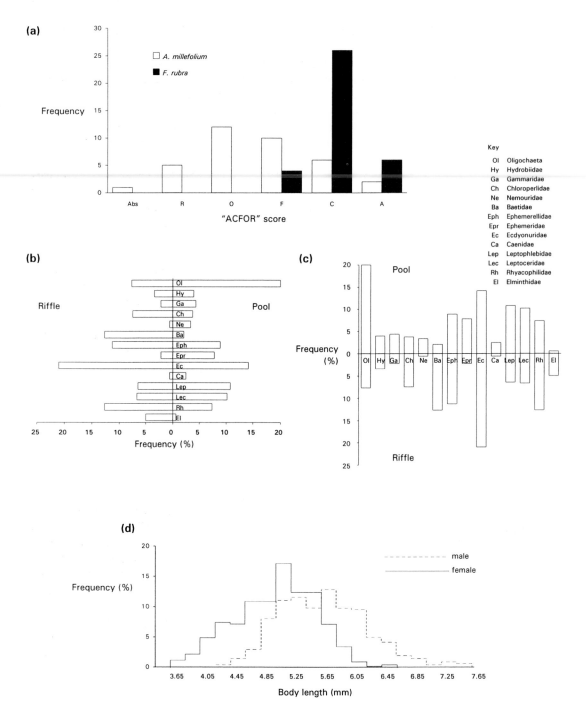

**Figure 4.3** Comparing frequency distributions. (a) Frequency of ACFOR scores recorded by 36 students for *Achillea millefolium* and *Festuca rubra*; (b) percentage composition by family of freshwater invertebrates in riffles and pools; (c) percentage composition by family of freshwater invertebrates in riffles and pools; (d) frequency distribution of body length in female and male isopods (*Sphaeroma rugicauda*).

quency is marked by a point directly above the class mark and adjacent points are joined by straight lines.

One feature of these frequency distributions is immediately apparent. They differ in how symmetrical they are. The histogram (Fig. 4.1d) is roughly symmetrical, by which we mean that the hump is about in the middle and the two tails are about the same size. The bar chart (Fig. 4.1b) is asymmetrical, because the hump is towards the left-hand side and one tail is much longer than the other. The distribution in (Fig. 4.1b) is said to be **skewed**. Because the right-hand tail is drawn out the distribution is described as skewed to the right or positively skewed.

Compared with a table, a graphical presentation has advantages and disadvantages. There would be little to be gained by presenting simple data, such as the number of damp and dry turns, as a bar chart, but when the data fall into many classes it is difficult to get the overall picture from a table. Under these circumstances, a graph is more appropriate. The same is true if you want to compare two (or more) sets of data. There are a number of ways of doing this, some of which are shown in Figure 4.3.

The drawbacks of graphs are that they take more time to produce than a table and they take up more space. Also you lose the actual numerical values which can only be read off approximately from the graph. As a general rule it is not worth presenting the same frequency distribution in both tabular and graphical form. The most efficient format for your purposes will depend on the complexity of the data, the message you are trying to convey and the readership. Do not forget to number all tables and graphs. You should also give them titles and if necessary a legend, which explains any symbols used.

## Summarizing data in a single number

The construction of frequency distributions is the first step in summarizing data so that we can begin to "see at a glance" what its main features are. However, we can carry this process further and reduce the information to a single number.

If we look at our frequency distributions, we can see that there are at least two different features that could be of interest and which therefore need to be described. The first feature is where along the horizontal axis the distribution is positioned or centred. This tells us what a typical value of the variable is. To describe this needs a "measure of location". This could be used to convey the information that a typical abundance score for *Achillea millefolium* is smaller than that for *Festuca rubra* (Fig. 4.3a) and that males are typically bigger than females (Fig. 4.3d).

The second feature of possible interest is the spread of the data around the typical value. This needs a "measure of variability or dispersion". For instance, we might want to convey the information that the scores for *Achillea millefolium* in Fig. 4.3c are more variable than those for *Festuca rubra*.

### Measures of location

Measures of location are called averages and three different ones are available: the **mode**; the **median;** and the **arithmetic mean**. These measures have different properties and uses.

## The mode

This is the simplest and the quickest measure of what constitutes a typical sampling unit. It is simply the category or the class of the variable with the most observations in it. Put another way, the modal category or class is the most frequently occurring category or value.

The mode is the only measure we can use with nominal variables. For example we could say that the modal type of turn in the woodlouse experiment is a turn to the damp alley of the chamber; typically woodlice turn to the damp. The modal family of invertebrates in riffles is the Ecdyonuridae, in pools it is the Oligochaeta. The mode can also be used with discontinuous and continuous variables. The mode of the distribution of the number of colonies is clearly one and the modal class of the body lengths of female isopods is 5.05 mm.

The mode as a quantitative measure, is rarely used with these two types of variables, because there are other, better, measures available. However, the word is used to describe the shape of a distribution. The distributions of colonies and body lengths are both **unimodal** because they have a single peak. Distributions with two fairly clear peaks (e.g. Fig. 5.9a) would be described as **bimodal**.

## The median

The median is simply the observation that is in the middle when the observations are arranged in ascending order. It is therefore the observation that has equal numbers of observations above and below it.

Box 4.5 gives some examples. In the case of the counts of the number of *Nucella lapillus* in 15 quadrats, we first arrange them in order. The middle observation is number 8 in the sequence, with seven observations above and below it. The median number of snails is therefore 1. If there is an even number of items as in (Box 4.5b) then no one item lies in the middle. We find the middle two observations, in this case observations 5 and 6 in the sequence, and the median is taken to lie halfway between them. So the median is $(17.2 + 18.0)/2 = 17.6$ mg/ml. Note that this value has equal numbers of observations above and below it.

One of the virtues of the median is that you do not have to deal with all the observations. To find the median only requires that you put the first half of the observations in order, so for the fungal colony data in Table 4.2a, you should easily be able to verify that the median is 1 (there are 90 observations, so the median is observation number 45.5; this observation lies in the class labelled 1).

Finding the median for a continuous variable which is grouped into classes is rather more tedious, but it works on the same principle. In the sample of isopod body lengths there are 400 observations, so the median lies between observations number 200 and 201; it is actually observation number 200.5. We first locate the class within which this observation lies and then calculate the median by following the steps below.

(a) Adding up the frequencies from the left-hand end of the distribution, we find that observation number 181 is the last observation in the class marked 4.9 mm.

(b) Observation number 214 is the last observation in the class marked 5.0 mm.

(c) Thus the median observation (number 200.5) lies somewhere in this class.

(d) Because there are 181 items in the preceding classes the median is observation number 19.5 in this class $(200.5 - 181 = 19.5)$.

(e) This class has implied limits at 4.95 and 5.05, so it has a class interval of 0.1 mm and it contains 33 items. We assume that the measurements are evenly distributed throughout this class so the median is located at a fraction $19.5/33 = 0.591$ of the way through the class.

**BOX 4.5  CALCULATING THE MEDIAN**

**(a)** Number of *Nucella lapillus* in 15 random 0.5 m × 0.5 m quadrats:

0,  15,  1,  0,  1,  2,  0,  7,  0,  6,  2,  0,  5,  1,  1.

Arrange in ascending order;  median is observation number $\left(\dfrac{n+1}{2}\right)$:

$n = 15$, so  median is observation number $\left(\dfrac{15+1}{2}\right)$ = 8th observation

| 1st | 2nd | 3rd | 4th | 5th | 6th | 7th | 8th | | | | | | | |
|---|---|---|---|---|---|---|---|---|---|---|---|---|---|---|
| 0 | 0 | 0 | 0 | 0 | 1 | 1 | 1 | 1 | 2 | 2 | 5 | 6 | 7 | 15 |

$\underbrace{\qquad\qquad}$ 7 values below median $\qquad$ median $\qquad$ $\underbrace{\qquad}$ 7 values above median

**(b)** Urea concentration (mg/ml)  in human urine (variates arranged in ascending order):

median is observation number $\left(\dfrac{n+1}{2}\right)$ = $\left(\dfrac{10+1}{2}\right)$ = observation number 5.5.

| 1st | 2nd | 3rd | 4th | 5th | 5.5 | 6th | | | | |
|---|---|---|---|---|---|---|---|---|---|---|
| 6.8 | 10.0 | 11.5 | 15.0 | 17.2 | | 18.0 | 20.2 | 31.8 | 36.0 | 38.1 |

5 values below median $\qquad\qquad\qquad$ 5 values above median

$$\text{Median} = \frac{17.2 + 18.0}{2} = 17.6$$

**(c)** Body length (mm) in a sample of 400 female isopods, *Sphaeroma rugicauda*:

$n = 400$, median is observation number $\left(\dfrac{400+1}{2}\right) = 200.5$

| Class mark | 3.6 | 3.7 | 3.8 | 3.9 | 4.0 | 4.1 | 4.2 | 4.3 | 4.4 | 4.5 | 4.6 | 4.7 | 4.8 | 4.9 | 5.0 | 5.1 | 5.2 | 5.3. . . |
|---|---|---|---|---|---|---|---|---|---|---|---|---|---|---|---|---|---|---|
| Frequency | 2 | 3 | 4 | 5 | 12 | 8 | 14 | 16 | 12 | 17 | 21 | 23 | 22 | 22 | 33 | | | |

(i)  Sum of frequencies to this point $\qquad\qquad\qquad\qquad$ = 181

(ii)  Sum of frequencies to this point $\qquad\qquad\qquad\qquad\qquad$ = 214

(iii)  Observation number 200.5 lies in this class, which contains 33 items

(iv)  Observation numer 200.5 is item number 19.5 in this interval (200.5 − 181 = 19.5)

(v)  Observation number 200.5 lies 19.5/33 = 0.591 of the way through this interval, which has a width of 0.1 mm.

(vi)  Median is 4.95 + (0.591 × 0.1) = 5.01 mm.

(f) The median therefore lies $0.591\,\text{mm} \times 0.1\,\text{mm} = 0.0591\,\text{mm}$ above the lower limit of the class which is 4.95. The median is therefore $4.95 + 0.0591 = 5.01\,\text{mm}$.

## The arithmetic mean

This is perhaps the best known average and is calculated by adding together all the observations and dividing by the number of observations (Box 4.6). The symbol $\Sigma$ (pronounced sigma) means "add together", so $\Sigma Y$ means add together all the values of $Y$. The sample mean is symbolized by $\overline{Y}$ which when spoken becomes "$Y$ bar". Incidentally, every observation has to be used, so if a value occurs twice, it has to be added in twice and you must also take account of zeros. Obviously the latter will not affect the quantity on the top line of the equation but they will be affect the quantity ($n$) on the bottom line. We shall be using equations a lot and it will help to know the proper terms for "the quantity above the line"

---

**BOX 4.6 CALCULATING THE MEAN**

$$\text{Mean} = \frac{\text{Sum of individual observations } (Y)}{\text{Total number of observations } (n)} = \frac{\Sigma Y}{n} = \overline{Y}$$

e.g. for a sample of five observations

$$\overline{Y} = \frac{Y_1 + Y_2 + Y_3 + Y_4 + Y_5}{5}$$

where $Y_1$ to $Y_5$ are individual variates.

(a) Activity of acid phosphatase at pH 3 ($\mu\text{M/min}$):

$$\overline{Y} = \frac{11.1 + 10.0 + 13.3 + 10.5 + 11.3}{5} = \frac{56.2}{5} = 11.24$$

(b) Number of dog whelks (*Nucella lapillus*) in $0.5\,\text{m} \times 0.5\,\text{m}$ quadrats:

$$\overline{Y} = \frac{(0 \times 5) + (1 \times 5) + (2 \times 5) + 5 + 6 + 7 + 15}{15} = 2.7333 = 2.7$$

(c) Number of colonies of micro-organisms per plate:

$$\overline{Y} = \frac{(0 \times 19) + (1 \times 34) + (2 \times 18) + (3 \times 14) + (4 \times 3) + (5 \times 1) + (6 \times 1)}{90}$$

$$= \frac{19 + 34 + 36 + 42 + 12 + 5 + 6}{90} = \frac{135}{90} = 1.5000 = 1.5$$

and the "quantity below the line". The first is known as the **numerator**, the second as the **denominator**. If, like me, you need a way to remember these, the "d" in denominator stands for "down below".

A calculator with a standard deviation function will calculate a sample mean easily and you should become familiar with how to do this. You will be using it a great deal! Although the details of how to do this vary between different models of calculator, it basically involves getting into standard deviation mode (abbreviated to SD or s.d.) and then simply entering each item of data. The calculator automatically adds them up and counts how many items there are. Pressing two buttons will give the mean which is usually symbolized by $\bar{x}$ on a calculator. Means for some of the data given in Box 4.1 have been calculated and are given in Box 4.6. If the data are in the form of a tabular frequency distribution and a value occurs more than once, then you should be able to cut down on the amount of button pushing. In Box 4.2a the value 0 occurs 19 times. Instead of entering these 19 values individually you should be able to enter them all at once by pressing the keys $0 \times 19$ and then whatever key you use to enter data. (This works on all the makes of calculators that I have come across with my students, but check it out on yours. It saves a great deal of time and reduces the chances of errors.)

As was the case with proportions, your calculator will give an answer to several decimal places and the same general arguments apply. To report the mean number of snails/quadrat as 2.73333 would be inappropriate because it gives a false impression of reliability. On the other hand, we want to use a sample mean as an estimate of the population mean, and we know that the larger the sample, the more reliable our estimate should be. By convention the mean is reported with one more decimal place than the original observations if the sample size is less than 100, and with two more if the sample size is between 100 and 1000. So the mean number of snails per quadrat is reported as 2.7 and the mean enzyme activity as 11.24 µM/min. Incidentally, although you obviously cannot ever have a quadrat containing two and a bit snails, it is quite in order to report the mean in this form. You should find that the mean body length using the data in Box 4.3 is 4.95125 mm, which we would report as 4.951 mm. Again, if you are going to use the mean in some further calculations, then you should not round off the value.

## Which measure to use

Although the mode, median and mean are all measures of location, they are not equivalent, and the situations in which they are employed are different. The mode is the only measure we can use with nominal variables; to talk of the median or mean family of organisms in a riffle does not make sense.

With an ordinal variable, we can use the mode, but it is rather inefficient because it only uses information from one value of the variable (the most frequent). The median is better because it uses information from all the values. It is particularly appropriate for an ordinal variable because it uses information on the order of all the values, but not the values themselves; remember the median merely specifies which observation is in the middle.

With discontinuous and continuous variables we can use the median or the mean. The mean uses all the information in the data and so is the most efficient of our three measures. Box 4.7 has a set of data which shows how these three measures give different answers. The mode is zero, the median is 2 and the mean is 4.3. In this case the mode does not give a very good idea of a typical observation, because well over half the quadrats have one or more flower spikes. The mean is misleading as well, because rather few quadrats have this number of spikes. For a skewed distribution the median is the best measure. It also has the virtue of being insensitive to (i.e. unaffected by) extreme values. If the largest number of spikes in a quadrat was 86 instead of 27, the median would remain the same. The mean on the other hand, would increase to 5.12.

---

**BOX 4.7 COMPARISON OF THE MODE, MEDIAN AND MEAN: NUMBER OF FLOWER SPIKES OF COMMON BROOMRAPE (*OROBANCHE MINOR*) IN 0.5 m × 0.5 m QUADRATS**

| Number of spikes ($Y$) | Frequency |
|:---:|:---:|
| 0 | 33 |
| 1 | 4 |
| 2 | 9 |
| 3 | 2 |
| 4 | 3 |
| 5 | 4 |
| 6 | 4 |
| 7 | 1 |
| 8 | 2 |
| 9 | 1 |
| 10 | 0 |
| 11 | 0 |
| 12 | 3 |
| 13 | 1 |
| 14 | 1 |
| 15 | 2 |
| 16 | 0 |
| 17 | 1 |
| 18 | 0 |
| 19 | 0 |
| 20 | 0 |
| 21 | 0 |
| 22 | 0 |
| 23 | 1 |
| 24 | 1 |
| 25 | 1 |
| 26 | 0 |
| 27 | 1 |
| Total | 75 |

Mode  =  0 flower spikes/quadrat

Median  =  observation number $(n+1)/2$ = 38th observation
     =  2 flower spikes/quadrat

Mean  =  4.3 flower spikes/quadrat

## Measures of variability

Measures of variability tell us how closely packed around the average, the values of the variable are. There are only three measures that we need concern ourselves with here. They are the **range**, the **standard deviation** and the **variance**. The last two are closely related and are by far the most important and useful.

The range is simply the difference between the highest and the lowest value and is quickly found. We would report it by simply stating these two values. For example, the range of body lengths in the sample of *Sphaeroma rugicauda* is 3.6–6.5 mm. The range only uses the two most extreme values, so it is wasteful of the information in the sample.

The variance and the standard deviation, on the other hand, use all the information in the sample and they have a number of mathematical properties which allow them to be used in various statistical tests. The first formula in Box 4.8 describes what the variance is and it is easy to see how this gives us a measure of variability or spread.

The bit inside the brackets tells us to take one observation $(Y)$, find out how far it is from the mean $(\bar{Y})$ by subtracting one from the other, and then square the answer. The $\Sigma$ means repeat this for every observation and add the answers together. An individual value of $(Y - \bar{Y})$ is called a **deviation from the mean** often abbreviated to a **deviate**. When squared, it is a **squared deviation from the mean**. The sum of all these values is called the **sum of squared deviations from the mean** or, more succinctly, the **sum of squares**. This is usually abbreviated to **SS**. To get the variance, we divide this by one less than the sample size, that is by $(n - 1)$.

Now let's see why this is a good measure of variability. At the centre of the equation is the distance between each observation and the mean (calculated as the difference). If there is little variability, all the observations will be near to the mean and these differences will be small. If there is a lot of variability the differences will be large. Note that all the observations are used. The rest of the equation is there to convert this information into a more useful form.

Some values of $Y$ will be larger than the mean and will yield positive deviations, others will be less than the mean and will give negative deviations. Squaring the deviations has the effect of making all the deviations positive, and as a result the sum of the squared deviations will always be positive. (If they were not squared the positive and negative deviations would cancel one another exactly and their sum would be zero.) You can see that the total of the squared deviations, that is the sum of the squares, will be large if the observations are a long way from the mean (i.e. widely spread) and small if they are close to the mean.

However, the size of the sum of squares will also depend on the number of observations. Other things being equal, the more observations in the sample the larger the total of the squared deviations will be. What we need is a sort of average squared deviation from the mean, which will tell us how far a typical observation is from the mean. To get this we have to relate the sum of squares to the sample size which is done by dividing by the number of observations minus one. This may seem a slightly odd way to get our average squared deviation but there is a good reason for it to which we will return in the final section.

Dividing the sum of squares by $(n - 1)$ gives us the **sample variance**, denoted by $s^2$. This measure of variability enters into many statistical tests but it is not used for describing and reporting variation. The units in which the variance is expressed are (original units)$^2$ which is conceptually awkward (what are $mg^2/ml^2$?). To get round this, the sample variance is converted to the standard deviation, $s$, by simply taking the square root. This has the practi-

## BOX 4.8 VARIANCE AND STANDARD DEVIATION

**(a) Formulae for describing the variance and the standard deviation**

$$\text{Sample variance } s^2 = \frac{\Sigma\left(Y - \bar{Y}\right)^2}{n-1} = \frac{\text{Sum of squares}}{n-1} = \frac{\text{SS}}{n-1}$$

$$\text{Sample standard deviation} = s = \sqrt{\text{Sample variance}} = \sqrt{\frac{\text{SS}}{n-1}}$$

**(b) Formulae for calculating a variance and standard deviation**

Calculate the sum of squares (sum of squared deviations):

$$\text{Sum of squares (SS)} = \Sigma Y^2 - \left(\Sigma Y\right)^2 \big/ n$$

$$\text{Variance} = \frac{\text{SS}}{n-1} = \frac{\Sigma Y^2 - \left(\Sigma Y\right)^2 \big/ n}{n-1}$$

**(c) Example**

Phosphate concentration $Y$ (mg/ml) measured by first year student.

| $Y$ | $Y^2$ |
|---|---|
| 1.25 | 1.5625 |
| 1.20 | 1.4400 |
| 1.22 | 1.4884 |
| 1.25 | 1.5625 |
| 1.30 | 1.6900 |
| 1.15 | 1.3225 |
| 1.40 | 1.9600 |
| 1.30 | 1.6900 |
| 1.21 | 1.4641 |
| $\Sigma Y = 11.28$ | $\Sigma Y^2 = 14.1800$ |

Sum of squares $= \Sigma Y^2 - (\Sigma Y)^2/n = 14.18 - (11.28)^2/9 = 0.0424$

Variance $= s^2 = $ Sum of squares$/(n-1) = 0.0424/8 = 0.0053$ mg$^2$/m$^2$

Standard deviation $= s = \sqrt{\text{variance}} = \sqrt{0.0053} = 0.072801 = 0.07$ mg / ml

**d) Obtaining the sum of squares from a calculator**

$$\text{Sum of squares (SS)} = \text{Variance} \times (n-1) = s^2 \times (n-1)$$

cal advantage that the measure of variability is now in the same units as the measure of location.

### An easier formula for calculating the standard deviation

The formula we have just been examining shows quite clearly what a standard deviation is, but it is not the best formula for calculating it. In particular, calculating the sum of squares by subtracting the mean from every observation is laborious and there is a much more convenient formula for the sum of squares. It is given in Box 4.8 along with some data.

This formula simply tells us to do the following. Square each value of $Y$ and add the answers to get $\Sigma Y^2$. This is the sum of the squared values of $Y$. Then add the values of $Y$ to get $\Sigma Y$ and square this to get $(\Sigma Y)^2$ which is the square of the sum of the values of $Y$. (Note that $\Sigma Y^2$ and $(\Sigma Y)^2$ are not the same thing.) Divide $(\Sigma Y)^2$ by $n$ and subtract the answer from $\Sigma Y^2$ to get the sum of squares. The variance is obtained as before by dividing the sum of squares by $(n-1)$.

### Using the standard deviation function on a calculator

Things are even easier if you have a calculator with a standard deviation function because you only need to enter the values of $Y$; any values which occur more than once can be entered using the quick method (p. 77). You can then obtain both the mean and the standard deviation at the press of one or two buttons. One potentially confusing point is that most calculators have the facility to calculate two slightly different standard deviations. These are retrieved using two different keys which, depending on the make of calculator, are either labelled $s$ and $\sigma$, or $\sigma_n$ and $\sigma_{n-1}$.

We shall see in the final section of this chapter why two different standard deviations can be calculated, however in practice you are only likely to want one of them. What is really confusing is that there seems to be no agreement amongst the manufacturers as to what the labels on the buttons mean nor do they conform to the standard symbols used in textbooks! The safest way to find out which is the correct button on your machine is to put in some data, press one button and note the answer and then press the other one. You should find that one answer is slightly larger than the other. You want the larger one. So for future use note which is the appropriate key or, even better, mark it in some way!

Most calculators also have buttons (labelled $\Sigma x^2$ and $\Sigma x$) which enable you to obtain values of what we are calling $\Sigma Y^2$ and $\Sigma Y$ and the latter can be squared to obtain $(\Sigma Y)^2$. As noted before these are not the same thing. The first is the sum of the individual values of $Y$ after they have been squared; the second is the sum of the individual (unsquared) values of $Y$, which is subsequently squared. These quantities are used in some statistical tests. Sums of squares are also used quite widely and, although most calculators do not produce them directly, you can easily calculate them from the value of $s$, the standard deviation. Remember that the variance is the sum of squares divided by $(n-1)$, so the sum of squares is equal to the variance multiplied by $(n-1)$. To calculate the sum of squares (Box 4.8d) all you need to do is to retrieve the value of $s$ (using the correct button), square it and multiply by $(n-1)$. You will be using these quantities a great deal in the rest of the book so I suggest that you always have an appropriate calculator to hand and that you practice with it so that these calculations become second nature. In some tests to be described later we have to calculate standard deviations from two samples. It is useful to find out whether you can store a calculated standard deviation in the ordinary memory; it does seem to work on many calculators.

## Sample values and population values

These numerical measures of location and spread based on samples are examples of **statistics**. As we saw in the previous chapter our real interest is in the populations from which the samples were taken, or more strictly, the values of these measures in the population. These have been referred to rather vaguely as the "true" values. These true values, which are unknowable in practice, are called the **parameters** of the population. The distinction between sample and population values is easy to remember. Statistic is to sample as parameter is to population. Statistics and parameters are identified by different symbols and wherever possible a roman letter is used for the statistic and a greek letter for the parameter. So, for example, the sample mean is $\bar{Y}$ and the population mean is $\mu$. We use $\bar{Y}$ as an estimate of $\mu$. Sometimes this system of symbols breaks down because a symbol is in use elsewhere. The sample proportion ought to be $p$ and the population proportion ought to be $\pi$, but $\pi$ is used elsewhere. In this case both statistic and parameter are given the same roman letter $p$, and the statistic is identified by putting a "hat" over it ($\hat{p}$, pronounced "p hat").

It is desirable that any measure which we use as an estimate has certain properties. One of them is that an estimate should be unbiased. This simply means that, if we imagine taking many samples, the average value of the estimates obtained should equal the population or parametric value. The sample mean $\bar{Y}$, calculated by dividing the sum of the values of $Y$ by the sample size ($n$) is an unbiased estimator of the population mean. However, if we had calculated the sample variance by dividing the sum of squares by $n$ it would have been a biased estimate of the population variance. The values would tend to come out too low and this is why we divide by ($n-1$) instead. The estimate calculated in this way is an unbiased estimate of the population variance. The expression ($n-1$) is known as the **degrees of freedom**, a term we will come across again.

You do not need to worry too much about why the degrees of freedom in this case are ($n-1$). It is probably one of those things that you will just have to accept. A simple explanation is that the population variance ($\sigma^2$) is a measure of the variability around the population mean ($\mu$). We have estimated this variance by calculating the variability around the sample mean ($\bar{Y}$), which is itself only an estimate of the population mean. In other words our estimate of the variance involves making use of another estimate! Dividing by ($n-1$) compensates for this.

# 5

# Three important models for the frequency distribution of variables

As we saw in Chapter 2, a statistical test enables us to decide, in an objective way, whether the result of an experiment is a likely or an unlikely event if the null hypothesis is true. To do this we had to know how the variable was distributed (assuming that the null hypothesis was true) and this required a mathematical model of the variable which incorporated the essential features of the biology. In the woodlouse experiment we used the binomial distribution, but this is only one of three commonly used models for the distribution of variables that you are likely to need at this stage. The other two are the Poisson distribution and the normal distribution. In this chapter we shall look at all three models in some detail, in terms of their biological meaning and relevance, their relationships to one another and their use as a basis for statistical procedures.

## The binomial distribution

We shall start by going back and taking another look at the binomial. In mathematical terms, a binomial distribution will arise whenever each trial has two possible, mutually exclusive outcomes, and when the outcomes of successive trials are independent. Adding this to our knowledge of the different types of variables described in Chapter 3, we can see that a binomial distribution should apply to nominal variables that fall into only two categories, such as simple behavioural choice experiments and simple genetic crosses. Provided that we know the probability of one of the outcomes occurring then we can find the associated probability of a given number ($Y$) of these outcomes in a sample of size $n$.

In the woodlouse experiment we had only one sample at our disposal and our subsequent statistical analysis was based on testing whether it was likely that this single sample could have been drawn from a specified binomial distribution. In other words, we assumed, on the basis of the biology, that a binomial distribution was applicable. This does not give us any idea about whether a binomial distribution of results would have occurred in practice, that is, whether the binomial was an appropriate model in the first place. The way to see whether this is so, is to take many samples, use them to produce a frequency distribution and ask whether this distribution is actually a binomial. This is tedious since it involves repeated sampling. However, the boredom factor can be alleviated by sharing the work between a class of students, as has been done in the following genetic example.

## Kernel colour in maize – is a binomial distribution appropriate?

This example (Box 5.1) concerns the inheritance of kernel colour in maize. Normally, the purpose of setting up a genetic cross is to use the frequencies of the different offspring phenotypes to deduce the genetic basis of the phenotypic differences, that is, the number of genes and the number of alleles involved. Here we use a phenotypic difference (yellow vs black kernels) with a known genetic basis involving just two alleles of one gene. The allele C gives yellow kernels, the allele c gives black kernels, with C dominant to c. This character difference is inherited according to Mendel's first law and produces a very clear-cut phenotypic difference which is conveniently displayed on the corn cob.

In a backcross between a cc and a Cc plant, all the ovules of the cc plant will carry a single c allele as a result of meiosis. The Cc plant will produce two types of pollen. Half the pollen grains will carry the C allele and half will carry the c allele. An ovule has an equal chance of being fertilized by a pollen grain of either type and it follows that the probability of a kernel being Cc (i.e. yellow) is 0.5. The probability of it being cc (i.e. black) is also 0.5. This gives rise to the familiar 1:1 ratio of offspring types.

From this we can see that there are only two possible outcomes for each kernel, black and yellow, and that they are mutually exclusive (i.e. a kernel cannot be both). Furthermore, whether a given kernel is black or yellow should be independent of the outcome for any other kernel. This is because pollen falls at random onto the tassel and so the type of pollen that fertilizes a given ovule should not influence the type of pollen that fertilizes some other ovule. Because the two events are mutually exclusive and independent, if we take repeated random samples of a given size ($n$) and count the number of yellow kernels ($Y$), then this variable should follow a binomial distribution. Since we also know what the probability of one outcome is (for a yellow kernel $p = 0.5$), we can use Table A2 to find out exactly what this distribution is expected to look like.

I asked each student in a class to count how many of the two colours occurred in a random sample of twenty kernels. Altogether 92 such samples were counted and the frequencies of samples with different numbers of yellow kernels are presented in the table in Box 5.1. In the second line these observed frequencies have been converted to relative frequencies based on the total number of samples. These are expressed as percentages. The third line has the percentages which we would expect if the variable followed a binomial distribution with $p = 0.5$ and $n = 20$. They were obtained from Table A2 by multiplying the expected relative frequencies by 100.

We can see that the two sets of figures are not identical and the discrepancies between them show up clearly when the observed and the expected frequencies are plotted as bar charts. For example, the observed distribution is not perfectly symmetrical, unlike the expected one, and there are rather more samples with 11 kernels than there should be in theory. On the whole the disparities are rather small and of course we would expect to see some because of sampling variation. After all, the expected distribution is what we would get if we took all possible random samples of size 20, while the observed distribution is based on only 92 samples. However, there is an alternative explanation for the differences – perhaps the population from which these samples were drawn does not follow this binomial distribution. At the moment we cannot discriminate between these two explanations, because to do so would require a statistical test. We will see how to test this in Chapter 6.

In fact, in the case of this particular variable which has been extensively studied, we know that the population does follow a binomial distribution with $p = 0.5$ so that sampling

## BOX 5.1 FREQUENCY DISTRIBUTION OF THE NUMBER OF YELLOW MAIZE KERNELS IN RANDOM SAMPLES OF TWENTY KERNELS

Cc × cc

| | Number of yellow kernels ($Y$) | | | | | | | | | | |
|---|---|---|---|---|---|---|---|---|---|---|---|
| | 0 | 1 | 2 | 3 | 4 | 5 | 6 | 7 | 8 | 9 | 10 |
| Observed frequency | 0 | 0 | 0 | 0 | 0 | 2 | 5 | 4 | 12 | 13 | 16 |
| Observed rel. freq. (%) | 0.00 | 0.00 | 0.00 | 0.00 | 0.00 | 2.20 | 5.40 | 4.30 | 13.60 | 14.10 | 17.40 |
| Expected rel. freq. (%) | 0.00 | 0.00 | 0.20 | 0.11 | 0.46 | 1.48 | 3.70 | 7.39 | 12.01 | 16.02 | 17.62 |

| | Number of yellow kernels ($Y$) | | | | | | | | | |
|---|---|---|---|---|---|---|---|---|---|---|
| | 11 | 12 | 13 | 14 | 15 | 16 | 17 | 18 | 19 | 20 |
| Observed frequency | 18 | 7 | 8 | 5 | 1 | 0 | 1 | 0 | 0 | 0 |
| Observed rel. freq. (%) | 19.60 | 7.60 | 8.70 | 5.40 | 1.10 | 0.00 | 1.10 | 0.00 | 0.00 | 0.00 |
| Expected rel. freq. (%) | 16.02 | 12.01 | 7.39 | 3.70 | 1.48 | 0.46 | 0.11 | 0.20 | 0.00 | 0.00 |

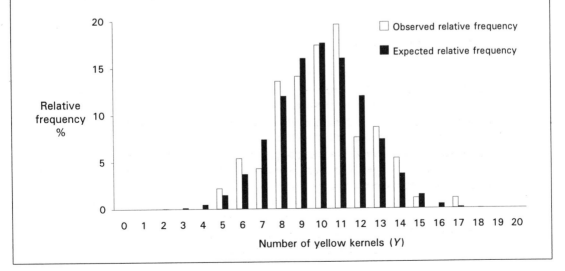

variation must be the explanation for the disparities. One indicator of this is given by examining more closely the pattern of the differences between the expected and observed percentages. For instance, cases in which the expected percentage is larger than the observed percentage are not concentrated in one particular part of the bar chart, rather they are scattered fairly haphazardly. This is just the pattern we would expect to arise by chance.

Note, however, that some students obtained extreme numbers of yellow kernels such as 5, 15 and 17. If we had set the (nominal) significance level at 5%, we can see from the expected frequencies that the critical values are 5 and 15 (these values cut of 4.14% of the distribution in the two tails which is why our significance level is nominally 5%). These particular results (of 5, 15 and 17) lie in the rejection region. What this tells us is that even when a variable does follow a specified distribution, it is possible to obtain, by chance, a sample that is so extreme that it would lead you to say that the variable does not follow this distribution. In other words, you will sometimes make the wrong decision.

This is an inevitable consequence of the chance element involved in sampling – you can sometimes get extreme values. Indeed, if getting results of five or fewer yellow kernels and 15 or more has a probability of 0.0414, then in a large number of random samples, we would expect to get extreme results of this sort in 4.14% of the samples. We obtained four such extreme results out of 92 (i.e. 4.3%) which, given the small number of samples involved, is fairly close to 4.14%. The important message here is that an extreme result can occur in any statistical test, so that there is always a possibility that we will come to the wrong conclusion. Statistical tests never tell us the answer with absolute certainty.

Finally, it is a useful exercise to try and think of possible reasons why this binomial might not have been an appropriate model for the distribution of this variable. Identifying these reasons can highlight issues of experimental design. You might ask:

(a) Were the kernels chosen representative? Left to their own devices, would students consciously or unconsciously choose samples that were atypical? We must never forget that all statistical procedures are based on the assumption that samples are taken at random. Clearly, if we choose samples on some other basis, then improbable results could occur more frequently than they should.

(b) Could some of the kernels have resulted from fertilization of ovules by pollen from a plant with a different genotype? Suppose a cc plant is exposed to a mixture of pollen from a Cc plant (intended) and a CC plant. What effect would this have?

(c) Could there be any confusion in assigning kernels to one of the two categories?

You should be able to see how the effect of some of these factors might show up in the pattern of positive and negative differences which we were considering earlier.

## Describing binomial distributions

A graphical presentation is one way of describing our two distributions but we can also use the numerical measures of location and spread (i.e. the mean and standard deviation) introduced in the last chapter. We can express the number of yellow kernels as a frequency, that is a number between 0 and 20, as we have done in Box 5.1. Alternatively we can use a proportion, based on the sample size of 20. The mean and the standard deviation can be expressed in terms of either.

Formulae for the mean and standard deviation can also be derived from a knowledge of the mathematical basis of the binomial (Box 5.2a). If we use these to calculate values they will be theoretical values which will be the same as the population parameters. They are the values which would expect to get if we took every possible sample. In our example (Box 5.2b) where $p = 0.5$ and $n = 20$, the mean expected frequency of yellow kernels is 10.0 and the mean expected proportion is 0.50. The expected standard deviation is 2.2 for the frequency and 0.112 for the proportion.

You can calculate the mean and standard deviation from the observed distribution using the formulae in Boxes 4.6 and 4.8 or, preferably, the standard deviation function on your calculator. You can use either the frequencies of yellow kernels (e.g. 5) or the proportions of yellow kernels (e.g. $5/20 = 0.25$). The answers (Box 5.2c) are sample statistics and you

---

**BOX 5.2 DESCRIBING BINOMIAL DISTRIBUTIONS**

(a) Formulae for expected (theoretical) values in sample of size $n$, when probability of one outcome is $p$.

| | Mean | Standard Deviation |
|---|---|---|
| Frequency in sample | $\mu = pn$ | $\sigma = \sqrt{pqn}$ |
| Proportion in sample | $p$ | $\sigma = \sqrt{pq/n}$ |

(b) Expected (theoretical) values for binomial distribution of number of yellow kernels ($p = 0.5, q = 0.5, n = 20$):

| | Mean | Standard Deviation |
|---|---|---|
| Frequency in sample | $\mu = 0.5 \times 20$ <br> $= 10.0$ yellow kernels | $\sigma = \sqrt{0.5 \times 0.5 \times 20}$ <br> $= 2.2$ yellow kernels |
| Proportion in sample | $p = 0.500$ | $\sigma = \sqrt{0.5 \times 0.5/20}$ <br> $= 0.112$ |

(c) Observed values for frequency distribution of the number of yellow kernels in 92 samples (values calculated using formulae in Boxes 4.6 & 4.8):

| | Mean | Standard Deviation |
|---|---|---|
| Frequency in sample | $\overline{Y} = 929/92$ <br> $= 10.1$ yellow kernels | $s = 2.3$ yellow kernels |
| Proportion in sample | $\hat{p} = 929/1840$ <br> $= 0.505$ | $s = 0.117$ |

---

will notice that the important theoretical distinction between statistics and parameters is denoted by the use of different symbols. We can use these statistics as estimates of the population parameters and you can see that the estimates are close to the theoretical values. This supports our earlier view that the distribution is a binomial with $p = 0.5$.

## Unequal probabilities

So far our attention has been focused on situations in which the probabilities of the two outcomes are equal, but there are plenty of situations in which this is not the case. Again, turning to genetics, we know that a cross between two individuals, both of whom are heterozygous (Aa) for a dominant allele, should produce two phenotypes of offspring in the ratio 3:1 (assuming that the characteristic obeys Mendel's first law). In other words the probability of a single offspring having the "A" phenotype is 0.75, while the probability of it having the "a" phenotype is 0.25.

For example, in tobacco plants *Nicotiana*, there is a recessive allele of a gene involved in chlorophyll production which, when homozygous, produces no chlorophyll and gives seedlings with white leaves. The frequency ($Y$) of white-leaved seedlings in random samples of 13 will follow a distribution that is a binomial with $p = 0.25$. This frequency distribution can be obtained from Table A2. We have to use the highlighted row of values of $p$ (which range from 0.01 to 0.50) found at the top of the page. The column of figures in the subtable for $n = 13$ vertically below the value of $p = 0.25$ are then the probabilities of getting samples with different numbers of the white-leaved phenotype. The 14 different possible numbers (0–13) are read off from the left-hand highlighted column of values of $Y$. Thus the

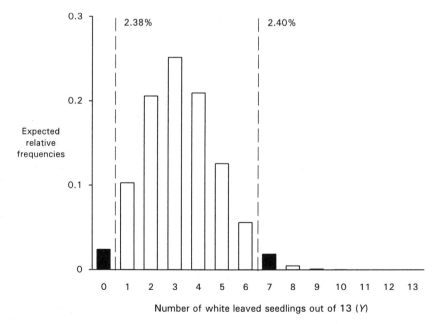

**Figure 5.1** Expected relative frequencies of the number of white leaved *Nicotiana* seedlings in samples of 13 seedlings from a cross of two heterozygous parents. Solid bars are critical values.

expected relative frequency of samples of 13 individuals, none of which is white-leaved, that is samples with $Y = 0$, is 0.0238. The complete distribution has been drawn in Figure 5.1 and we can see that now the distribution is asymmetrical. Samples with a few white-leaved individuals are more likely to occur than samples with many white-leaved individuals and this merely reflects the lower frequency with which this phenotype occurs in the population. We can still get critical values at the nominal 5% significance level as before by working out what values of $Y$ cut off 2.5% in each tail, but we find in this case that they are asymmetrical. They are 0 and 7.

The complementary distribution for the green-leaved phenotype is obtained from the column of figures vertically above the value of $p = 0.75$ in the highlighted row at the very bottom of the page. These are read off against the values of $Y$ in the highlighted column at the right-hand side of the page which you will see run in the opposite direction to the values in the column on the extreme left. We end up using the same set of figures as we used for $p = 0.25$ but in the reverse order. This makes sense because the probability of getting a sample with 13 green-leaved individuals is the same as the probability of getting a sample with no white-leaved individuals.

## Calculating the expected frequencies of a binomial distribution

Binomial distributions can have any values of $n$ and $p$ so there is an infinite number of different binomial distributions. Table A2 gives those more commonly required. If you wanted to produce the binomial distribution for other values of $p$ or for sample sizes larger than 20 you would have calculate the frequencies with which each value of $Y$ would occur using a formula. This can be easily done on a calculator (Box 5.3).

The biology may be familiar. The abundance of a plant species can be measured using a point quadrat, which involves randomly placing a frame of pins and then lowering the pins onto the vegetation. Often, as in this example, we use a frame of 10 pins. The pin is the sampling unit and for each one we record whether or not it touches the species of interest, so this is a nominal variable because the pin either hits or it misses. Does the frequency of hits in frames of 10 pins follow a binomial distribution?

As a preliminary, we calculate the proportion of hits ($\hat{p}$) and misses ($\hat{q}$) in the sample. We shall also need the value of $\hat{p}/\hat{q}$, denoted by $Q$, which should be stored in the memory because we shall need to use it several times. None of these values should be rounded at this stage. The expected frequency of frames with no hits ($Y = 0$) is obviously the same as that for frames with 10 misses. If the chance of one pin missing is 0.768, then the chance of all 10 pins missing is $0.768^{10}$, so this gives us the expected frequency of frames with $Y = 0$. To get the expected frequency of frames with $Y = 1$, we multiply the previous answer by 10 and then by $Q$, which is in the memory. For $Y = 2$, we multiply the previous answer by 9, divide by 2 and multiply by $Q$. You can see how this sequence continues.

With a calculation like this you do need to be careful, because if you make a mistake with one frequency it will affect all subsequent frequencies. Use Table A2 to check that the answers are about right and also check that the frequencies add up to one. Obviously for other sample sizes the calculations would have to be modified. If we had 20 pins in a frame the first term would be $0.768^{20}$, the second term would involve multiplying by $Q$ and 20/1, the third term by $Q$ and 19/2, etc. You can find a general formula in Sokal & Rohlf (1981).

We can see that in this example the observed and expected frequencies are quite different

BOX 5.3 CALCULATION OF THE EXPECTED FREQUENCIES OF A BINOMIAL DISTRIBUTION: FREQUENCY OF PIN HITS ON RESTHARROW (*ONONIS REPENS*) IN 50 RANDOMLY PLACED FRAMES OF 10 PINS

| Number of hits in frame of 10 pins ($Y$) | Observed frequency of frames with $Y$ hits | Observed relative frequency of frames with $Y$ hits | Expected relative frequency of frames with $Y$ hits | |
|---|---|---|---|---|
| 0 | 10 | 0.20 | $= \hat{q}^{10} = 0.768^{10}$ | $= 0.0714$ |
| 1 | 17 | 0.34 | $= 0.714 \times 10/1 \times Q$ | $= 0.2156$ |
| 2 | 8 | 0.16 | $= 0.2156 \times 9/2 \times Q$ | $= 0.2931$ |
| 3 | 2 | 0.04 | $= 0.2931 \times 8/3 \times Q$ | $= 0.2361$ |
| 4 | 0 | 0.00 | $= 0.2931 \times 7/4 \times Q$ | $= 0.1248$ |
| 5 | 6 | 0.12 | $= 0.1248 \times 6/5 \times Q$ | $= 0.0453$ |
| 6 | 5 | 0.10 | $= 0.0453 \times 5/6 \times Q$ | $= 0.0114$ |
| 7 | 0 | 0.00 | $= 0.0114 \times 4/7 \times Q$ | $= 0.0020$ |
| 8 | 1 | 0.02 | $= 0.0020 \times 3/8 \times Q$ | $= 0.0002$ |
| 9 | 1 | 0.02 | $= 0.0002 \times 2/9 \times Q$ | $= 1.5 \times 10^{-5}$ |
| 10 | 0 | 0.00 | $= 1.5 \times 10^{-5} \times 1/10 \times Q$ | $= 4.5 \times 10^{-7}$ |
| | Total = 50 | | | |

Observed proportion of hits $= \hat{p} = 116/500 = 0.232$

Observed proportion of misses $= \hat{q} = 1 - 0.232 = 0.768$

$$\hat{p}/\hat{q} = 0.3020833 = Q$$

and that there is also a pattern to the differences. Frames with a small number of pin hits and frames with a large number are more frequent than we would expect. Frames with intermediate numbers are less frequent than expected. These two facts suggest that perhaps a binomial distribution is not appropriate, although we would need to carry out a statistical test to confirm this. The reason why it is not appropriate is because the events are not independent. The pins are in a frame of 10 and restharrow occurs in patches, so if one pin in a frame has hit a plant, there is a very high chance that the other pins will too.

## Extreme probabilities

Earlier we saw how a binomial distribution becomes asymmetrical when the probabilities of the two outcomes are unequal. This becomes increasingly pronounced as one of the outcomes becomes less and less likely. You can easily see this by looking in Table A2 at the probabilities in each column to the left of the one for $p = 0.25$. A series of these distributions for different values of $p$ and $n = 10$ are shown in Figure 5.2 which clearly demon-

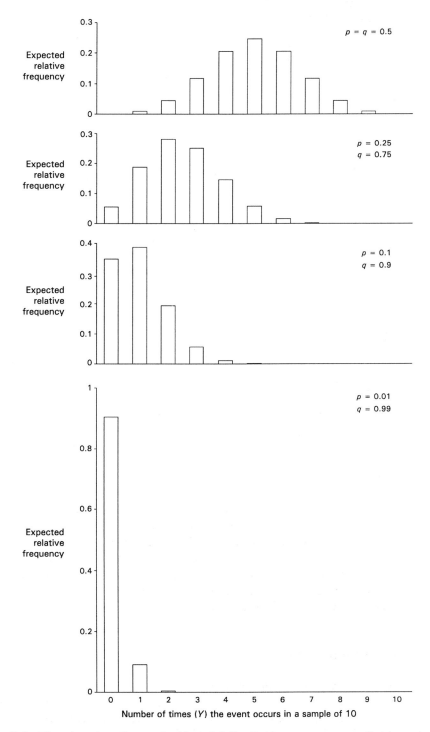

**Figure 5.2** The changing shape of a binomial distribution as *p* gets smaller (sample size = 10).

strates how the modal value of $Y$, that is, the most frequently occurring value, becomes progressively smaller as $p$ decreases. The hump in the distribution, which was in the middle when $p = 0.5$, gradually moves to the left-hand end of the axis. At the same time the frequency of the modal value increases.

Genetic data do not usually involve such extreme values of $p$, but these asymmetrical distributions become important in a rather different type of situation. For example, a seed merchant may claim that the germination success of seeds of a particular variety is 99.9%, which simply means that in the long run 999 out of every thousand seeds will germinate and only one will fail to do so. The proportion of non-germinating seeds is claimed to be 0.001. Clearly we could test this claim by germinating a batch of seeds and counting the number which did not germinate. Because there are only two mutually exclusive, independent outcomes, this variable should follow a binomial distribution. We could use this sampling distribution to test whether the number not germinating was the sort of result that was too extreme to be attributed to chance. The only problem that we would have, is that it is tedious to calculate the terms of a binomial distribution when $n$ is large and $p$ is small. In circumstances like this we can use a different model, which is also very useful for describing the frequency distribution of events in space and time.

## The Poisson distribution

We saw in the last section how the binomial could be applied to situations where $p$ (or $q$) was small and the sample size was large, but that there were some practical difficulties associated with the calculations. An alternative model for this type of situation is the Poisson distribution.

### From the binomial to the Poisson distribution

The basis of the Poisson distribution can be explored by thinking about the spatial distribution of events, such as our example of micro-organism colonies on nutrient plates. At first sight, this appears to be a topic quite unrelated to the problem of the proportion of non-germinating seeds but, as we shall see, there are some similarities.

Each plate can be thought of as consisting of a large number of very small points arranged in a two-dimensional array. When we count the number of colonies on a plate we are simply counting how many of these points have been occupied by a spore which subsequently grew into a colony. In our example of germinating seeds the sample units were individual seeds, we assigned each one to a category, either germinated or not germinated, and then counted up how many sampling units fell into each category. With the micro-organisms, the sampling units are points on a surface and we can imagine classifying them on the basis of whether or not they have been occupied by a spore.

The sample now consists of all the points on a plate and, in theory at any rate, we could imagine counting the number of both occupied points (colony present) and unoccupied points (colony absent). If we were to repeat this for a large number of plates, we could find out how many plates had no occupied points, how many had one occupied point, two occupied points, etc., and present the information as a frequency distribution. This distribution would be a binomial, because for each point there are only two, mutually exclusive, independent outcomes.

The outcomes are mutually exclusive because a point can be either occupied or not. The outcomes are independent because what happens at one point should not affect what happens at any other point. In other words, the fact that a particular point is, say, already occupied by a spore should not affect the chance of some other point being occupied. This latter condition is quite reasonable. The spores have no control over where they land, so a given spore is just as likely to land near a point that is already occupied as it is to land elsewhere. As a result every point has an equally likely chance of being occupied and so the pattern of dispersion of the colonies is said to be **random**. The word has the same meaning here as it did in Chapter 3 where we described sampling as being random.

What would this binomial distribution look like? Because the spores are not numerous and there are a large number of points available for them to occupy, the probability ($p$) that any particular point is occupied by a spore is small (i.e. this outcome is rare). The probability ($q$) that a point is not occupied, is correspondingly high, so $p$ and $q$ will be very unequal. As a result the binomial distribution for an event of this type would be very asymmetrical. If we wanted to describe this distribution in more detail we would need to know the sample size ($n$) and the probability of one of the outcomes ($p$), but we know neither. This is because, in practice, we can only count the number of occupied points, so we do not know what the total number of available points is. In the absence of this information we cannot work out what $p$ (and $q$) are.

However, we can make use of a Poisson distribution to describe the frequency distribution of events such as these. Its theoretical basis, which need not concern us is different from that of the binomial. The important points are that a Poisson distribution is a frequency distribution of the number of times that a rare and random event occurs in a sample and that it can be mathematically described using a relatively simple formula. It can be used as an approximation to the extreme sorts of binomial distributions which we have been looking at and also as a model for the distribution of events in space and time. It is of particular value in the analysis of the pattern of spatial dispersion of organisms.

## Properties of Poisson distributions

The formula is given in Box 5.4a and, although it looks rather complicated, it turns out to be easy to use. It enables us to find the expected relative frequency of samples (i.e. plates) with a given number of colonies ($Y$). You can see that to get the answer from the formula, we need three pieces of information, $e$, $Y$ and $\mu$. Note that the formula, unlike that for the binomial, does not contain a $p$ or a $q$, that is, we do not need to know the probability of a point being occupied or unoccupied, nor does it require a knowledge of $n$. We know $e$, it is a constant (actually 2.71828, the base of natural logarithms). $Y$ is specified by the question we are asking because it is the number of occurrences in the sample (e.g. one colony on a plate) whose expected relative frequency we wish to know.

The third item, $\mu$, is the population mean. This is the mean number of colonies on a plate which we would get if we were to expose all possible plates. The population mean will obviously be different in different situations, as it will depend on the number of spores in the air, the size of the plate and the length of time for which it is exposed. So, just as for the binomial, there is a huge number of different Poisson distributions, each one specified by a

## BOX 5.4 THE POISSON DISTRIBUTION

### (a) Formula for expected relative frequencies.

Expected relative frequency of a plate having $Y$ colonies $= \dfrac{e^{-\mu}\mu^{Y}}{Y!}$

### (b) Expected relative frequencies for Poisson distributions (part of Table A3)

| $\mu =$ | 1.0 | 1.1 | 1.2 | 1.3 | 1.4 | 1.5 | 1.6 | 1.7 | 1.8 | 1.9 | 2.0 | 2.1 | | |
|---|---|---|---|---|---|---|---|---|---|---|---|---|---|---|
| $Y=0$ | .3679 | .3329 | .3012 | .2725 | .2466 | .2231 | .2019 | .1827 | .1653 | .1496 | .1353 | .1225 | ... | 0 |
| 1 | .3679 | .3662 | .3614 | .3543 | .3452 | .3347 | .3230 | .3106 | .2975 | .2842 | .2707 | .2572 | ... | 1 |
| 2 | .1839 | .2014 | .2169 | .2303 | .2417 | .2510 | .2584 | .2640 | .2678 | .2700 | .2707 | .2700 | ... | 2 |
| 3 | .0613 | .0738 | .0867 | .0998 | .1128 | .1255 | .1378 | .1496 | .1607 | .1710 | .1804 | .1890 | ... | 3 |
| 4 | .0153 | .0203 | .0260 | .0324 | .0395 | .0471 | .0551 | .0636 | .0723 | .0812 | .0902 | .0992 | ... | 4 |
| 5 | .0031 | .0045 | .0062 | .0084 | .0111 | .0141 | .0176 | .0216 | .0260 | .0309 | .0361 | .0417 | ... | 5 |
| 6 | .0005 | .0008 | .0012 | .0018 | .0026 | .0035 | .0047 | .0061 | .0078 | .0098 | .0120 | .0146 | ... | 6 |
| 7 | .0001 | .0001 | .0002 | .0003 | .0005 | .0008 | .0011 | .0015 | .0020 | .0027 | .0034 | .0044 | ... | 7 |
| 8 | .0000 | .0000 | .0000 | .0001 | .0001 | .0001 | .0002 | .0003 | .0005 | .0006 | .0009 | .0011 | ... | 8 |
| 9 | .0000 | .0000 | .0000 | .0000 | .0000 | .0000 | .0000 | .0001 | .0001 | .0001 | .0002 | .0003 | ... | 9 |
| 10 | .0000 | .0000 | .0000 | .0000 | .0000 | .0000 | .0000 | .0000 | .0000 | .0000 | .0000 | .0001 | ... | 10 |

### (c) Observed and expected relative frequencies of plates with different numbers of colonies.

| | Number of colonies on plate ($Y$) | | | | | | | |
|---|---|---|---|---|---|---|---|---|
| | 0 | 1 | 2 | 3 | 4 | 5 | 6 | 7 or more |
| Observed frequency | 19 | 34 | 18 | 14 | 3 | 1 | 1 | 0 |
| Observed relative frequency | 0.2111 | 0.3778 | 0.2000 | 0.1556 | 0.0333 | 0.0111 | 0.0111 | 0.0000 |
| Expected relative frequency | 0.2231 | 0.3347 | 0.2510 | 0.1255 | 0.0471 | 0.0141 | 0.0035 | 0.0009 |

different value of $\mu$. The only problem in using this formula is that we are unlikely to know the value of $\mu$. However, we do know the sample mean ($\overline{Y}$) so, in practice, we use this as an estimate of $\mu$.

Although it is useful to see what the formula "does" and what has to be put into it, you will not necessarily have to use this formula. A selection of some of these frequency distributions are given in Table A3. If you do have to calculate the terms of a Poisson distribution, there is an alternative formula which is computationally easier and which we shall look at later.

## Using tables of Poisson distributions

In Table A3, part of which is reproduced here (Box 5.4b), you will find expected relative frequencies for a number of Poisson distributions, each corresponding to a different value of $\mu$. These tables are set out in exactly the same way as the tables for the expected relative frequencies for binomial distributions, which we have already looked at. Along the top of each table is a range of values of $\mu$, the mean number of occurrences, and down the side of each table a range of values of $Y$, where $Y$ is the number of times the event occurs in a sample. The values in the body of the table are the relative frequencies with which the values of $Y$ are expected to occur in the long run. These, of course, are the same thing as the probabilities of obtaining a specified value of $Y$ in a single sample. It is clear that the resulting distributions are asymmetrical. Samples with zero occurrences or with low numbers of occurrences predominate and samples with high numbers are rare, which is what we would expect with a rare event.

We can see how to use the table by taking our example of the fungal colonies, in which the mean number of colonies per plate conveniently happened to be 1.5. We use this sample mean $(\overline{Y})$ as an estimate of $\mu$ and, reading down the column of figures under $\mu = 1.5$ in the table, we find the expected relative frequency of plates having 0, 1, 2, etc., colonies. The expected frequencies for seven and eight colonies have been added together because they are very small. These expected frequencies are presented in Box 5.4c along with the observed frequencies and you can see that the two sets are similar, but not identical.

There are two possible explanations for the discrepancies which are the same as those which applied to the observed and expected frequency distributions of maize kernels. Perhaps the frequency distribution of the colonies does follow a Poisson distribution with a mean of 1.5 and the discrepancies between observed and theoretical frequencies are due to sampling variation. Don't forget that the frequency distribution from the tables is the distribution that you would get if the population mean actually was 1.5 and if you took every possible sample. Clearly we do not know the true mean so we might be looking in the wrong column of the table. Even if somehow we did know that the mean was 1.5, so that we were using the correct column in the table, we only have a sample of plates. By chance, the make-up of this sample is likely to depart from the theoretical values.

The alternative explanation is that the discrepancies are not due to chance and that the distribution does not follow a Poisson. You may already have thought of a reason. In developing the model, I talked about spores occupying points at random, that is, not influencing one another. While this is reasonable we have to remember that we have not observed the spores but the colonies that they produced. The growth of one colony may well influence the growth of adjacent colonies, by overgrowing or killing them, and this could lead to a distribution which was not a Poisson. We will see how to distinguish between these two explanations in the next chapter.

Finally, there is one important property of any Poisson distribution: the mean is always numerically equal to the variance. If you wish you can try to verify this by working out the standard deviations of some of the Poisson distributions in Table A3. (You can use the relative frequencies of each value of $Y$.) Since this is one of those rare instances where we are using the whole population rather than a sample, the standard deviation is calculated by dividing the sum of squares by $n$ rather than by $(n - 1)$ (see Chapter 4). To get this standard deviation you will need to use the other button on the calculator and don't forget you will need to square the result to get the variance! (The answer may not be exactly the same as the mean because the relative frequencies in the table have been rounded.) Quite why the variance should always equal the mean is something you will just have to accept as there is no easy explanation of why it should be so. However, it turns out to be very useful, as we shall see.

## Calculating expected frequencies for a Poisson distribution

Because there are an infinite number of different Poisson distributions, the chances are that the exact value of our sample mean ($\bar{Y}$) will not correspond to a value of $\mu$ in the tables. If this happens, then we have to use the formula to calculate the expected frequencies and it is easier to do this using a rearranged version (Box 5.5). We first calculate the expected relative frequency of plates with no colonies (i.e. $Y = 0$). We then use this answer to calculate the expected frequency of plates with one colony ($Y = 1$), which in turn is used in the calculation for $Y = 3$, etc. This formula cuts down the work because the answer to the preceding calculation is already in the calculator ready to be used in the current calculation. The disadvantage is that if you make a mistake early on in the sequence then all the subsequent answers will be wrong as well, so be careful! You can always check that the answers are roughly right, either by referring to the nearest value of $\mu$ in the tables or by checking that your frequencies add up to one.

To get some practice with this type of calculation, you might like to use the two sets of data on spatial dispersion in Box 5.6. The two sample means are given. In both sets of data you will find that the expected relative frequencies in the upper tail of the distribution are very small and it is too tedious to calculate them separately. We can obtain the frequency of quadrats containing 12 or more flower spikes much more simply by using the fact that the sum of all the frequencies is 1. To do this we simply add the calculated frequencies for quadrats containing between zero and eleven spikes (0.998) and subtract it from one.

With these two examples, you can see that the expected frequencies depart quite markedly from the observed frequencies and, what is more, there is something of a pattern to the differences. This suggests that chance variation is not the only thing at work here and that perhaps a Poisson distribution is not a very good model for these two variables. This would happen if there are features of the biology that do not match the assumptions which underlie the Poisson distribution. This is something we shall examine in the next section.

---

**BOX 5.5  CALCULATING THE EXPECTED RELATIVE FREQUENCIES FOR A POISSON DISTRIBUTION: FREQUENCIES OF PLATES WITH DIFFERENT NUMBERS OF COLONIES OF MICRO-ORGANISMS**

Mean $= \bar{Y} = 1.5$ colonies/plate

| Number of colonies on plate ($Y$) | Expected relative frequency of plates with $Y$ colonies | |
|---|---|---|
| 0 | $= 1/e^{\bar{Y}} = 1/4.482$ | $= 0.2231$ |
| 1 | $= 0.2231 \times \bar{Y}$ | $= 0.3347$ |
| 2 | $= 0.3347 \times \bar{Y}/2$ | $= 0.2510$ |
| 3 | $= 0.2510 \times \bar{Y}/3$ | $= 0.1255$ |
| 4 | $= 0.1255 \times \bar{Y}/4$ | $= 0.0471$ |
| 5 | $= 0.0471 \times \bar{Y}/5$ | $= 0.0141$ |
| 6 | $= 0.0141 \times \bar{Y}/6$ | $= 0.0035$ |
| 7 | $= 0.0035 \times \bar{Y}/7$ | $= 0.0009$ |

## BOX 5.6   CALCULATION OF THE EXPECTED FREQUENCIES FOR  A POISSON DISTRIBUTION

(a)  **The number of flower spikes of common broomrape (*Orobanche minor*) in 0.5 m × 0.5 m quadrats.**

(b)  **The number of stomata per field of view  (× 40 magnification).**

1.  Calculate mean:   flower spikes/quadrat, $\bar{Y} = 4.333$
     stomata/field of view, $\bar{Y} = 6.8667$

2.  Use formula:   expected frequency with $(Y = 0) = 1/e^{\bar{Y}} = f_0$
     expected frequency with $(Y = 1) = f_0 \times \bar{Y} = f_1$
     expected frequency with $(Y = 2) = f_1 \times \bar{Y}/2 = f_2$
     etc.

| No. of spikes $(Y)$ | Observed frequency | Observed relative frequency | Expected relative frequency | No. of stomata $(Y)$ | Observed frequency | Observed relative frequency | Expected relative frequency |
|---|---|---|---|---|---|---|---|
| | | Flower spikes | | | | Stomata | |
| 0 | 33 | 0.440 | 0.013 | 0 | 0 | 0.000 | 0.001 |
| 1 | 4 | 0.053 | 0.057 | 1 | 0 | 0.000 | 0.007 |
| 2 | 9 | 0.120 | 0.123 | 2 | 0 | 0.000 | 0.025 |
| 3 | 2 | 0.027 | 0.178 | 3 | 0 | 0.000 | 0.056 |
| 4 | 3 | 0.040 | 0.193 | 4 | 0 | 0.000 | 0.096 |
| 5 | 4 | 0.053 | 0.167 | 5 | 2 | 0.133 | 0.132 |
| 6 | 4 | 0.053 | 0.121 | 6 | 2 | 0.133 | 0.152 |
| 7 | 1 | 0.013 | 0.075 | 7 | 8 | 0.533 | 0.149 |
| 8 | 2 | 0.027 | 0.040 | 8 | 2 | 0.133 | 0.128 |
| 9 | 1 | 0.013 | 0.020 | 9 | 1 | 0.067 | 0.098 |
| 10 | 0 | 0.000 | 0.008 | 10 | 0 | 0.000 | 0.067 |
| 11 | 0 | 0.000 | 0.003 | 11 | 0 | 0.000 | 0.042 |
| 12 | 3 | 0.040 | | 12 | 0 | 0.000 | 0.024 |
| 13 | 1 | 0.013 | | 13 | 0 | 0.000 | 0.013 |
| 14 | 1 | 0.013 | | 14 | 0 | 0.000 | 0.006 |
| 15 | 2 | 0.027 | | 15 | 0 | 0.000 | |
| 16 | 0 | 0.000 | | 16 | 0 | 0.000 | |
| 17 | 1 | 0.027 | | 17 | 0 | 0.000 | 0.005 |
| 18 | 0 | 0.000 | | 18 | 0 | 0.000 | |
| 19 | 0 | 0.000 | | 19 | 0 | 0.000 | |
| 20 | 0 | 0.000 | 0.002 | 20 | 0 | 0.000 | |
| 21 | 0 | 0.000 | | | | | |
| 22 | 0 | 0.000 | | $n = 15$ | | 0.999 | |
| 23 | 1 | 0.013 | | | | | |
| 24 | 1 | 0.013 | | | | | |
| 25 | 1 | 0.013 | | | | | |
| 26 | 0 | 0.000 | | | | | |
| 27 | 1 | 0.013 | | | | | |
| $n = 75$ | | 1.000 | | | | | |

## Other patterns of spatial dispersion

As we saw in the last section, it looks as though frequency distributions relating to patterns of spatial dispersion may not necessarily follow a Poisson distribution. These situations are biologically just as interesting as situations where the variable does follow a Poisson distribution. A Poisson distribution results when there is a pattern of random dispersion such that every point has an equal chance of being occupied. Figure 5.3b shows a random dispersion of individuals "on the ground". Note that a random pattern is not an even pattern. By chance, it has patches where individuals are common and patches where individuals are rare or absent, in just the same way that the units in a random sample are unevenly distributed. If we were to take a large number of quadrats of the size shown and produce a frequency distribution of numbers/quadrat, it would follow a Poisson distribution.

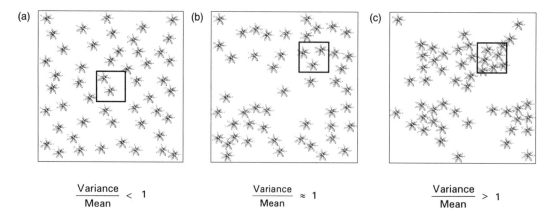

Figure 5.3  Patterns of spatial dispersion: (a) regular; (b) random; (c) contagious.

Random spatial dispersion is in fact not all that common in the living world and the reason is that events are not usually independent. This lack of independence can take two forms which we can see by contrasting the landing of spores with the settling behaviour of the larvae of an aquatic invertebrate and the egg laying behaviour of an insect. The key difference is that larvae can exercise a choice of where to settle and the insects can choose where to lay eggs; spores have no choice. For a larva and an insect this choice may be in one of two directions. Individuals may choose an area where other individuals or eggs are absent or rare and this will tend to produce a more evenly spaced pattern than a random pattern (Fig. 5.3a). Such a dispersion pattern, produced when one event impedes or decreases the probability of a similar event occurring again, is called **regular**. Alternatively, individuals may choose an area where other individuals or eggs are already present and avoid areas from which they are absent. This behaviour will lead to clumps of individuals and empty spaces which are more pronounced than would occur with random dispersion (Fig. 5.3c). A pattern like this which occurs when one event encourages the occurrence of further events is called **contagious** (or **clumped**) dispersion. Plates 8–10 show actual examples of these patterns.

The biological reasons for these two types of choice are varied and were briefly examined in Chapter 1; for example, the need to avoid competition or predation may favour a

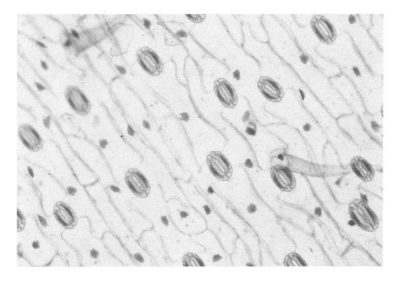

**Plate 8** A regular pattern of spatial dispersion; stomata on leaf surface.

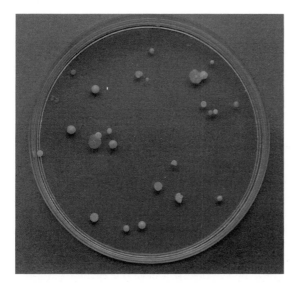

**Plate 9** A random pattern of spatial dispersion;
colonies of micro-organisms on plates.

regular pattern, while patchiness in the habitat or the need for mating partners may produce a contagious pattern. We can get patterns like these even when choice is not involved, as we saw in Chapter 1. For example, unequal survival rates due to competition could change a random pattern to one that is more regular.

Because these different patterns tell us something about the biology of the organism, it is useful to be able to distinguish between them, always remembering that sampling variation is going to obscure the picture somewhat. One obvious way is to look at the pattern of differences between the expected frequency distribution which is based on independent

**Plate 10**   A contagious pattern of spatial dispersion; dog whelks *Nucella lapillus* on a rocky shore.

events and the observed frequency distribution. You can see from Figure 5.3 that a regular pattern should have very few quadrats with very high and very low numbers when compared with a random pattern. If, on the other hand, the individuals have a contagious pattern of dispersion then quadrats with a very high and very low numbers will be more common compared with the random pattern. The two examples in Box 5.6 illustrate these two patterns; the flower spikes have a contagious dispersion, while the stomata have a more regular dispersion.

Another way we can distinguish between these situations is to use the ratio of the variance to the mean. As we saw earlier the mean of a Poisson distribution is always numerically equal to its variance and so the variance/mean ratio of a random dispersion pattern is 1. In contrast, regular and clumped dispersion patterns have variance/mean ratios of less than 1 and greater than 1 respectively, and it is easy to see why.

The total number of organisms in each area in Figure 5.3 is the same – there are actually 50. If we imagine placing quadrats over the whole of the area in Figure 5.3b and counting the number of organisms in each, there will be some variation in numbers between quadrats. Some quadrats will have none, some will have one or two and a few will have three or four and to describe this variation we would use the variance. Because this is a random dispersion the frequency distribution would follow a Poisson distribution and the variance of these values will equal the mean. If we repeat this for Figure 5.3a then there will be less variation in the numbers in each quadrat on account of the regular pattern. Using a quadrat of the size indicated, we can see that most quadrats would have 0 or 1 individuals so there will be little variation. The variance would be low and because the mean is the same as in Figure 5.3b, the variance/mean ratio will now be less than 1. In Figure 5.3c, the opposite will happen and there will be a lot of variability between quadrats. Some quadrats will have no individuals but some will have six or seven, leading to a high variance. The variance/mean ratio will be greater than in Figure 5.3b, that is, greater than 1.

The variance/mean ratio is thus a convenient measure of the pattern of spatial dispersion, in fact, it is called the **index of dispersion**, denoted by $I$. You can work it out for the two examples in Box 5.6 and also see some other examples in Box 5.7 – but there is a problem! When we are working with a sample, the values of the mean and the variance are both subject to chance variation, which means that we will only get an estimate of the true variance/mean ratio. If the distribution really is random it follows that we are unlikely to get a value of exactly 1. Conversely, a value greater than 1 does not necessarily imply that the pattern is clumped. This is, of course, nothing new – it is the persistent problem that we have met before and the solution as ever will be to employ a statistical test. It will tell us whether the ratio is too far away from 1 to be reasonably attributable to chance. We shall see how to do this in the next chapter.

---

**BOX 5.7 VARIANCE: MEAN RATIOS AS AN INDEX OF SPATIAL DISPERSION**

| Number of organisms per sampling unit ($Y$) | Observed frequency Simulium larvae | Tetrahymena | Nucella lapillus |
|---|---|---|---|
| 0 | 2 | 1 | 5 |
| 1 | 9 | 4 | 4 |
| 2 | 1 | 9 | 2 |
| 3 | | 2 | 0 |
| 4 | | 5 | 0 |
| 5 | | 3 | 1 |
| 6 | | 3 | 1 |
| 7 | | 1 | 1 |
| 8 | | 2 | 0 |
| 9 | | | 0 |
| 15 | | | 1 |
| $n$ | 12 | 30 | 15 |
| $\bar{Y}$ | 0.9 | 3.5 | 2.7 |
| $s^2$ | 0.3 | 4.7 | 16.8 |
| Variance/mean ($s^2/\bar{Y}$) or Index of dispersion ($I$) | 0.29 | 1.37 | 6.14 |

---

## The Poisson distribution as an approximation to the binomial distribution

We started our exploration of the Poisson distribution by thinking about a binomial. However, a Poisson distribution is only an approximation to the appropriate binomial and we ought to check on the validity of this approximation. In Figure 5.4 you will find some binomial and Poisson distributions compared.

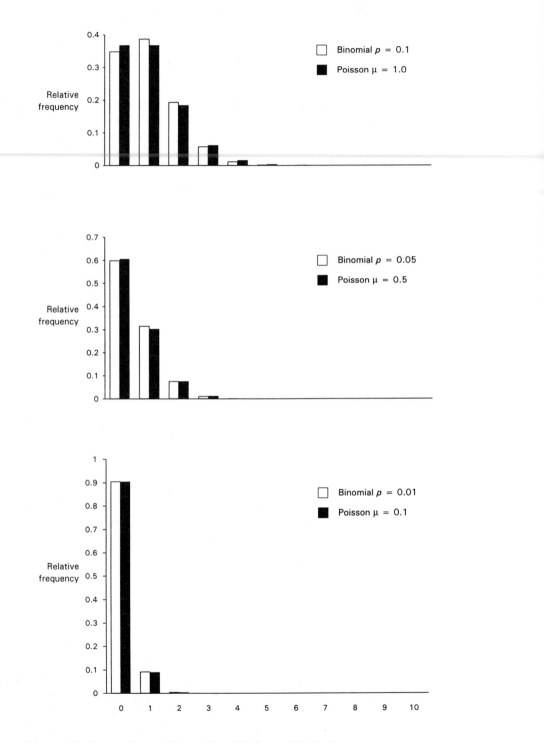

**Figure 5.4** Comparison of binomial and Poisson distributions.

The three binomial distributions are for different, extreme, values of $p$ (0.1, 0.05 and 0.01), but all are for a sample size of 10. We already know that the mean of a binomial distribution is given by the formula $\mu = np$, so we can calculate the means of these three distributions as 1.0, 0.5 and 0.1. If we use these values of $\mu$ in the Poisson tables (Table A3), we can find the three Poisson distributions which have the same means as the three binomials.

These are plotted alongside the relevant binomial distributions and you can see that the similarity between the two distributions increases as $p$ becomes smaller. The difference is hardly detectable in the case where $p = 0.01$, that is, where one outcome is very rare and the difference would be even smaller if we further decreased $p$ and/or increased the sample size.

This means that we could use a Poisson distribution (as an approximation) for the sampling distribution in our examination of the seed merchants claim of a germination rate of 99.9%. Suppose we find three non-germinating seeds in a batch of 100, what should we conclude? The probability of a seed not germinating is claimed to be 0.001. If we germinate a batch of 100 seeds the mean number of non-germinating seeds is expected to be 0.1 ($= np$). We can use this information to formulate a null hypothesis that the number of non-germinating seeds (0, 1, 2, etc.) in random samples of 100 would be expected to follow (approximately) a Poisson distribution with $\mu = 0.1$. Using Table A3 you should be able to verify that the probability of finding three or more non-germinating seeds is 0.0002. This would lead us to doubt the claim made.

## The normal distribution

The binomial and the Poisson distributions both apply to situations in which the variable is discontinuous. As we saw in Chapter 3, many variables are continuous so we need to find a mathematical model for their distribution. Once more, we can get some way towards the model by starting from the binomial and to see why this is so we shall have to look at the way in which continuous variation arises. It is easiest to do this by thinking about the genetic basis of phenotypic differences between individuals which is something we touched on in Chapter 1. Later, we shall see how the same line of reasoning can be applied to the two other ways in which continuously varying characteristics are produced. These are the action of the environment on phenotypes and the occurrence of experimental variation at various stages in a procedure.

### Genes, alleles and phenotypes

A gene is a sequence of bases in the DNA which carries the information to produce a specific protein. This protein may be an enzyme, a structural protein or a regulatory molecule (which switches other genes on and off) and it is through the production of these proteins at the correct time and place during development that the phenotype of the individual emerges. Changes in the sequence of bases, arising through mutation, give rise to variant forms of each gene called alleles. Different alleles of the same gene produce slightly different forms of the specific protein, which may differ in their properties and so produce different pheno-

103

types. During meiosis, independent assortment of chromosomes and crossing over between homologous chromosomes produces gametes carrying different combinations of alleles. The amount of genetic variability in most populations is such that every gamete is likely to be unique, so that when gametes fuse at fertilization to form zygotes, each new individual is likely to be genetically unique.

Some phenotypic differences are produced by alleles of just one gene and this tends to produce discontinuous phenotypic variation of the nominal type in which individuals can be assigned to one of a small number of categories. For instance, the ABO blood group pheno-types (A, B, AB and O) arise through the different combinations (in pairs) of three alleles of one gene. In contrast, phenotypic differences in variables such as height, weight and other morphological and physiological characteristics are controlled by many genes, each with a small effect. They are said to be polygenic. Because a large number of genes are involved, there are a very large number of possible combinations of the alleles of those genes. This leads to a large number of phenotypes with small differences between them, that is, a con-tinuously varying characteristic. It is relatively simple to show how this happens and how the resulting frequency distribution relates to the binomial.

## From the binomial to the normal distribution

Consider first a simple situation (Fig. 5.5a) which you should have come across in genetics where a phenotypic difference, such as red versus white flower colour, is controlled by one gene with two alleles (A and a). A is the allele for redness and it is incompletely domi-nant to the a allele. This simply means that both alleles will contribute equally to the char-acteristic and the heterozygote (Aa) will be pink rather than red. The effects of the two alleles are said to be additive. If these two alleles are at equal frequencies (0.5) in the (bio-logical) population, then the probability of an offspring receiving one A allele from each parent via a gamete (i.e. being AA) is $0.5 \times 0.5 = 0.25$. Likewise, the probability of an individual being aa will also be 0.25. The probability of an individual being a heterozygote will be $2 \times 0.5 \times 0.5 = 0.5$, because there are two different ways to get an Aa combina-tion. This is shown in the Punnett square on the left, which also indicates by the shading what the colour of each genotype will be. This gives us a familiar ratio of 1:2:1 and you may also recognize these probabilities as the terms of a binomial distribution with $n = 2$ and $p = 0.5$.

Why do the frequencies of the three genotypes follow a binomial distribution? It is be-cause each zygote is the result of two independent events and each event has two mutually exclusive outcomes. The event in this case relates to the formation of a gamete. Each dip-loid zygote results from the fusion of two gametes, a pollen grain carrying one allele and an ovule carrying one allele. Each parent has two of each allele so for each gamete one allele has to be "chosen". Each "choice" has two possible outcomes which are mutually exclu-sive, that is, the allele which ends up in the gamete can be A or a. The genotype of a zygote depends on which allele is carried by each of the two gametes involved and these will be in-dependent of one another; a pollen grain carrying one particular allele is equally likely to fuse with an ovule carrying either A or a. On this simple genetic model the characteristic will be variable and will have a discontinuous frequency distribution as shown in Figure 5.5a.

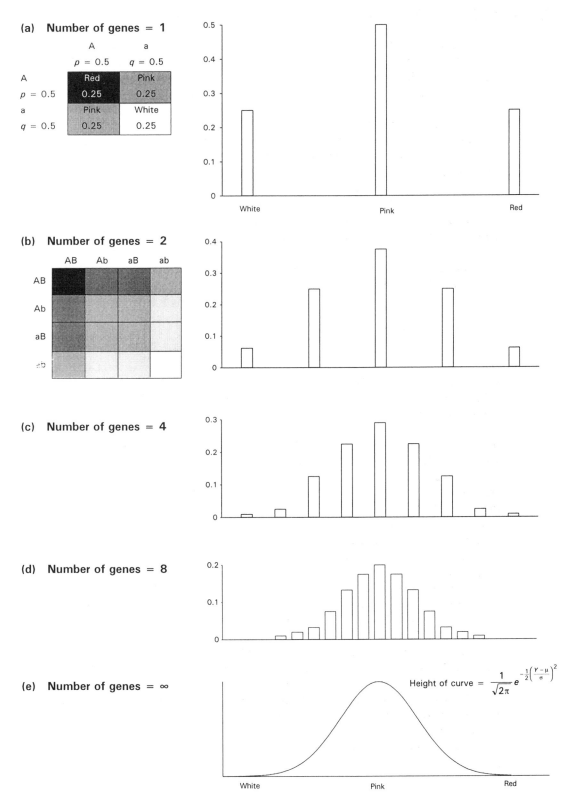

**Figure 5.5** From a binomial distribution to a normal distribution: a genetic model for a continuously varying characteristic (flower colour).

Now let's assume that a second gene, the B gene, is involved, and that this contributes equally to coloration in the same way as the A gene, i.e. there are two alleles B and b at equal frequencies and with additive effects. The different possible combinations of these alleles in diploid individuals are given in the Punnet square in Figure 5.5b. Now there are four alleles to be "chosen" for each individual and the probabilities of having 0, 1, 2, 3 and 4 alleles for redness will be given by the binomial distribution for $n = 4$ and $p = 0.5$. AABB individuals with four alleles for redness will still be red, aabb individuals will be white and individuals with two alleles for redness (AAbb, aaBB, AaBb) will be pink. But now there are two additional classes of genotypes, those with three alleles for redness and those with only one. These produce two new intermediate phenotypes, dark and light pink. The resulting frequency distribution of the variable is as shown in Figure 5.5b.

Having established that the frequency distribution of the different phenotypes follows a binomial distribution it is an easy matter to find out what happens when we increase the number of genes involved. We simply consult the binomial tables. The frequency distributions for four genes ($n = 8$ alleles) and eight genes ($n = 16$ alleles) are shown in Figure 5.5c,d. As the number of genes increases, so we increase the number of phenotypes with different, intermediate shades of colour. Since the extreme phenotypes are always the same (red and white), the steps or discontinuities between the intermediates must get progressively smaller as the number of genes increases. Eventually, in theory, with an infinite number of genes, the gaps disappear altogether and the resulting binomial distribution is a continuous distribution (Fig. 5.5e). This is known as a **normal distribution**. It can be described mathematically using the fairly complicated formula in Figure 5.5e, and it forms the basis of many statistical tests.

Now, by this stage, you might be saying to yourself that the model is not very realistic. It needs an infinite number of genes, each contributing equally to the variable, with only two alleles of each gene, at equal frequencies and with additive effects. In fact, it can be shown mathematically that these assumptions can be substantially relaxed without unduly affecting the validity of the model. In practice, a continuous distribution is likely to be produced with quite a small number of genes because of environmental effects on the phenotypes and errors in measurement. Both these sources of variation will tend to blur the distinction between the phenotypes. For example, in a classic demonstration of the genetic basis of continuous variation, ear colour in wheat was shown to be controlled by only three genes, yet the distribution was continuous.

The model can also be modified to account for discontinuous variables such as clutch size which are also controlled by a number of genes. All we need to assume is that some underlying characteristic (e.g. amount of energy available for reproduction) has a continuous distribution but that, in some way, this gets divided up into discrete packages.

## Environmental and experimental variation

We can construct similar models for the effects of environmental factors and experimental variation. A variable such as plant height is likely to be affected by many environmental factors, for instance, light, water and nutrient availability. Each environmental factor can be thought of as occurring in a variety of different states (high light, low light, intermediate light). Some of these states (e.g. high light) will lead to an increase in the variable; we could label these as "+" states. Others (e.g. low light) will lead to a decrease and could be labelled "–" states and the effects of different factors are likely to be additive. This means that if high light increases plant height by three units and low nutrient level decreases height by one units then plants in high light and low nutrients will increase in height by two units.

Most individuals will experience, by chance, a fairly even mixture of (+) and (–) states for the different factors, just as most individuals in the genetic example end up with a fairly balanced mixture of alleles for redness and whiteness. It is less likely that an individual will experience an extreme environment in which most factors are either in the positive or the negative state. As a result, most individuals will have intermediate heights and, the more extreme the height, the rarer it will be. Once again, if there are several factors involved, each with several states, the resulting distribution of plant height will be a normal distribution even if all individuals are genetically the same.

The same argument will explain how the cumulative effect of experimental errors will produce a normal distribution of measurements. In the example in "Machines and techniques" in Chapter 1 we saw how each source of error can be thought of as either increasing (+) or decreasing (–) the final result. By chance, repeated measurements will end up with a different mixture of (+) and (–) errors, although the numbers of (+) and (–) errors are likely to be more or less equal. Again, it will be rather unlikely that all the errors will be in one direction. Since the effect of errors at different stages will be additive, a normal distribution of measurements will result.

Although we have treated these three sources of variation separately, it should be obvious that, in many situations of biological interest, they will act together. In our own species, it can be deduced that the continuous variation in skin pigmentation in the descendants of marriages between people with different skin colour is due to just three or four genes, but skin colour is also affected by the environment and measurements of skin colour are subject to experimental variation.

## Properties of normal distributions

A normal distribution is a frequency or probability distribution, which is smooth, symmetrical and bell-shaped as shown in Figure 5.5e. The formula, which can be used to work out the height of the curve, is not something that the average biologist would want to get involved with, although it does allow the probabilities of occurrences to be calculated. These can then be used in the same way we used the probabilities associated with the binomial distribution – that is to decide whether an event is likely or unlikely.

There is, however, one major difference which arises from the fact that this is a continuous distribution. With the binomial and the Poisson distributions, we could calculate the probability of a particular value of the variable occurring, for instance eight yellow kernels in a sample of 20 or a plate with two colonies. We cannot do this with a continuous distribution because (theoretically) there is an infinite number of different values which the variable could take and each one therefore has a zero probability of occurring. In practice, this is not a problem because, as we saw in Chapter 2, we are usually interested in the probability of obtaining a value as extreme as, or more extreme than, some particular value. This is a cumulative probability which is equal to an area under the curve and this can be calculated for a continuous distribution.

From what has been said in the previous sections, it is obvious if we think of all the different variables that could be measured, that there are many, many different normal distributions. However, if we look at the formula it turns out that only two parameters are necessary to define any particular distribution. The equation gives us the height of the curve for a given value of the variable $Y$ which appears on the righthand side of the equation along with two constants, $\pi$ and e. The other two symbols, $\mu$ and $\sigma$, should also be familiar; they are the mean and the standard deviation, respectively, and they will be characteristic of the particular variable being studied. Two things of interest follow from this formula. First, differences in either the mean or the standard deviation or both are the only two things which can account for the huge variety of different normal distributions which occur. Secondly, any normal distribution can be described exactly, provided that we know the values of the mean and the standard deviation.

Figure 5.6 illustrates this point with a hypothetical example in which biological factors produce different normal distributions of the same variable, namely the height of individual plants of one species. In (a) the plants have been growing in their natural environment, while in (b) fertilizer has been applied to the plants, also in their natural environment. In (c) the plants have been grown under controlled, optimal, conditions in a greenhouse, while the plants in (d) have all been produced from one plant by taking cuttings and growing them under the same conditions as (c). They are all normal distributions and in each one the mean is indicated by the solid line, while the two dotted lines on either side indicate the standard deviation. The size of the standard deviation is a measure of how spread-out the distribution is. We can see that the distributions differ from one another in their location, i.e. their means ($\mu$), in their "spread-outness", i.e. their standard deviations ($\sigma$), or in both parameters.

The biological reasons for these differences are straightforward. In Figure 5.6a the plants are both genetically different from one another and, because the natural environment is spatially heterogeneous, different individuals are exposed to different environments. As a result of these genetic and environmental differences, plant height is variable and the distribution has a wide spread, i.e. a large standard deviation. In Figure 5.6b there are still genetic and environmental differences between individuals but the environment is improved by the addition of fertilizer. Overall, plants grow taller under these conditions so that the mean of the distribution is now greater, but the standard deviation is unchanged. In Figure 5.6c the mean is even higher than in Figure 5.6b, because the greenhouse environment is better for growth, while the variation in height between individuals has been reduced because all the plants are exposed to a similar environment. In Figure 5.6d the plants, on average, grow very well because of the good environment, but now the standard deviation of the distribution is even smaller than in Figure 5.6c. This is because not only is the environment similar for all the plants but they are also genetically identical, because they originated as cuttings from a single plant.

Despite the fact that these distributions differ as described above, they share one very important feature, which is common to all normal distributions. Put simply, the proportion of the distribution between the two vertical, dotted lines is the same for all four distributions. The vertical lines on either side of the mean, mark off values of plant height ($Y$) which are one standard deviation below the mean ($\mu - \sigma$) and one standard deviation above the mean ($\mu + \sigma$). In Figure 5.6a, the mean is 20 cm and the standard deviation is 5 cm, so the values of $Y$ would be 15 cm and 25 cm, respectively. The equivalent lines on the other three distributions obviously correspond to different plant heights, because the means and/or the standard deviations of the distributions are not the same. Irrespective of these differences in $\mu$ and $\sigma$, it is a mathematical property of any normal distribution that the propor-

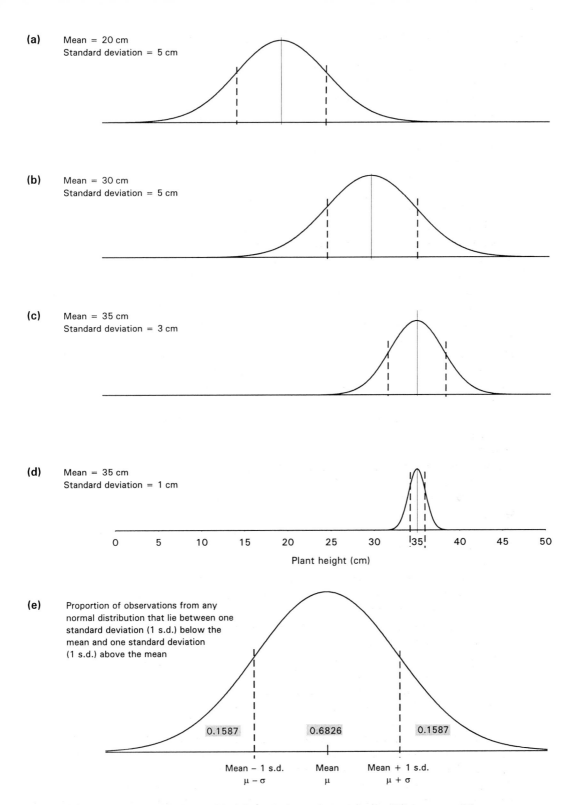

**Figure 5.6** Normal distributions of heights of plants grown under different conditions.

tion of the distribution which lies between these two limits is always 0.6826 (Fig. 5.6e). In other words 68.26% of all the observations in any normal distribution have values that are larger than one standard deviation below the mean and smaller than one standard deviation above the mean. In the case of Figure 5.6a, this would mean that about 68 out of every 100 plants would have heights in the range 15–25 cm.

It follows from this that if we take one observation at random from this normal distribution, then the probability that it will lie within this range is 0.6826. Looked at another way, the proportion of observations which are outside these limits is 0.3174 and, because of the symmetrical nature of the distribution, 0.1587 lies in each tail (Fig. 5.6e). Thus, for example, the probability that a plant taken at random from the natural environment (Fig. 5.6a) is as short as, or shorter than, 15 cm is 0.1587. Likewise, the probability that a random plant is as tall as, or taller than, 25 cm is 0.1587. This is simply putting values to what is obvious from the graph, namely that very tall and very short plants are less likely to occur than plants with intermediate heights.

The same principle applies to any other multiple of the standard deviation. Moving further out in both directions to the points where $Y = \mu - 1.96\sigma$ and $Y = \mu + 1.96\sigma$, the proportion of the distribution between these limits is 0.95 or 95%, the proportion outside is 0.05 or 5% (Fig. 5.7). Again, because the distribution is symmetrical 2.5% (0.025) of the distribution will lie below the lower limit and 2.5% (0.025) will lie above the upper limit. If you are beginning to think that this looks rather like the acceptance and rejection regions that we used with the binomial distribution, then you are correct. What it means is that for any normal distribution most of the observations (actually 95%) lie within the range $\mu \pm 1.96\sigma$ and only 5% lie outside these limits. So, if we were to take any one observation ($Y$) at random from such a distribution, the probability that it would lie outside these limits is only 0.05.

You can see from all this that normal distributions are not only reasonable models for certain sorts of variables, but that they also have useful properties which allow us to make decisions about whether an observation is likely or unlikely.

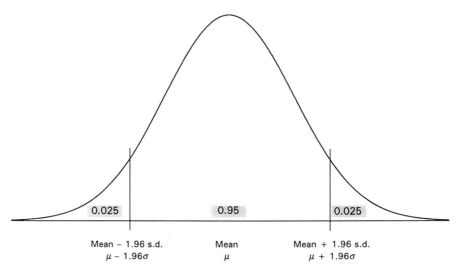

**Figure 5.7** Proportion of the observations from any normal distribution which lie between 1.96 standard deviations below the mean and 1.96 standard deviations above the mean.

## Standardization

Having grasped this theoretical property of all normal distributions we have to cope with a slight practical problem. Suppose that we have an observation, $Y$, taken at random from some normal distribution whose mean and standard deviation we know. We wish to find out the associated probability, that is, the chance of getting a value which is as extreme as, or more extreme than, the one we have obtained. How could we find this probability? One way would be to calculate the probability from first principles, using the formula for the normal distribution, but this would involve integration to get the relevant area under the curve. Another way would be to have a set of tables (as for the binomial distribution) which gave us the probability directly, but this would be totally impractical because, as I said earlier, there are literally millions of different normal distributions. Each one would require its own table. (Incidentally, exactly the same is true of binomial distributions, because there are so many different values of $n$ and $p$. However, here there are some values of $p$, corresponding to very common situations which are frequently needed.)

To get over this problem and still use tables, what we do is convert our particular normal distribution into a standard form which can be looked up in a table. How to do this follows simply and directly from the ideas discussed in the previous section.

We can see from Figure 5.6 that what is important is not the absolute value of $Y$ on its own, but rather the difference between $Y$ and the mean. A very small value of $Y$ is not unusual if the mean of the distribution is very small, but it is unusual if the mean of the distribution is large. If we now focus on this difference $(Y - \mu)$, again the absolute value on its own is meaningless, it has to be seen in the context of how widely spread out the distribution is, that is, in relation to $\sigma$. A large value of $(Y - \mu)$, that is a value of $Y$ which is a long way from the mean, is not unusual if the distribution has a large standard deviation, but it is unusual if the distribution has a small standard deviation. So what is important is the size of the $(Y - \mu)$ relative to $\sigma$ and the question we really need to answer is "How many standard deviations away from the mean of the distribution $\mu$ is our observation $Y$?". Thus we need to calculate $(Y - \mu)/\sigma$.

This answer is a value of something called $z$ and it can be either positive or negative, depending on whether our observation is larger or smaller than the mean. Converting a value of $Y$ to a value of $z$ is called **standardization**. Its effect is to convert any normal distribution to one standard normal distribution with a mean of zero and a standard deviation of 1. Then all we need to have is a table of probabilities relating to this one standard normal distribution. A value of $z$ obtained in this way can therefore be used as a test statistic, because its sampling distribution is known.

If you find standardization difficult to visualize, I have illustrated it in action in Figure 5.8 by considering a plant of height 32.4 cm taken from our hypothetical distribution with a mean of 20 cm and a standard deviation of 5 cm. First, we imagine taking the mean away from every value in the distribution which will shift the whole distribution to the left and give it a mean of zero. Our observation of 32.4 now has a value of 12.4. We then divide by the standard deviation, to convert from the original units to units of standard deviation. Now, the distribution has a standard deviation of 1 instead of 5. This converts our value of 12.4 to 2.48 and tells us how many standard deviations above the mean our original value is. If you are still finding it difficult, then remember that all it is doing is getting rid of the absolute values of $Y$, $\mu$ and $\sigma$ (which will be different for different distributions) by converting them into a single value $z$, which is simply the number of standard deviations that the observation is away from the mean.

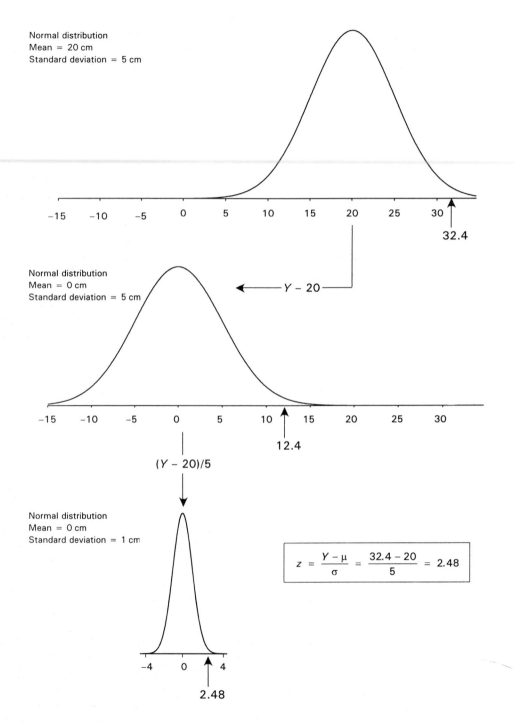

**Figure 5.8** Standardizing a normal curve.

You may be a bit suspicious of this manoeuvre since it looks like a sleight of hand to make things that are different look the same! If that is a worry think of it as exactly the same sort of procedure as you would use to convert temperatures from degrees Fahrenheit or degrees Kelvin to degrees Celsius. In practice, it means that one table is sufficient to describe the probabilities associated with any normal distribution.

## Tables of the standard normal distribution

Tables A4 and A5 both relate to the standard normal distribution and are used to find the associated probability for a given value of $z$. Remember that this is not the probability of a particular value of $z$, but the probability of that value and all the more extreme values.

Table A4 (reproduced in part as Table 5.1) lists a range of values of z (to two decimal places) and for each one, gives the proportion of the standard normal distribution which lies to the left of (i.e. below) that value. The values of $z$ (to one decimal place) are given in the shaded column down the left-hand side of the table, while the second decimal place is given in the shaded row along the top of the table. The proportions of the distribution are given in the body of the table.

Let's see how these tables work by asking a question to which we already know the answer. What proportion of a normal distribution lies beyond a point 1.96 standard deviations below the mean? Look down the column of values of z until you find the value –1.9 and then look across that row. The figure in the column headed 6, gives us the required

Table 5.1   Proportions of the area under the standard normal curve.

| z | 0.00 | 0.01 | 0.02 | 0.03 | 0.04 | 0.05 | 0.06 | 0.07 | 0.08 | 0.09 |
|---|---|---|---|---|---|---|---|---|---|---|
| -3.0 | 0.0013 | 0.0010 | 0.0007 | 0.0005 | 0.0003 | 0.0002 | 0.0002 | 0.0001 | 0.0001 | 0.0000 |
| -2.9 | 0.0019 | 0.0018 | 0.0017 | 0.0017 | 0.0016 | 0.0016 | 0.0015 | 0.0015 | 0.0014 | 0.0014 |
| -2.8 | 0.0026 | 0.0025 | 0.0024 | 0.0023 | 0.0023 | 0.0022 | 0.0021 | 0.0020 | 0.0020 | 0.0019 |
| -2.7 | 0.0035 | 0.0034 | 0.0033 | 0.0032 | 0.0031 | 0.0030 | 0.0029 | 0.0028 | 0.0027 | 0.0026 |
| -2.6 | 0.0047 | 0.0045 | 0.0044 | 0.0043 | 0.0041 | 0.0040 | 0.0039 | 0.0038 | 0.0037 | 0.0036 |
| -2.5 | 0.0062 | 0.0060 | 0.0059 | 0.0057 | 0.0055 | 0.0054 | 0.0052 | 0.0051 | 0.0049 | 0.0048 |
| -2.4 | 0.0082 | 0.0080 | 0.0078 | 0.0075 | 0.0073 | 0.0071 | 0.0069 | 0.0068 | 0.0066 | 0.0064 |
| -2.3 | 0.0107 | 0.0104 | 0.0102 | 0.0099 | 0.0096 | 0.0094 | 0.0091 | 0.0089 | 0.0087 | 0.0084 |
| -2.2 | 0.0139 | 0.0136 | 0.0132 | 0.0129 | 0.0126 | 0.0122 | 0.0119 | 0.0116 | 0.0113 | 0.0110 |
| -2.1 | 0.0179 | 0.0174 | 0.0170 | 0.0166 | 0.0162 | 0.0158 | 0.0154 | 0.0150 | 0.0146 | 0.0143 |
| -2.0 | 0.0228 | 0.0222 | 0.0217 | 0.0212 | 0.0207 | 0.0202 | 0.0197 | 0.0192 | 0.0188 | 0.0183 |
| -1.9 | 0.0287 | 0.0281 | 0.0274 | 0.0268 | 0.0262 | 0.0256 | 0.0250 | 0.0244 | 0.0238 | 0.0233 |
| -1.8 | 0.0359 | 0.0352 | 0.0344 | 0.0336 | 0.0329 | 0.0322 | 0.0314 | 0.0307 | 0.0300 | 0.0294 |
| -1.7 | 0.0446 | 0.0436 | 0.0427 | 0.0418 | 0.0409 | 0.0401 | 0.0392 | 0.0384 | 0.0375 | 0.0367 |
| -1.6 | 0.0548 | 0.0537 | 0.0526 | 0.0516 | 0.0505 | 0.0495 | 0.0485 | 0.0475 | 0.0465 | 0.0455 |
| -1.5 | 0.0668 | 0.0655 | 0.0643 | 0.0630 | 0.0618 | 0.0606 | 0.0594 | 0.0582 | 0.0570 | 0.0559 |
| -1.4 | 0.0808 | 0.0793 | 0.0778 | 0.0764 | 0.0749 | 0.0735 | 0.0722 | 0.0708 | 0.0694 | 0.0681 |
| -1.3 | 0.0968 | 0.0951 | 0.0934 | 0.0918 | 0.0901 | 0.0885 | 0.0869 | 0.0853 | 0.0838 | 0.0823 |
| -1.2 | 0.1151 | 0.1131 | 0.1112 | 0.1093 | 0.1075 | 0.1056 | 0.1038 | 0.1020 | 0.1003 | 0.0985 |
| -1.1 | 0.1357 | 0.1335 | 0.1314 | 0.1292 | 0.1271 | 0.1251 | 0.1230 | 0.1210 | 0.1190 | 0.1170 |
| -1.0 | 0.1587 | 0.1562 | 0.1539 | 0.1515 | 0.1492 | 0.1469 | 0.1446 | 0.1423 | 0.1401 | 0.1379 |
| -0.9 | 0.1841 | 0.1814 | 0.1788 | 0.1762 | 0.1736 | 0.1711 | 0.1685 | 0.1660 | 0.1635 | 0.1611 |
| -0.8 | 0.2119 | 0.2090 | 0.2061 | 0.2033 | 0.2005 | 0.1977 | 0.1949 | 0.1922 | 0.1894 | 0.1867 |
| -0.7 | 0.2420 | 0.2389 | 0.2358 | 0.2327 | 0.2297 | 0.2266 | 0.2236 | 0.2206 | 0.2177 | 0.2148 |
| -0.6 | 0.2743 | 0.2709 | 0.2676 | 0.2643 | 0.2611 | 0.2578 | 0.2546 | 0.2514 | 0.2483 | 0.2451 |
| -0.5 | 0.3085 | 0.3050 | 0.3015 | 0.2981 | 0.2946 | 0.2912 | 0.2877 | 0.2843 | 0.2810 | 0.2776 |

(continued overleaf)

| z | 0.00 | 0.01 | 0.02 | 0.03 | 0.04 | 0.05 | 0.06 | 0.07 | 0.08 | 0.09 |
|---|---|---|---|---|---|---|---|---|---|---|
| -0.4 | 0.3446 | 0.3409 | 0.3372 | 0.3336 | 0.3300 | 0.3264 | 0.3228 | 0.3192 | 0.3156 | 0.3121 |
| -0.3 | 0.3821 | 0.3783 | 0.3745 | 0.3707 | 0.3669 | 0.3632 | 0.3594 | 0.3557 | 0.3520 | 0.3483 |
| -0.2 | 0.4207 | 0.4168 | 0.4129 | 0.4090 | 0.4052 | 0.4013 | 0.3974 | 0.3936 | 0.3897 | 0.3859 |
| -0.1 | 0.4602 | 0.4562 | 0.4522 | 0.4483 | 0.4443 | 0.4404 | 0.4364 | 0.4325 | 0.4286 | 0.4247 |
| -0.0 | 0.5000 | 0.4960 | 0.4920 | 0.4880 | 0.4840 | 0.4801 | 0.4761 | 0.4721 | 0.4681 | 0.4641 |
| 0.0 | 0.5000 | 0.5040 | 0.5080 | 0.5120 | 0.5160 | 0.5199 | 0.5239 | 0.5279 | 0.5319 | 0.5359 |
| 0.1 | 0.5398 | 0.5438 | 0.5478 | 0.5517 | 0.5557 | 0.5596 | 0.5636 | 0.5675 | 0.5714 | 0.5753 |
| 0.2 | 0.5793 | 0.5832 | 0.5871 | 0.5910 | 0.5948 | 0.5987 | 0.6026 | 0.6064 | 0.6103 | 0.6141 |
| 0.3 | 0.6179 | 0.6217 | 0.6255 | 0.6293 | 0.6331 | 0.6368 | 0.6406 | 0.6443 | 0.6480 | 0.6517 |
| 0.4 | 0.6554 | 0.6591 | 0.6628 | 0.6664 | 0.6700 | 0.6736 | 0.6772 | 0.6808 | 0.6844 | 0.6879 |
| 0.5 | 0.6915 | 0.6950 | 0.6985 | 0.7019 | 0.7054 | 0.7088 | 0.7123 | 0.7157 | 0.7190 | 0.7224 |
| 0.6 | 0.7257 | 0.7291 | 0.7324 | 0.7357 | 0.7389 | 0.7422 | 0.7454 | 0.7486 | 0.7517 | 0.7549 |
| 0.7 | 0.7580 | 0.7611 | 0.7642 | 0.7673 | 0.7703 | 0.7734 | 0.7764 | 0.7794 | 0.7823 | 0.7852 |
| 0.8 | 0.7881 | 0.7910 | 0.7939 | 0.7967 | 0.7995 | 0.8023 | 0.8051 | 0.8078 | 0.8106 | 0.8133 |
| 0.9 | 0.8159 | 0.8186 | 0.8212 | 0.8238 | 0.8264 | 0.8289 | 0.8315 | 0.8340 | 0.8365 | 0.8389 |
| 1.0 | 0.8413 | 0.8438 | 0.8461 | 0.8485 | 0.8508 | 0.8531 | 0.8554 | 0.8577 | 0.8599 | 0.8621 |
| 1.1 | 0.8643 | 0.8665 | 0.8686 | 0.8708 | 0.8729 | 0.8749 | 0.8770 | 0.8790 | 0.8810 | 0.8830 |
| 1.2 | 0.8849 | 0.8869 | 0.8888 | 0.8907 | 0.8925 | 0.8944 | 0.8962 | 0.8980 | 0.8997 | 0.9015 |
| 1.3 | 0.9032 | 0.9049 | 0.9066 | 0.9082 | 0.9099 | 0.9115 | 0.9131 | 0.9147 | 0.9162 | 0.9177 |
| 1.4 | 0.9192 | 0.9207 | 0.9222 | 0.9236 | 0.9251 | 0.9265 | 0.9278 | 0.9292 | 0.9306 | 0.9319 |
| 1.5 | 0.9332 | 0.9345 | 0.9357 | 0.9370 | 0.9382 | 0.9394 | 0.9406 | 0.9418 | 0.9430 | 0.9441 |
| 1.6 | 0.9452 | 0.9463 | 0.9474 | 0.9484 | 0.9495 | 0.9505 | 0.9515 | 0.9525 | 0.9535 | 0.9545 |
| 1.7 | 0.9554 | 0.9564 | 0.9573 | 0.9582 | 0.9591 | 0.9599 | 0.9608 | 0.9616 | 0.9625 | 0.9633 |
| 1.8 | 0.9641 | 0.9648 | 0.9656 | 0.9664 | 0.9671 | 0.9678 | 0.9686 | 0.9693 | 0.9700 | 0.9706 |
| 1.9 | 0.9713 | 0.9719 | 0.9726 | 0.9732 | 0.9738 | 0.9744 | 0.9750 | 0.9756 | 0.9762 | 0.9767 |
| 2.0 | 0.9772 | 0.9778 | 0.9783 | 0.9788 | 0.9793 | 0.9798 | 0.9803 | 0.9808 | 0.9812 | 0.9817 |
| 2.1 | 0.9821 | 0.9826 | 0.9830 | 0.9834 | 0.9838 | 0.9842 | 0.9846 | 0.9850 | 0.9854 | 0.9857 |
| 2.2 | 0.9861 | 0.9864 | 0.9868 | 0.9871 | 0.9874 | 0.9878 | 0.9881 | 0.9884 | 0.9887 | 0.9890 |
| 2.3 | 0.9893 | 0.9896 | 0.9898 | 0.9901 | 0.9904 | 0.9906 | 0.9909 | 0.9911 | 0.9913 | 0.9916 |
| 2.4 | 0.9918 | 0.9920 | 0.9922 | 0.9925 | 0.9927 | 0.9929 | 0.9931 | 0.9932 | 0.9934 | 0.9936 |
| 2.5 | 0.9938 | 0.9940 | 0.9941 | 0.9943 | 0.9945 | 0.9946 | 0.9948 | 0.9949 | 0.9951 | 0.9952 |

value, which is 0.025. Should we need it, we can also find (by subtraction) the proportion lying above a given value of $z$. When $z = +1.96$, the table tells us the proportion of the distribution below this value is 0.975 which means that the proportion above $z = +1.96$ is $1 - 0.975 = 0.025$.

We can use the table the other way round to find the value of $z$ which cuts off a specified proportion of the curve. To do this we find the appropriate proportion in the body of the table and read off the value of $z$ in the margins. For example, what value of $z$ cuts off 0.005 in the upper tail of the distribution? This will be the same value of $z$ which has $1 - 0.005 = 0.995$ of the distribution below it. The nearest we can get to 0.995 in the table is either 0.9949 or 0.9951, so $z$ obviously lies between 2.57 and 2.58. Should we wish to give an additional decimal place in our value of $z$ we can do this approximately using what is called linear interpolation. This works on the basis that, because (in this case) the required proportion (0.995) is halfway between the tabled proportions (0.9949 and 0.9951), the required value of $z$ will be halfway between the two tabled values of $z$, i.e. 2.575. This should be accurate enough for most purposes; the correct value in this case is actually 2.5758.

## Calculating the expected frequencies for a normal distribution

Table A4 is most useful because it enables us to compare an actual frequency distribution with a normal distribution which has the same mean and standard deviation. We can use it to find out how many observations would be expected to fall into each class interval, just as

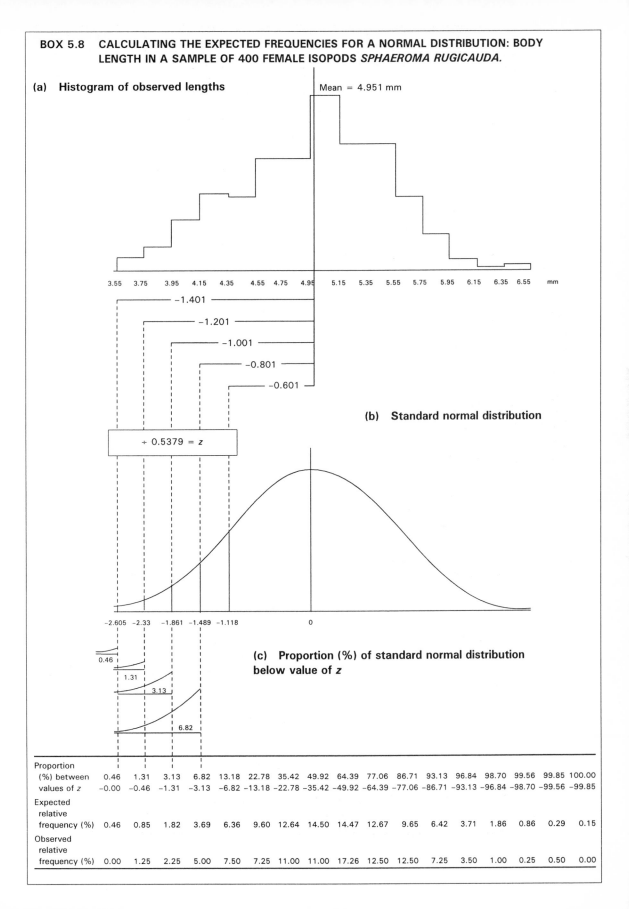

BOX 5.8 CALCULATING THE EXPECTED FREQUENCIES FOR A NORMAL DISTRIBUTION: BODY LENGTH IN A SAMPLE OF 400 FEMALE ISOPODS *SPHAEROMA RUGICAUDA*.

we did for binomial and Poisson distributions. These expected values can be compared with those observed. The example in Box 5.8 shows you how to do this. The top histogram is the actual frequency distribution of the 400 body lengths which we have used before (Box 4.3), with a minor change made to the horizontal axis. The values given are not the class marks but the upper class limits. For example, in Box 4.3, the first class had a class mark of 3.65 mm, but the upper class limit was 3.75 mm. The horizontal lines below this and the figures on the lines simply tell us how far (in millimetres) each class limit deviates from the mean of the distribution (which is indicated by the vertical line), these values being obtained by subtraction.

These deviates have then been divided by the standard deviation of the distribution to convert them to values of $z$, which are shown marked on the standard normal distribution. From Table A4 we can find what proportion of the normal curve lies to the left of each of these values of $z$, and these proportions are given as percentages. The percentage of the normal curve lying between two adjacent values of $z$ can be obtained by subtraction. For example, 0.46% of the distribution lies to the left of $z = -2.605$ and 1.31% of the distribution lies to the left of $z = -2.233$. So the proportion between these two values of $z$ is $1.31 - 0.46 = 0.85\%$. The proportion lying between $z = -2.233$ and $z = -1.861$ is $3.13 - 1.31 = 1.82\%$, etc. This value is, of course, the same thing as the relative frequency of observations expected in this interval which can be compared to the observed relative frequency in the final line. The two sets of figures are not identical, although there is reasonable agreement and once again there are two possible explanations for the discrepancies. Either the population of body lengths is normally distributed and sampling variation accounts for the disparities or they arise because body length is not normally distributed. We shall see how to distinguish between these two explanations in the next chapter.

## Critical values of $z$

Table A5, which is reproduced in Table 5.2, is a table of critical values of $z$. The values in it have been taken from Table A4, but it is much simpler because in many applications there

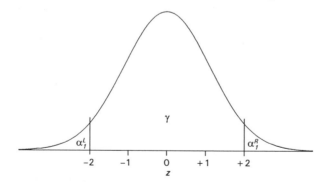

**Table 5.2  Critical values of $z$ (Table A5).**

| $\alpha_1^R$ | 10% | 5% | 2.5% | 1% | 0.5% | 0.1% | 0.05% |
|---|---|---|---|---|---|---|---|
| $\alpha_2$ | 20% | 10% | 5% | 2% | 1% | 0.2% | 0.1% |
| $\gamma$ | 80% | 90% | 95% | 98% | 99% | 99.8% | 99.9% |
| $z$ | 1.2816 | 1.6449 | 1.9600 | 2.3263 | 2.5758 | 3.0902 | 3.2905 |

are only a few commonly used values of $z$ that you need to be aware of. These values are the critical values which relate to the different significance levels commonly used in biological work. If you cast your mind back to the woodlouse experiment, you will remember that we divided the binomial distribution up into a central portion, the acceptance region comprising likely results, and a rejection region in the two tails comprising the more extreme, less likely results. The size of the rejection region was set by the significance level ($\alpha$), which in that case was 5% (0.05). We were then able to identify an upper and a lower critical value which marked the borderlines between the acceptance region and the two rejection regions. We now need to do exactly the same thing with the distribution of $z$ so that we have a handy list of critical values for different significance levels. Values of $z$ calculated from data can then be compared with these critical values.

The notation involved all looks rather complicated but its meaning is illustrated on the diagram of the distribution at the top. An equivalent notation is used in the other tables of critical values so you will only need to learn it once! The areas in the left-hand (or lower) tail and the right-hand (or upper) tail of the distribution, are labelled $\alpha_I^L$ and $\alpha_I^R$ respectively. The first row at the head of the table, gives selected values of $\alpha_I^R$, the percentage of the distribution, which lies in the right-hand tail. Taken together with the equivalent left-hand tail they make up the proportion in both tails, labelled $\alpha_2$, values of which are given in the second row at the head of the table. The highlighted values of $\alpha$ correspond to the commonly used significance levels.

In the body of the table, in the appropriate column, are the values of $z$ corresponding to the values of $\alpha$. For example, we can pick out the familiar value of $z = 1.96$ as the value which cuts off 2.5% in each tail ($\alpha_I^R = 2.5\%$) and 5% in both tails ($\alpha_2 = 5\%$). Strictly, when we are considering the left-hand, lower tail the value of $z$ is $-1.96$ and so for $\alpha_2$, there are two values of $z$ involved, an upper and a lower one. In the example just used, these are $-1.96$ and $+1.96$. However, because the distribution is symmetrical the absolute values (i.e. the values ignoring the sign) are always the same. To save space only one value without a sign is given and, as we shall see, in many situations we can ignore the sign anyway. This table shows clearly what we already know which is that the more extreme the value of $z$, the less of the distribution lies beyond it.

The portion of the distribution which lies between an upper positive value of $z$ and a lower negative value, is called the **confidence level** and is labelled $\gamma$. A selection of values of $\gamma$ (as percentages) is given in the third row at the head of the table, highlighted values being those most commonly used. You can see that the value of $\alpha_2$ and the value of $\gamma$ add up to 100% because, together, these two areas account for the whole of the area under the curve. Confidence levels relate to the reliability of estimates, something we shall deal with in Chapter 7.

## Beak size in Darwin's finches: a simple example of the use of $z$

Table A5 can be put to a variety of uses which we shall come across later. For the moment I shall use a simple example (Box 5.9) to illustrate what has so far been a rather theoretical treatment.

The large cactus finch (*Geospiza conirostris*) is one of Darwin's finches and is found on several of the Galapagos islands. Suppose that mean beak depth of males is known to be 10.76 mm, with a standard deviation of 0.799 mm. An individual is captured with a beak

---

**BOX 5.9  CALCULATING A VALUE OF $z$: COMPARISON OF BEAK DEPTH IN AN INDIVIDUAL LARGE CACTUS FINCH (*GEOSPIZA CONIROSTRIS*) WITH THE MEAN OF A POPULATION**

Mean beak depth = 10.76 mm

Standard deviation of beak depth = 0.799 mm

Beak depth of individual = 13.0 mm

Calculated value of $z = \dfrac{13.0 - 10.76}{0.799} = 2.803$

Critical value of $z = 1.9600$

Probability of obtaining value of $z$ as extreme as, or more extreme than, 2.803 is less than 0.01.

---

depth of 13.0 mm – does it belong to this population? Note that the "it" in the question is the measurement, not the bird, and that population here means "population of variates", that is "population of beak measurements" rather than a biological population of *G. conirostris*.

We can rephrase this question as follows. "How many standard deviations away from the mean of the population is this one observation and what is the associated probability?". To do this we need to calculate a value of $z$ and refer it to Table A5 where we find the critical values at the 5% significance level to be –1.96 and +1.96 ($\pm 1.96$ for short). Our calculated value of 2.803 is more extreme than 1.96. Indeed we can see from the table that the critical value of $z$ at the 1% level is 2.5758, which means that values of $z$ as or more extreme than this will only occur one time in 1000. Our value of $z = 2.803$ is clearly an extreme, unlikely value. So, although there is always the chance that this variate might belong to the specified distribution, a more reasonable conclusion would be that it did not. This is the statistical interpretation of what we have found, but what is the biological explanation? A beak of this size is intermediate between the mean of *G. conirostris* and the mean of another species *G. magnirostris*, the large ground finch, which lives on the same island. It could be that the individual concerned is a hybrid between these species.

In order to carry out this analysis, we have had to assume that both $\mu$ and $\sigma$ are known, whereas in practice we are only likely to have estimates based on a sample. If the samples were small, then these estimates could be far from the true values and this type of calculation would lead to unreliable conclusions. In this case the estimates of $\mu$ and $\sigma$ were based on quite large samples ($n = 246$) so that the procedure followed, although an approximation, is fairly reasonable. We'll see later how to modify things to deal with small samples. We must also remember that the analysis is based on the variable having a normal distribution, so this is a convenient point at which to look at some of the reasons why variables might not be normally distributed.

## How usual are normal distributions?

The normal distribution looks as though it ought to apply to a wide range of continuous variables and, as such, should form a useful model for making decisions about likely and unlikely occurrences. However, we need to be careful because the real world of the biologist

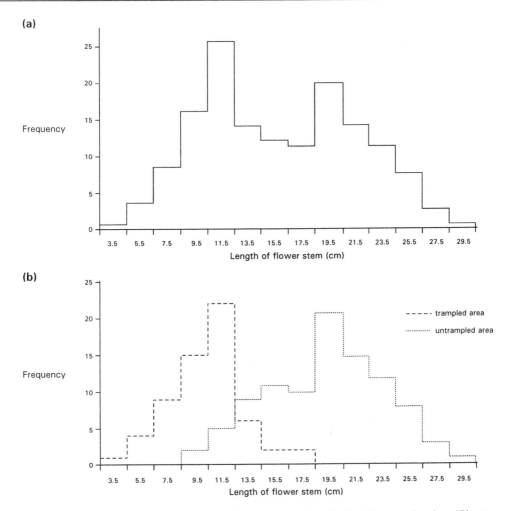

**Figure 5.9** Frequency distribution of flower stem lengths in ribwort plantian (*Plantago lanceolata*).

is often much more messy than the neat world of the mathematician, and normal distributions are not necessarily that usual.

Three examples with interesting biological interpretations will show the pitfalls. Figure 5.9a is a frequency distribution of the length of the flower stems of ribwort plantain (*Plantago lanceolata*). This does not look like a normal distribution, in fact, it is bimodal with two peaks instead of one. This is because two different populations have been sampled and treated as though they were one and the distribution is the result of adding together the two separate distributions shown in Figure 5.9b. The distribution with the smaller mean relates to plants growing in an area subject to trampling, the other to a non-trampled area. This sort of effect, produced by inadvertently taking a single sample from two different environments, can easily occur unless care is taken. A similar effect will arise if there are two (or more) different sorts of individuals, for example, different sexes or different generations. In many species, males are larger on average than females, so a frequency distribu-

**(a) Area**

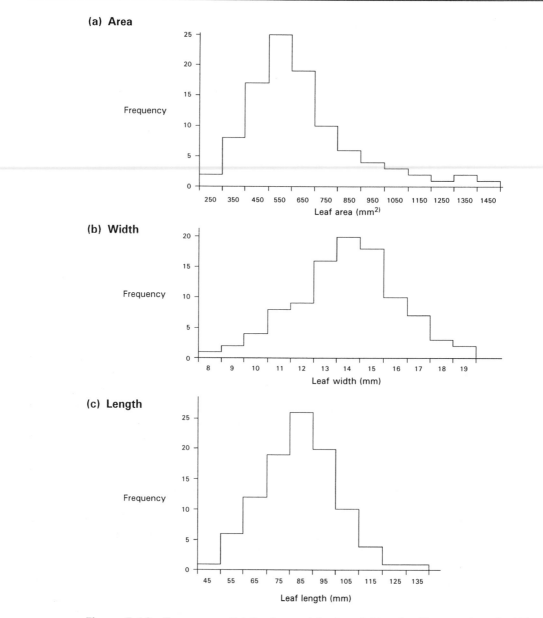

**(b) Width**

**(c) Length**

**Figure 5.10** Frequency distributions of leaf variables in ribwort plantain (*Plantago lanceolata*).

tion of body length in a sample of both sexes may be bimodal. The same will be true if our sample contains adults and juveniles.

Figure 5.10a shows a frequency distribution of leaf areas, again in ribwort plantain. This is a continuous variable, controlled by many genes and affected by the environment, but it is clearly not normally distributed. It is skewed to the right, that is, the right-hand tail is much more extended than the left-hand tail. At first sight this seems curious especially as both leaf length and leaf width are more or less normally distributed (Fig. 5.10b & c). The reason

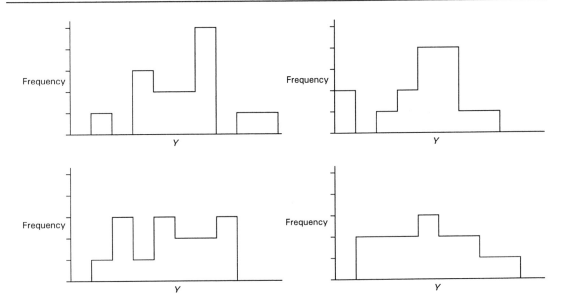

**Figure 5.11**  Four random samples of fifteen observations from a normal distribution.

why area does not follow a normal distribution is due to the mathematical relationship between the area and the linear dimensions (length and width). The area is proportional to the length multiplied by the width, that is, to the linear dimensions squared, which means that the relationship between area and length (or width) is not linear. We can see the effect that this will have by assuming, for the sake of simplicity, that the leaves are rectangular. The smallest leaf (45.0 mm long and 8.0 mm wide) would have an area of 360 mm$^2$ and an average size leaf (85.0 mm × 14.0 mm) would have an area of 1190 mm$^2$. The area of the largest leaf (135 mm × 19.0 mm) would be 2565 mm$^2$. Although large and small leaves are more or less equidistant from the mean in terms of their lengths and widths, this is not true for their areas. This is what produces the skewed distribution.

We could also consider a situation in which a variable has a normal distribution, but this is modified by either mortality or a bias in sampling. If, for example, individuals who are smaller than average survive dry conditions less well than large individuals, then the left-hand, lower tail of the distribution will be reduced relative to the upper tail. If the distribution was normal to begin with, it will not be now. The same thing will happen if small individuals tend to escape capture because, for example, they can swim more easily through the holes in the net or because they tend to differ in behaviour or habitat requirements.

Finally, all the preceding examples have involved large samples. Each distribution in Figure 5.11 is based on a sample of size 15 taken at random from a normal distribution. You can see that when sample sizes are small, the shape of the distribution can be very misleading. It is impossible to tell from the sample, whether or not the variable has a normal distribution. In these situations the only solution is to rely on our knowledge of the biology of the variable.

# 6

# Tests on a single sample: do the data fit the model?

In the previous chapter, we examined the details of three different models which could be used to describe variables. We also saw that the theoretical results produced by the model do not necessarily match the actual results obtained from the real world – that is, there is usually a discrepancy between what we find in a sample and what the model predicts. There were always two alternative explanations for this. The model could be right and the discrepancies could be due to the chance effects that always accompany the use of samples. Alternatively, it could be that the model is inappropriate. Remember that the models are based on assumptions which are supposed to be biologically realistic. If some of the assumptions were not valid for the particular situation being considered then the model would be inappropriate.

There are two reasons why we need to be able to discriminate between these two alternatives. First, as we have seen, the assumptions of any model relate to important biological features of the variable. If one or more of the assumptions is incorrect, then it means that there is something about the biology of the situation that we have not taken into account. By identifying it we can enhance our understanding of the biology. Secondly, statistical procedures are also based on assumptions about the form of the frequency distribution of the variable being tested. For example, some statistical tests assume that the variable is normally distributed and so it is important to be able to see whether or not this is the case.

In this chapter we shall see how to use statistical tests to distinguish between the "sampling variation" and the "invalid model" hypotheses and, at the same time, we shall use the examples to examine several other features common to all statistical tests. These tests are called **hypothesis tests** or **significance tests**. We have already met and used the simplest of these tests on our woodlouse data, when we asked this question. Assuming that the variable follows a binomial distribution with $n = 17$ and $p = 0.5$, how likely are we to get a result as big as or bigger than 14, or as small or smaller than 3? For obvious reasons, this test is called the **binomial test**. Let's look at it again.

## The binomial test

As you have probably gathered from Table A2 this test can be used in a variety of situations. The conditions under which it would be appropriate are those situations in which we have a nominal variable falling into just two categories. There are two situations in biology where this is likely to be the case. First, in laboratory and field studies of behaviour in which individuals have a choice of two courses of action and, secondly in simple genetic crosses where we expect offspring to be of two types. All we have to be able to produce is a null hypothesis which specifies what the probabilities of the two outcomes should be, although these need not be equal as they were in our woodlouse example. You may have used the chi-squared test in these circumstances, but it is less useful when the sample sizes are small and the underlying theory is less easy to understand. We shall deal with it later in the chapter.

## Unequal probabilities

We already know that when the probabilities of the two outcomes are unequal, the resulting binomial distribution is asymmetrical. This does not affect the principle of using it to make decisions about likely and unlikely events – we can still divide the distribution up into a central acceptance region and rejection regions in the two tails. All that happens is that the critical values become asymmetrical as we saw with the white-leaved tobacco plants in Chapter 5. So if our null hypothesis was that the ratio was 3:1 the sampling distribution, would be as shown in Figure 5.1.

The left-hand rejection region is delimited by a critical value of 0, which cuts off 2.380% of the distribution, while the critical value for the right-hand tail is 7 which cuts off 2.403%. Again we cannot get a rejection region of exactly 2.5% in each tail, so we err on the side of caution and make the rejection region smaller rather than larger. What should we say if we obtained one white seedling in a random sample of 13? Clearly this is greater than the lower critical value. Since it lies in the acceptance region we should accept the null hypothesis that the ratio is 3:1.

If we want to know how likely we are to get a result which is equal to, or more extreme than, the one we obtained, we need the associated probability. We can calculate this from the table. The probability of getting a number of white-leaved plants which is equal to or less than one, is $0.0238 + 0.1029$. We need to add to this the equivalent probabilities from the other tail of the distribution, corresponding to extreme results of 12 or more white-leaved plants. These probabilities are both 0.0000, so the associated probability is 0.1267. The reason why we use both tails is dealt with in the next section.

## One-tailed and two-tailed tests

I want now to use the binomial test on some data from another simple experiment in animal behaviour to illustrate an important principle which applies to all statistical testing. First of all we shall need some biological background.

### Wing size and mating efficiency in fruit-flies

In the fruit fly (*Drosophila melanogaster*), there is a recessive allele called vestigial which produces small, deformed wings when homozygous (Box 6.1) – you may have come across it in your genetics practicals. If we set up a culture of vestigial and normal winged flies in a bottle and let the flies reproduce, we find that the vestigial flies become less and less common in succeeding generations. One possible explanation is that the vestigial winged flies transmit

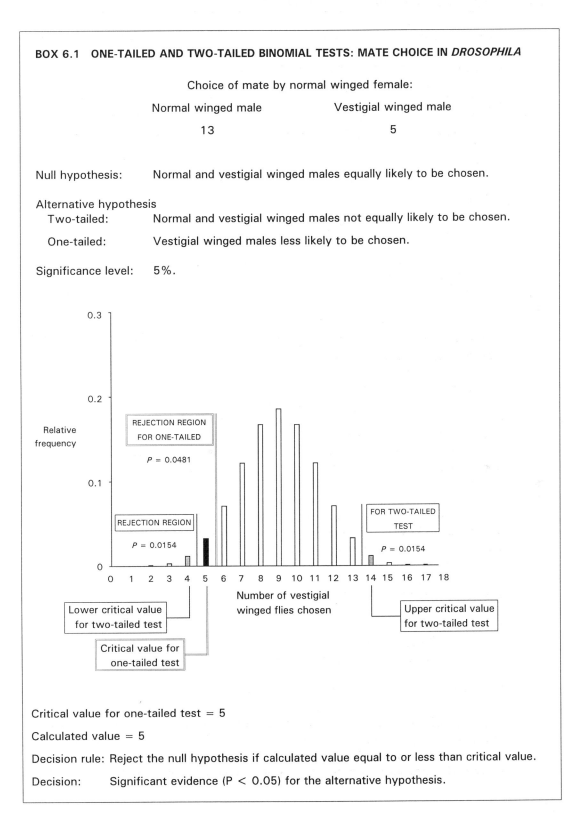

**BOX 6.1   ONE-TAILED AND TWO-TAILED BINOMIAL TESTS: MATE CHOICE IN *DROSOPHILA***

Choice of mate by normal winged female:

| Normal winged male | Vestigial winged male |
|:---:|:---:|
| 13 | 5 |

Null hypothesis:   Normal and vestigial winged males equally likely to be chosen.

Alternative hypothesis
 Two-tailed:   Normal and vestigial winged males not equally likely to be chosen.

 One-tailed:   Vestigial winged males less likely to be chosen.

Significance level:   5%.

REJECTION REGION
FOR ONE-TAILED

$P = 0.0481$

REJECTION REGION

$P = 0.0154$

FOR TWO-TAILED
TEST

$P = 0.0154$

Relative frequency

Number of vestigial winged flies chosen

Lower critical value for two-tailed test

Upper critical value for two-tailed test

Critical value for one-tailed test

Critical value for one-tailed test = 5

Calculated value = 5

Decision rule: Reject the null hypothesis if calculated value equal to or less than critical value.

Decision:    Significant evidence ($P < 0.05$) for the alternative hypothesis.

their genes to the next generation less efficiently than do normal winged flies. If this happens, then the vestigial allele will become less and less common and as a result fewer flies will display this characteristic in later generations. Natural selection would be occurring.

One very important process which can affect the efficiency of genetic transmission is mating behaviour, because a genotype which mates inefficiently will pass its genes on inefficiently and this could well be the explanation for the decline in frequency observed. This is because in fruit flies mating involves a complex courtship ritual in which the male signals to the female by vibrating his wings, so reduced wing size could well affect mating efficiency. We can examine this possibility by giving a female fly a choice of two males – one with normal wings, the other with vestigial wings. To do this, individual females (with normal wings) were placed in a glass tube with both sorts of male. The male chosen by the female was recorded (Box 6.1).

### The null and alternative hypotheses

The parallels between this experiment and the woodlouse experiment should be obvious and the data are clearly suitable for testing with the binomial test. The null hypothesis, set up in advance of doing the experiment, would be that the wing size of males has no effect on the female's preference, so that both sorts of male have an equal probability of being chosen. If this is true we would expect approximately equal numbers of the two sorts of males to be chosen – any departures from equality being due to sampling variation. The alternative hypothesis would, of course, be that wing size does affect female preference, so that the two sorts of male do not have equal probabilities of being chosen. It is here that we encounter another issue.

In the woodlouse example, to keep things simple, we assumed that if the difference in humidity did have an effect, then it could be in either direction. Woodlice might have preferred the damp or the dry environment. On this basis, extreme results in either direction, that is 14 (or more damp) turns and three (or fewer damp turns) would have been evidence against the null hypothesis. The rejection region lay in both tails of the sampling distribution and we carried out what is called a **two-tailed** or a **two-sided test**.

In the case of the mate choice experiment, there is a good reason for having a different form of the alternative hypothesis. We already know, before making the observations, that wings are important in courtship and it is reasonable to assume that evolution will have adjusted the mating behaviour to fit in with the normal wing size. We could therefore argue that a small, deformed wing could not possibly be better than the normal wing in terms of courtship, although it could be worse.

If we are prepared to take this position, then we can incorporate it into our alternative hypothesis by stating not only that wing size has an effect but also what the direction of that effect will be. In this case, it would be that small wing size reduces mating efficiency (i.e. **decreases** the probability of a male fly being chosen). The logical consequence of specifying the direction is that the only evidence which would lead us to reject the null hypothesis would be extreme results in that one, predicted, direction (i.e. very low numbers of vestigial winged flies chosen). This follows because we have already argued from our existing biological knowledge that having small wings, if it does anything, can only reduce the probability being chosen, it cannot increase the chance. Under these circumstances if we were to find that more vestigial winged flies were chosen (an extreme result in the other direction), it could not have a biological explanation because it could not possibly be an effect of the wing size. An extreme result like this would have to be attributed to sampling variation.

A test like this, in which the alternative hypothesis predicts the direction of the effect, is called a **one-tailed** or a **one-sided test** because the rejection region now lies in only one tail of the distribution. The critical value at the 5% significance level is now the value of the test statistic which cuts of 5% of the distribution in that one tail.

We shall return to a further discussion of one-tailed and two-tailed tests in the next section, but there is an important point to be made at this stage. In general one-tailed and two-tailed tests carried out on the same data will have different critical values. We can see this with the our fly data.

### Critical values

A total of 18 different females were given a choice, with the result that 13 of the females mated with normal winged males and five mated with vestigial winged males. Is there any evidence that females are discriminating between the two types of male? Now in practice, we should have decided on whether this was to be a one-tailed or a two-tailed test before we obtained the data, but for our present purpose we need to examine both types of test so that we can see the difference in terms of critical value (Box 6.1). For a two-tailed test with a sample size of 18, the critical values are four and 14 at a nominal significance level of 5% (the actual size of rejection region is 3.08%). For a one-tailed test at the same nominal significance level, the critical value is 5 (actual size of rejection region 4.81%).

If we were doing a one-tailed test, then we would be able to reject the null hypothesis at the 5% significance level because our value of the test statistic is equal to the critical value. We would conclude that females do have a preference for males with normal wings. However, if we were doing a two-tailed test, then with this result we would have had to accept the null hypothesis, because the result is larger than the critical value. In other words, the same data can lead to different conclusions. It follows that we must ensure that we are doing a test with the correct number of tails and that we have used the appropriate critical value (and of course the appropriate significance level).

Critical values for one-tailed and two-tailed tests are usually different, but when the sampling distribution is discontinuous (as it is in the binomial test), the critical values can be the same for both tests. This is because under these conditions, the rejection regions cannot be made exactly the right size, for example, exactly 5%. You might like to verify that if we had used a sample size of 20 flies, the one-tailed and two-tailed critical values at the 5% significance level would both have been 5.

### One-tailed or two-tailed tests – how to choose which one to use?

As we saw above, a one-tailed test allows us to reject the null hypothesis with a less extreme result than that required by a two-tailed test. This advantage is, however, balanced by a disadvantage, because a one-tailed test can only detect a difference in the predicted direction. If, for example, we had decided on a one-tailed test and the result had been that 18 vestigial flies and no normal flies were chosen, we would have had to accept the null hypothesis, despite the extreme nature of the result. This is because it is not in the direction predicted by the alternative hypothesis. In setting up our one-tailed alternative hypothesis we have said that there can be no possible biological explanation for an extreme result of this type. Therefore, the only explanation must be that it is one of those unlikely events that will occasionally happen when the null hypothesis is true. It would be invalid at this stage to change our minds and carry out a two-tailed test.

It follows from this that the decision about whether to do a one-tailed or a two-tailed test must be made before the experiment is carried out. Students sometimes think that you make the decision after doing the experiment and examining the results and that if the results "look" as if they are clearly in one direction, then you test this with a one-tailed test. This is not correct. A one-tailed test is used if you think you know enough about the biology of the system to be sure that extreme results in one direction can only be due to chance, that is, they cannot have any biological meaning. If you do decide to carry out a one-tailed test, then you must specify in the alternative hypothesis what the direction of the difference will be.

There is clearly something of a risk in deciding to do a one-tailed test, because there could be good but totally unexpected reasons why vestigial flies might be chosen. For example, the gene which produces vestigial wings might have some other effect, such as increased pheromone output, which makes vestigial flies more attractive than normal winged flies. Or the female flies could be abnormal in some way so that they genuinely prefer vestigial-winged males. You could argue that a one-tailed test would be appropriate in the woodlouse experiment, after all woodlice are known to be poorly adapted to terrestrial life and to be frequently found in places that are obviously damp. It may seem inconceivable that they might have a preference for dry conditions – but some species do live in dry grasslands and others live in deserts!

One-tailed tests would be appropriate if we were examining whether an exercise programme affected physical fitness or whether vaccination altered infection rates. In both cases we know enough about the biology to say what the direction of any effects would be. Other possible applications arise in the context of very well supported general biological laws. For example, it is well known from a very large number of studies on a wide range of species that inbreeding (that is breeding between close relatives) leads to inbreeding depression, that is, low production and survival of offspring. In the light of this general finding, we would be justified in applying a one-tailed test to a set of data from a particular species. It would be appropriate too in a field experiment on the effects of grazing molluscs on seaweed abundance, in which the molluscs had been excluded from areas of the shore as described in Chapter 1. If molluscs do have an effect it will surely be to reduce seaweed abundance. Finally, we could use a one-tailed test where a claim has been made such as the seed supplier's claim of a 99.9% germination rate. We would only want to detect whether the rate is less than this.

If you are not particularly convinced by the above then don't worry; some statisticians think that one-tailed tests should never be done. The best general guideline is "If in doubt, carry out a two-tailed test". If you carry out a two-tailed test and reject the null hypothesis when you should really have done a one-tailed test, then you would have come to the same decision with the correct one-tailed test. This is because two-tailed tests have larger critical values than one-tailed tests and so represent a more conservative (i.e. safer) procedure. If, on the other hand, you carry out a one-tailed test and reject the null hypothesis when you should have done a two-tailed test, then you could end up rejecting the null hypothesis when properly you should not.

### Back to the flies – what does the result tell us?

Before leaving this example there are one or two interesting issues which relate to the fact that this is an observational study, not an experiment. We are not manipulating the factor of interest (wing size) but are making use instead of naturally occurring variation in the factor. This distinction means that care has to be taken with the biological interpretation of the result. Remember that in an experiment in which we manipulate the factor of interest, the aim is to arrange things so that on average the only difference between the two treatments is in this factor; with care this should be possible. It is much more difficult (perhaps impossible) to achieve this using naturally occurring variation in something like wing size. One reason is that genes often affect more than one characteristic; suppose, for example, that the gene for vestigial wings reduces body size. If vestigial winged flies are smaller than normal winged flies, then their reduced mating success could be due to their small body size rather than their small wings. We could not tell which is responsible.

Now you could solve this problem by choosing flies of equal size, but what if the vestigial gene has an effect on something like pheromone production which you are not aware of, but which is very important in mating? The message here is that if we merely want to know whether the two wing types have different mating efficiencies then the method used above is adequate. If, however, we want to know what the difference in mating efficiency is due to, then we would have to use a manipulative experiment. This means that before we decide on the method to be employed, we have to have a clear idea about the question we want to answer.

## Levels of significance

Now that we have carried out a few simple statistical tests and grasped the basics, we need to re-examine our ideas on the relationship between significance levels, critical values and the decisions which we then make.

### The test procedure

Most of the examples so far have dealt with simple statistical tests based on the binomial distribution, however they have also illustrated the standard procedure used in any statistical test. Let's just review this. We first find an appropriate model which describes the distribution of "results" under the null hypothesis. We may be able to use the results directly (as in the binomial test), but in other situations we may have to do some calculations on the results to produce some other measure whose distribution is known, for instance a value of $z$. In general, whatever the measure, we call it a test statistic. Then we set the significance level at an appropriate value (e.g. 5%) producing rejection regions containing those results which we consider to be unlikely to occur if the null hypothesis is true. There may be one or two such regions, depending on whether we are doing a one-tailed or a two-tailed test. If we cannot make the regions exactly the right size, then we err on the side of caution and make them smaller than the nominal significance level. We can then identify the critical value (or values) of the test statistic which mark off these rejection regions. We then compare our calculated value of the test statistic with the critical values to see whether it lies in the acceptance region or the rejection region, that is, whether it rates as a likely or an unlikely result. This determines our decision with respect to the null hypothesis and it is the nature of this decision which needs clarification.

## Why a significance level of 5%?

This is a question which comes to mind immediately. Obviously the level needs to be quite low, remembering that it is the area of the sampling distribution which is supposed to include only the less likely values of the test statistic. These, of course, account for only a small proportion of the sampling distribution. In addition, 5% is a convenient number to remember.

However, a 5% significance level does have a drawback which was apparent in the example of the maize kernels. Sometimes, actually in 5% of random samples, we will, by chance, get a value of the test statistic which falls in this 5% of the distribution. If we happen to pick such a sample our decision will be to reject the null hypothesis even though it is true. Our conclusion will be wrong and we will have made what is known as a **type I error**. The chance of making a type I error is obviously equal to the significance level, so the logical way to reduce the chance of coming to the wrong decision would be to make the size of the rejection region smaller. Critical values will then be more extreme and this will reduce the chances of our calculated value being equal to or more extreme than the critical value. While this may sound comforting, there is a real problem in that on those occasions when the alternative hypothesis is true, these more extreme results should be leading us to reject the null hypothesis. As we make the rejection region smaller and reduce the chance of a type I error so we increase the chance that we will accept the null hypothesis when it is false. Erroneously accepting the null hypothesis when it is false is called a **type II error**.

Ideally, the chance of making both types of error should be as small as possible. For most applications in biology the 5% level is a reasonable compromise between these two conflicting requirements. This is borne out by experience, because, on the whole, our understanding of biology has progressed a great deal. This would not have happened if our statistical tests were giving us answers which were often wrong.

## A more sophisticated approach to decision rules

In Chapter 2 we used a simple rule for deciding whether to accept or reject the null hypothesis. It was made simple so that the basic idea came across clearly. By this stage you should be familiar with the basic idea so what I want to do now is to develop a more sophisticated approach.

### Should we ever accept the null hypothesis?

Our statistical test is based on the sampling distribution which would occur if the null hypothesis were true and to produce this we have to specify the distribution exactly. This means making an exact statement about the null hypothesis. For example, in the case of the green and white seedlings, to get the sampling distribution, we had to specify that $p$, the probability of getting a white-leaved seedling, was exactly 0.25. We then divided the distribution up into acceptance and rejection regions. When the value of the test statistic fell in the acceptance region we accepted the null hypothesis, but what does this mean? Logically it means that we are accepting that $p$ is exactly 0.2500000..., which is almost certainly not true. Perhaps $p$ is very near to this but is actually 0.250008. Also, thinking back to the woodlouse example, it is clear that if we take a small enough sample (e.g. $n = 4$) there is then no possible result which is so extreme as to be judged improbable. All the possible results fall into the acceptance region. On the basis of such limited evidence, it would not be reasonable to accept that the null hypothesis was true. So, bearing these two points in mind, it is more sensible to think of the outcome of a statistical test in terms of the evidence it supplies in favour of the alternative hypothesis.

We can do this using our simple rule. This involved a distinction between likely and unlikely, made on the basis of the relationship between the calculated and the critical value of the test statistic. A likely event, as in our example of the white-leaved seedling, lies within the critical values, has an associated probability of occurrence of greater than 0.05 and would give us no significant evidence in favour of the alternative hypothesis. The result is said to be statistically **non-significant**, a conclusion we can denote by the abbreviation **n.s.** An unlikely value as in the woodlouse experiment, is equal to or more extreme than the critical value and has a probability of occurrence under the null hypothesis of less than 0.05. This does give us the required evidence in favour of the alternative hypothesis and is a statistically **significant** result. When a result is declared non-significant a common fault is to state that this proves the null hypothesis. The preceding discussion should have made it clear that this conclusion is incorrect.

This approach, though simple, is misleading. It is easy to think, particularly when we have put our data into an impressive equation and reduced it to a single number, that this number must enable us to make a clear decision with respect to the alternative hypothesis. Unfortunately it is more complicated than this.

### Making more use of the associated probabilities

Imagine doing the same test on two sets of data and finding the probabilities associated with the calculated values of the test statistic to be 0.051 in one case and 0.049 in the other. Formally the first is not significant and the second is, but there is not that much to choose between them. The 5% level is arbitrary and it has no magical properties that enable us to draw a clear line. It is merely a convenient point on a continuum from very probable to very improbable. In addition, as we shall see, most statistical tests involve assumptions and approximations which will be more or less valid for the particular data being analysed. This in turn means that any associated probabilities calculated are not likely to be strictly accurate. For both these reasons it is best to give an indication of how non-significant the result is rather than just reporting it as non-significant. (The way to do this is described at the end of this section.) The reader can make up their own mind. However, you ought to stick to your chosen significance level and the decision which flows from it, otherwise an element of subjectivity will creep into your conclusions.

The same applies to the reporting of significant results, because here we can usefully take into account the strength of the evidence for the alternative hypothesis, based on the degree of improbability of the result. As we saw with the maize kernels, unlikely events can occur, even if the alternative hypothesis is not true, but the more unlikely the result, the stronger must be the evidence for the alternative hypothesis. So in the woodlouse example, results as extreme as the one obtained have an associated probability of occurrence of just less than 0.05. We would describe this as **statistically significant** evidence for the alternative hypothesis, or a **significant** result. In the case of beak size in the finches we saw that the probability associated with a value of $z = 2.803$ is less than 0.01. We would call this a **highly significant** result. Finally, consider the case of our test of seed germination rates in which we found three non-germinating seeds out of 100. The associated probability (probability of three or more) is 0.0002 which is less than 0.001. This is **very highly significant evidence** for the alternative hypothesis.

---

### BOX 6.2 REPORTING THE RESULTS OF STATISTICAL ANALYSES

| Example | Test statistic | Calculated value | Associated probability (P) | Shorthand notation | Verbal |
|---|---|---|---|---|---|
| *Nicotiana* seedlings | Number of white-leaved plants | 1 | > 0.10 | n.s. | Non-significant |
| Woodlouse habitat choice | Number of damp turns | 14 | < 0.05 | * | Significant |
| Beak size in *Geospiza* | z | 2.803 | < 0.01 | ** | Highly significant |
| Seed germination success | Number of non-germinating seeds | 3 | < 0.001 | *** | Very highly significant |

We should note this information in our conclusion to a test and it can be done shorthand in a number of ways which are illustrated in Box 6.2. We can state that the associated probability is less (or more) than some value, e.g. $P < 0.01$, or use a system of asterisks to denote the probability level. The most usual system is * for $P < 0.05$, ** for $P < 0.01$ and *** for $P < 0.001$, but this is not a universal convention, so you ought to state somewhere what your symbols stand for. Alternatively, you can give the exact probabilities; many computer statistics packages calculate these as a matter of routine.

It is easy to get confused when reporting results in this form. When you write that you have significant evidence for the alternative hypothesis ($P < 0.05$), the two pieces of information can seem contradictory. It can seem as though it is the alternative hypothesis which has the low associated probability, which would suggest that there is little evidence for it. So what is this probability? Well it is not the probability that the null hypothesis is true; if you think about it for an experiment done under specified conditions the null hypothesis is either true or it is not. It cannot be true on only some occasions! The probability given is the probability associated with obtaining the value of the test statistic were the null hypothesis to be true.

## The chi-squared test

You will probably have realized by this stage that the binomial test has some limitations. One of these is that the binomial tables given here only go up to $n = 20$, so that we cannot conveniently deal with a sample that is larger than this. More restrictive is the fact that it can only be used in situations where there are two outcomes, yet, as we have seen, nominal variables often fall into more than two categories. How would we deal with these situations? The solution is to use a test which almost every biologist has come across, usually in the context of data from genetic crosses namely the **chi-squared ($\chi^2$) test**. The word chi ($\chi$) is pronounced in the same way as the "ki" in kite.

## Where does $\chi^2$ come from?

This first section is one which you may prefer to miss out on a first reading because it tells you why the test works, rather than how to carry it out. If you do miss it out you may need to return to it later.

To begin with, we need to go back to a normal distribution with mean $\mu$ and standard deviation $\sigma$ (Fig. 6.1a). You will remember that if we had a single random value, $Y$, we could obtain a standardized measure of how deviant or unusual it was by asking how far away it was from the true mean $\mu$ in units of standard deviation. The answer was a value of $z$, which could be positive or negative. Now imagine extending this idea to a sample of several random values of $Y$ in order to get an idea of how far away from the mean all the items in the sample are, that is a standardized measure of the total variability. This is illustrated in Figure 6.1b, where each random value of $Y$ is represented by a filled circle. We could obtain our measure by simply calculating a value of $z$ for each observation and then adding them together to get a total $z$ for the sample, but this would not be very satisfactory because the positive and negative values of $z$ will tend to cancel one another out when they are added together. This is easy to overcome with the same device we used when calculating a variance in Chapter 4. We simply square each deviation and then divide this by $\sigma^2$ before adding them. The sum of these squared standardized deviations is a value of something called $\chi^2$ which has a known sampling distribution and which can be used as a test statistic.

The mathematics of this distribution is too complex for us to examine, but we can develop an understanding of it as follows. The size of an actual value of $\chi^2$ will be determined by the variation of the individual values of $Y$ in the sample. A typical sample drawn from the population, with most observations quite near to the mean and a few more extreme values (Fig. 6.1c) would yield a medium sized value of $\chi^2$. If, by chance, the observations in our sample were all very close to the mean, i.e. there was very little variability in the sample (Fig. 6.1d), then the value of $\chi^2$ would be small, but you will appreciate that this is not a likely occurrence. If, on the other hand, there was a lot of variability in the sample, that is, the observations were very far from the mean (Figs 6.1e & f), then the values of $\chi^2$ would be large. Samples like these would not be likely either. Note too how the calculation takes account of the underlying variability, that is, the measure is standardized. In Figure 6.1g, although the observations have the same values as in Figure 6.1f they have been drawn from a more variable distribution, but this is taken into account by dividing by $\sigma^2$. The value of $\chi^2$ will be smaller than in Figure 6.1f. You should be able to see from this how $\chi^2$ provides a comparative measure of the variability in a sample.

Now let's do a thought experiment of a type that you will have to do at several other stages in this book. Imagine taking every possible sample of a given size from the population and calculating all the possible values of $\chi^2$ which result and then plotting these values as a frequency distribution. This will in fact be a sampling distribution of $\chi^2$. To describe the exact shape of this distribution, we would need some complex mathematics which we shall not worry about, but the formula for $\chi^2$ gives us some hints. First, $\chi^2$ has a minimum value of zero, which will happen when all the items in the sample have the same value as the mean. Secondly, since $\chi^2$ is the sum of the deviations of each value of $Y$ from the mean, for a given population, the larger the sample the larger the values of $\chi^2$ will tend to be. This means that there will be a different sampling distribution of $\chi^2$ for every sample size. It turns out that this is not quite correct because what determines the shape of the distribution is something called the degrees of freedom which are closely related to the sample size, but more of that later.

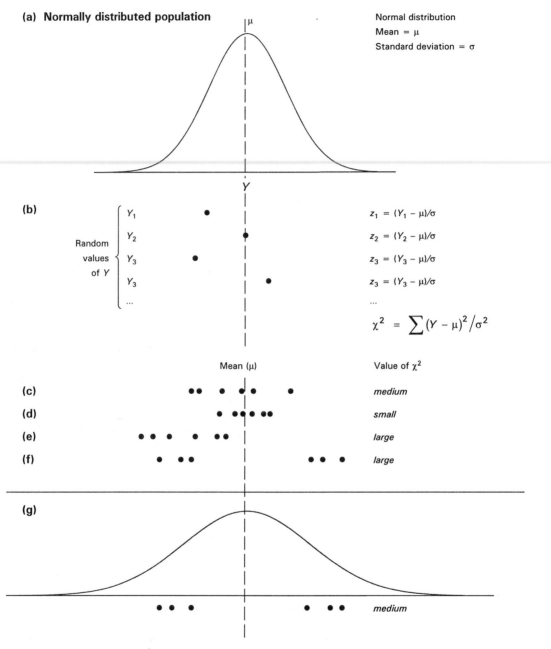

**Figure 6.1** Where does $\chi^2$ come from?

Figure 6.2 shows the distributions of $\chi^2$ for three and eight degrees of freedom. They are continuous distributions and we can see from their shape that both very small and very large values of $\chi^2$ are unlikely to occur. This simply reflects the point we considered earlier that with random samples the values of Y are unlikely to be either all very close to the mean or all very far from the mean. The most likely type of sample will be a mixture of typical and

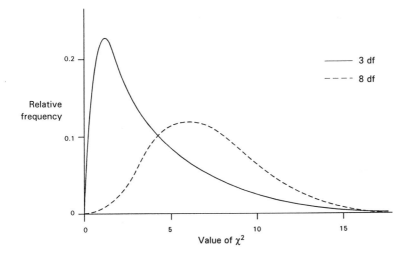

**Figure 6.2** Frequency curves of $\chi^2$ distributions for 3 and 8 degrees of freedom (df).

atypical observations, yielding an intermediate value of $\chi^2$. Because these distributions can be described mathematically, $\chi^2$ can be used as a test statistic to answer questions about the likely extent of chance variation among the observations in a sample. In particular we can use these distributions to set critical values and to find the probability associated with a calculated value of $\chi^2$.

You may be wondering how all this helps, given that we started off looking for a test for use on nominal variables and the mathematical model for $\chi^2$ is based on a continuous variable. In fact mathematicians have shown that similar calculations using other types of variables produce something that closely approximates to $\chi^2$. This finding enables us to use $\chi^2$ distributions in a wide range of applications, including tests on nominal variables. The most common of these are the so-called **goodness-of-fit tests** which, as their name suggests, test how well the observed frequencies in a single sample fit with the frequencies expected on the basis of a model.

## Chi-squared goodness-of-fit test for a 1:1 ratio

In this section we shall look at some simple genetic data on spore colour in the fungus *Sordaria fimicola* (Box 6.3) in order to see how to use this test. Spore colour is controlled by two alleles of a single gene, producing either white or black spores. To cross a black- and a white-spored strain, all we need to do is inoculate a sterile plate of medium with a small amount of fungal mycelium from each of the two strains. The hyphae grow and where they come into contact with one another, they fuse and form diploid cells. These undergo meiosis and eventually produce eight haploid spores which are held in a sausage-shaped structure called an ascus. There are equal numbers of the two colours of spores because of the segregation of the alleles at meiosis and their subsequent replication during mitosis. The spores can be arranged in a number of different patterns in the ascus, two of which are shown. An ascus with the four black spores at the top (type B) was produced from a diploid cell in which the pair of homologous chromosomes at metaphase was arranged so that the

135

**BOX 6.3   CHI-SQUARED GOODNESS-OF-FIT TEST: SPORE COLOUR PATTERNS IN THE FUNGUS *SORDARIA FIMICOLA*.**

Null hypothesis:          Ratio of the two patterns in the population is 1:1.

Alternative hypothesis:   Ratio of the two patterns in the population is not 1:1.

Significance level:       5%.

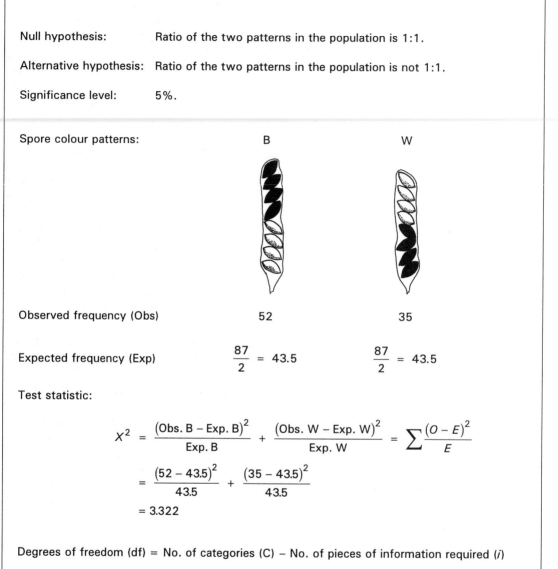

Spore colour patterns:                    B                    W

Observed frequency (Obs)                 52                   35

Expected frequency (Exp)         $\dfrac{87}{2}$ = 43.5       $\dfrac{87}{2}$ = 43.5

Test statistic:

$$X^2 = \frac{(\text{Obs. B} - \text{Exp. B})^2}{\text{Exp. B}} + \frac{(\text{Obs. W} - \text{Exp. W})^2}{\text{Exp. W}} = \sum \frac{(O - E)^2}{E}$$

$$= \frac{(52 - 43.5)^2}{43.5} + \frac{(35 - 43.5)^2}{43.5}$$

$$= 3.322$$

Degrees of freedom (df) = No. of categories (C) − No. of pieces of information required ($i$)

$$= 2 - 1 = 1$$

Critical value of $\chi^2$ = 3.841

Decision rule: Reject null hypothesis if calculated value is equal to, or larger than, critical value.

Decision:      No evidence for alternative hypothesis ($P < 0.10$), result non-significant.

chromosome carrying the allele for black colour was directed towards the "top" of the cell. If the pair is orientated so that the chromosome carrying the allele for white is in this position, then an ascus with four white spores at the top (type W) results.

Now it is a fundamental principle of Mendelian genetics that the orientation of a pair of homologous chromosomes at metaphase should be random. This means that the chromosome carrying the black allele is just as likely to be directed to the "top" of the cell as to the "bottom". It follows that the two types of spore pattern should occur in a 1:1 ratio in the population, a hypothesis which we can test by simply taking a random sample of asci and counting up how many of them are of each type. Suppose we count 52 asci with the four black spores at the top and 35 with the four black spores at the bottom. Does this provide any evidence (at the 5% significance level) that the orientation is not random?

Our variable, spore pattern, is clearly a nominal variable because each sampling unit (an ascus) falls into one of two categories, so we can use chi-squared. We start, as always, from the null hypothesis, which is that the ratio in the population is 1:1. On this hypothesis any deviations from this ratio in the sample are due to sampling variation. We next calculate for each of the two categories, the number which should occur if the null hypothesis were true. These values are called the **expected frequencies**. The use of the word "expected" here is rather confusing because if the null hypothesis is true then a range of different, whole numbers of the two patterns can be "expected" to occur because of sampling variation. The word "expected" actually stands for the value which we would get if we were to take the mean of all possible samples of 87 asci. So if the ratio were 1:1, then the expected frequency of either pattern in a sample of size 87 would be 43.5. Note that this expected (mean) value need not be a whole number.

Comparing these expected frequencies with those observed, it is clear that there are more asci with black spores at the top than expected and fewer with white spores but, as before, there are two different explanations for this disparity. It could be that the ratio in the population is actually 1:1 and that, by chance, we have obtained these rather unequal numbers. Alternatively, the population ratio could be different from 1:1. So the problem comes down to answering the question "How likely are we to get observed values which are as extreme as, or more extreme than, these if the null hypothesis is true?". This is the same sort of question we had in the binomial test. To answer this we need to produce, from the sample data, a test statistic, whose sampling distribution under the null hypothesis is known.

Mathematicians have shown that an appropriate test statistic called $X^2$ is produced by taking the observed and expected numbers for each category in turn and calculating the following:

$$\frac{\left(\text{Observed frequency} - \text{Expected frequency}\right)^2}{\text{Expected frequency}}$$

We then add together the answers for each category. This can all be written in shorthand as a formula (see Box 6.3), where O and E are the observed and expected frequencies and $\Sigma$ simply means add the answers for each category.

We can get an idea of what $X^2$ is doing by looking at the formula. It is clear that the formula gives a measure of how far each observed frequency deviates from the mean (expected) frequency based on the null hypothesis. These values are later added together to provide a measure of the total deviation in the sample. Because some of these deviations are positive and some are negative, we employ the mathematical convenience of squaring them all before adding them – if this was not done, the sum would be zero. If the null hypothesis is true, then the observed and expected frequencies will only differ because of sampling variation and the differences are likely to be small. If the null hypothesis is not true, then the observed and expected values are likely to be very different. So a small value for this sum would be likely if the null hypothesis is true, while a large value would be likely if the null hypothesis were not true.

However, the absolute values of the deviations are not much use on their own, because they need to be seen in the context of the sample size, which after all governs the likely magnitude of sampling variation. A difference of 8.5 between an observed and an expected value is striking if the expected value is 43.5, but is not so remarkable if the expected value is 435. Dividing the squared deviation by the expected number, measures the magnitude of the deviation with respect to the sample size – it is, if you like, a way of standardising the deviations to take into account the number of observations involved. This final step will not alter the basic feature of $X^2$, which is that large values are more likely to occur when the null hypothesis is not true.

To find out how large a value of $X^2$ needs to be before it will count as evidence for the alternative hypothesis, we shall need to consult a set of tables. Mathematicians have shown that the test statistic $X^2$ has a distribution which closely approximates a $\chi^2$ distribution. (You can probably see that this whole line of reasoning follows quite closely to the one used in the preceding section on the origin of $\chi^2$. In fact the formula for $\chi^2$ does resemble the one for $X^2$ in its general appearance.) This means that we can refer our calculated value of $X^2$ to a table of critical values of $\chi^2$. This will enable us to see whether the value calculated from our sample is so large as to fall into the rejection region.

## Using tables of critical values of $\chi^2$

In our example the calculated value of the test statistic is 3.322, which we now need to compare to the appropriate critical value of $\chi^2$ in Table A6, part of which is reproduced in Table 6.1. We have already set our significance level at 5%, the conventional value for work of this type, so what is the critical value?

To find this out we need to take into account the number of categories involved, because this will alter the shape of the distribution. In our example we have two categories of data, but there is no reason why we could not have more categories in a test of this type. For

Table 6.1  Critical values of $\chi^2$ (see Table A6).

| $\alpha_1^R$ | | | | 10% | 5% | 2.5% | 1% | 0.5% | 0.1% |
|---|---|---|---|---|---|---|---|---|---|
| $\alpha_2$ | 1% | 5% | 10% | 20% | 10% | 5% | 2% | 1% | 0.2% |
| df = 1 | 0.000 | 0.001 | 0.004 | 2.706 | 3.841 | 5.024 | 6.635 | 7.879 | 10.828 |
| 2 | 0.010 | 0.051 | 0.103 | 4.605 | 5.991 | 7.378 | 9.210 | 10.597 | 13.816 |
| 3 | 0.072 | 0.216 | 0.352 | 6.251 | 7.815 | 9.348 | 11.345 | 12.838 | 16.266 |
| 4 | 0.207 | 0.484 | 0.711 | 7.779 | 9.488 | 11.143 | 13.277 | 14.860 | 18.467 |
| 5 | 0.412 | 0.831 | 1.145 | 9.236 | 11.070 | 12.833 | 15.086 | 16.750 | 20.515 |
| 6 | 0.676 | 1.237 | 1.635 | 10.645 | 12.592 | 14.449 | 16.812 | 18.548 | 22.458 |
| 7 | 0.989 | 1.690 | 2.167 | 12.017 | 14.067 | 16.013 | 18.475 | 20.278 | 24.322 |
| 8 | 1.344 | 2.180 | 2.733 | 13.362 | 15.507 | 17.535 | 20.090 | 21.955 | 26.124 |
| 9 | 1.735 | 2.700 | 3.325 | 14.684 | 16.919 | 19.023 | 21.666 | 23.589 | 27.877 |
| 10 | 2.156 | 3.247 | 3.940 | 15.987 | 18.307 | 20.483 | 23.209 | 25.188 | 29.588 |

example, if we wanted to test the goodness-of-fit of some genetic data to a 9 : 3 : 3 : 1 ratio, we would have four categories. It should be obvious from the formula for $X^2$ that, in this case, the value of the test statistic would be larger even if the sizes of the deviations were the same. We would after all be summing four deviations rather than two. From this it follows that the critical value will increase as the number of categories increases, so we need to take this into account when using the table. In fact, the critical value, although related to the number of categories ($C$), is determined by something called the **degrees of freedom**, a term that we have already met when discussing the calculation of the variance.

The degrees of freedom (df) are always a whole number which is less than $C$, and there is a simple rule for working out what the number is. We simply count up how many separate pieces of information ($i$) from the data were used to calculate the expected values and subtract this from the number of categories ($C$). In our example, to obtain the expected values, we only had to use the total sample size (we divided 87 up in the ratio 1:1), so this counts as one piece of information. The degrees of freedom are therefore $C - i = 2 - 1 = 1$. Beware, it is always the number of categories ($C$) that are used to calculate the degrees of freedom, not the sample size ($n$).

The table of critical values of $\chi^2$ is set out in much the same way as the table of $z$ values and it uses the same notation. The only difference is the highlighted column at the left, which gives the degrees of freedom. The critical value is found in the row for the appropriate degrees of freedom and the column for the appropriate significance level. As we saw earlier, large values of $X^2$ are unlikely to occur if the null hypothesis is true, but the direction of the departure of the observed values from the expected values makes no difference to the value of $X^2$. You can easily see this by imagining that the figures were 35 and 52 rather than 52 and 35 in which case the value of $X^2$ would be unchanged. As a result, the rejection region for a test like this always lies in the right-hand tail of the sampling distribution. The decision rule is to reject the null hypothesis if the calculated value is equal to or larger than the critical value at the chosen significance level.

So, with the rejection region comprising 5% of the sampling distribution in the right-hand tail, that is $\alpha_I^R = 5\%$ and with df = 1, we find that the critical value is 3.841. Our calculated value of $X^2$ is smaller than this, so we have a non-significant result. We have no evidence for the alternative hypothesis of non-random orientation of bivalents at metaphase; in other words, the behaviour of chromosomes at meiosis parallels that of an unbiased coin when tossed. However, note that the calculated value is not that much smaller than the critical value. In fact we can see that the probability of getting a value of 3.332 or more lies between 0.1 and 0.05. Bearing in mind the earlier discussion on the meaning of significance levels we ought to draw this to the reader's attention. They can then make up their own minds and might consider it worth repeating the observations.

## Further points about $\chi^2$

It is worth mentioning here that the most common error made when carrying out this test is to use proportions or percentages, rather than frequencies when doing the calculations. This happens because we often convert frequencies to proportions or percentages for comparative purposes, but you can probably see why they are inappropriate for this test. Proportions and percentages lose all the information on sample size. For example, observed frequencies of 6 : 4, 60 : 40 and 600 : 400 would all produce the same proportions of 0.6 : 0.4. If these were used in the calculation of $X^2$ for a goodness-of-fit test to a 1:1 ratio the expected frequencies would be 0.5 and the $X^2$ value calculated would be the same very low value for all three sets of data. Yet common sense tells us that 6:4 is a likely result and 600:400 is not.

As we saw earlier a $\chi^2$ distribution is a theoretical distribution relating to samples of a continuous variable. Our example has been concerned with a discontinuous variable and, as has been stressed, the test statistic we have calculated is not actually a value of $\chi^2$. That is why we have labelled it as $X^2$ rather than using the greek letter. However, the distribution of our test statistic $X^2$ does approximate closely enough to a $\chi^2$ distribution under most circumstances for us to be able to test the significance of $X^2$ by using the distribution of $\chi^2$. This approximation can be improved by incorporating a modification known as **Yates' correction** which is a **correction for continuity**. As this name suggests, it is designed to compensate for the fact that we are using a distribution based on a continuous variable in a test for a discontinuous variable. Box 6.4 shows you how to do it, but it is only used when there is one degree of freedom. Each value of $(O - E)$ is reduced by 0.5, but you have do this using the absolute value of $(O - E)$. If $(O - E)$ is negative then ignore the minus sign, which is what the two vertical lines stand for. Since this correction makes $(O - E)$ smaller the effect will be to reduce the calculated value of $X^2$. Thus, values which are significant without the correction may not be significant if the correction is used.

$\chi^2$ tests of this type have a wide range of uses, some of which we shall look at in this chapter. We could use it to test our woodlouse data. With a total of 17 trials the expected frequency of the two types of turn under the null hypothesis 8.5. The calculated value of $X^2$ is 7.118 without Yates' correction, while the corrected value is 5.882. There is only one

---

**BOX 6.4   YATES' CORRECTION FOR GOODNESS-OF-FIT TEST WITH ONE DEGREE OF FREEDOM: FREQUENCY OF SPORE COLOUR PATTERNS IN *SORDARIA FIMICOLA*. (DATA FROM BOX 6.3)**

Null hypothesis:   Ratio of the two patterns in the population is 1:1.

Alternative hypothesis:   Ratio of the two patterns in the population is not 1:1.

Significance level:   5%.

$$X^2 = \frac{\left(\left|\text{Obs. B} - \text{Exp. B}\right| - 0.5\right)^2}{\text{Exp. B}} + \frac{\left(\left|\text{Obs. W} - \text{Exp. W}\right| - 0.5\right)^2}{\text{Exp. W}}$$

$$= \frac{\left(\left|52 - 43.5\right| - 0.5\right)^2}{43.5} + \frac{\left(\left|35 - 43.5\right| - 0.5\right)^2}{43.5}$$

$$= \frac{\left(8.5 - 0.5\right)^2}{43.5} + \frac{\left(8.5 - 0.5\right)^2}{43.5}$$

$$= 2.943$$

Critical value of $\chi^2 = 3.841$

Decision:   No evidence for the alternative hypothesis.

degree of freedom so the appropriate critical value will be the same as before (3.841). Our conclusion is the same as with the binomial test – there is significant evidence for the alternative hypothesis.

One drawback of $\chi^2$ tests is that they become inaccurate when sample sizes are small or, more precisely, when the expected numbers in any category become small. What constitutes small can be decided by the following rule of thumb. *No expected frequency should be less than 1 and no more than one-fifth of expected frequencies should be less than 5.*

If a set of data breaks this rule, we can sometimes overcome the problem by combining categories, something which we will see in action in some of the later examples. It is much better, wherever possible, to try and avoid the problem by taking a sample which is sufficiently large. This is one example of a more general issue, namely that we have to design our experiments and sampling programmes with a view to the limitations of the statistical test to be used. Many statistical tests have minimum sample sizes.

Finally, there is one slightly confusing point about this use of the $\chi^2$ test with two categories. It relates to the number of tails involved. The problem arises because the formula for $X^2$ produces a large value when there are large departures of observed from expected values in either direction. As was pointed out in the last example, the value of $X^2$ will be the same whether the observed numbers of the two spore types are $52 : 35$ or $35 : 52$. As it stands, the test is a two-tailed test because it detects deviations in either direction, but since these large values of $X^2$ only occur in the right-hand tail of the sampling distribution, we only use this one tail of the distribution for determining the critical values.

The complication arises if we want to do a one-tailed test. We are not likely to want to do this on our spore pattern data because there is no rational basis for predicting the direction of any difference, but the mate choice data in Box 6.1 could be analysed by $X^2$ and would need a one-tailed test. Should we want to carry out a one-tailed test, we proceed with the calculation of $X^2$ as before, but the critical value has to be changed. We simply look up the value in the column with a significance level twice that which is desired. So for a one-tailed test at the 5% significance level we use the critical value in the $\alpha_I^R = 10\%$ column, i.e. 2.706. You might like to verify that the mate choice data gives a corrected $X^2$ value of 2.772, which is sufficiently large to reject the null hypothesis on the basis of a one-tailed test but not a two-tailed test. Incidentally, this is only a problem where there are two categories; with more than two categories we cannot frame a sensible one-tailed alternative hypothesis (see the end of the next section for an explanation of why this is so).

## Chi-squared goodness-of-fit tests with more than two categories

This type of test can be used whenever each observation is assigned to only one of a number of categories, no matter how many categories there are. We could, for example, use it to analyse a behavioural experiment in which individuals had been offered a choice of four food types in equal frequencies and it is widely applied in genetics for testing the fit of observed phenotype frequencies to those expected on the basis of Mendelian inheritance.

Box 6.5 has a familiar example from the $F_2$ of a dihybrid cross in tomato plants in which we expect four categories of offspring in the familiar ratio of $9 : 3 : 3 : 1$. Of the offspring, 9/16 should have both dominant traits, 3/16 have only one dominant trait, 3/16 the other dominant trait, and 1/16 should have the double recessive phenotype. The total number of offspring is 172 so the expected absolute frequencies in the four categories are obtained by

## BOX 6.5 CHI-SQUARED GOODNESS-OF-FIT TEST WITH MORE THAN TWO CATEGORIES: PHENOTYPIC RATIOS IN THE $F_2$ OF A DIHYBRID CROSS

Null hypothesis:   Ratio of the four phenotypes in the population is $9:3:3:1$.

Alternative hypothesis:   Ratio of the four phenotypes in the population is not $9:3:3:1$.

Significance level:   5%.

| | Phenotype | | | | |
| | Purple stems | | Green stems | | |
| | cut leaves | uncut leaves | cut leaves | uncut leaves | Total |
|---|---|---|---|---|---|
| Observed frequencies ($O$) | 110 | 29 | 30 | 3 | 172 |
| Expected frequencies ($E$) | 96.75 | 32.25 | 32.25 | 10.75 | 172 |

$$X^2 = \frac{(100-96.75)^2}{96.75} + \frac{(20-32.25)^2}{32.25} + \frac{(30-32.5)^2}{32.5} \frac{(3-10.75)^2}{10.75}$$
$$= 7.886$$

Degrees of freedom = Number of categories ($C$) – Number of pieces of information required ($i$)

$$= 4 - 1 = 3$$

Critical value of $\chi^2$  = 7.815

Decision rule:   Reject null hypothesis if calculated value is equal to, or greater than, critical value.

Decision:   Significant evidence for alternative hypothesis, associated probability ($P < 0.05$), significant difference between observed and expected values.

**Quick formula**

$$X^2 = \sum \frac{O^2}{E} - n$$
$$= \left( \frac{100^2}{96.75} + \frac{29^2}{32.25} + \frac{30^2}{32.25} + \frac{3^2}{10.75} \right) - 172 = 7.886$$

multiplying 172 by 9/16, 3/16, 3/16 and 1/16, respectively, to give 96.75 : 32.25 : 32.25 : 10.75 – note again that these need not be whole numbers. There are four categories ($C = 4$) and we have used one piece of information from the data (the total number of offspring) to work out the expected values, so $i = 1$. Our calculated value of $X^2 = 7.886$ has $4 - 1 = 3$ degrees of freedom. Although the formula we have been using shows what $X^2$ is, it is not very convenient for doing the calculations. There is an easier to use version at the bottom of Box 6.5.

The critical value at the 5% significance level is 7.815. Our calculated value just exceeds this and we can conclude that we have evidence for the alternative hypothesis although, given the marginal nature of the result, we might want to repeat the observations. This significant result tells us that overall, there are differences between the observed and expected frequencies which are too large to be likely to be due to sampling variation. Unfortunately, it does not tell us whether this is due to differences in a particular category or categories.

We can obtain some idea about this by inspecting the observed and expected frequencies in each category and noting which are most divergent. The most marked discrepancy is that there are fewer double recessives than expected, but it would be wrong to conclude that it was this deficiency which was statistically significant. If we want to show whether this phenotype occurs less often than expected, then we ought to repeat the cross and test the data with a modified hypothesis. The null hypothesis would be that the ratio of double recessives to all other phenotypes was 1 : 15.

The alternative hypothesis in a test like this which involves more than two categories, has to be two-tailed because there is no one logical one-tailed alternative hypothesis. As we saw in the last section, in a test with only two categories there are only two possible directions for the differences between observed and expected values. The expected values in one category can be either above or below the observed values in that category, while those in the other category have to be in the opposite direction. It is possible to specify a direction in a one-tailed alternative hypothesis. When there are more than two categories then there are many possible alternative hypotheses. For instance, if we were to predict that the double dominant phenotype will be more common than expected, then some or all of the other three categories must be less common than expected. There are clearly many combinations of results which will meet this condition and as a result we cannot formulate a one-tailed alternative hypothesis.

## Extrinsic and intrinsic hypotheses

The preceding examples have all been tests of what is called an **extrinsic hypothesis**. It is called an extrinsic hypothesis because the expected ratios (e.g. 1 : 1, 3 : 1 and 9 : 3 : 3 : 1) arise from a hypothesis which is external to the data. To obtain the expected frequencies, we only need one piece of information from the sample – the total sample size, and, as a result, the degrees of freedom are one less than the number of categories. In other goodness-of-fit tests, the expected ratios do not arise from an external hypothesis and the sample size on its own is insufficient to calculate the expected frequencies. They have to be calculated using additional information from the sample itself and as a result the hypothesis being tested is said to be an **intrinsic hypothesis**.

We have already come across this idea in Chapter 5 where we visually compared the observed frequency distribution of colonies on plates with the expected frequency distribution based on a Poisson distribution which had the same mean as our sample. To calculate the expected frequencies, we had to use the sample size and the mean value from the sample (as an estimate of the parametric mean of the population). If we now want to test the goodness-of-fit of the observed and the expected distribution we shall be testing an intrinsic hypothesis. Another common situation for a test of an intrinsic hypothesis is in population genetics, when we wish to test whether the frequencies of genotypes in a sample are in agreement with those expected on the basis of the Hardy–Weinberg principle. If there are two alleles and three genotypes present, then to work out the expected genotype frequencies we first have to find the frequency ($p$) of one of the alleles. This is calculated from the observed genotype frequencies (i.e. from the sample). This piece of information, along with the sample size, enables us to calculate the expected values.

Once we have obtained the expected frequencies the actual calculations are carried out as for an extrinsic hypothesis, but the degrees of freedom are reduced because we have had to use an extra piece (or pieces) of information from the sample to calculate the expected values. This means that the critical value of $\chi^2$ for an intrinsic hypothesis is going to be lower than that for an extrinsic hypothesis with the same number of categories. Let's see how this works in practice, using our data on colony distribution.

## Chi-squared goodness-of-fit to a Poisson distribution

The figures in the first three columns in Box 6.6 are based on those produced in the previous chapter with only two changes. First, the expected relative frequencies (based on the hypothesis that the distribution is a Poisson) have been converted to frequencies by multiplying them by the sample size. Remember we must not use proportions when calculating $X^2$. Secondly, because the expected numbers of plates with four or more colonies are low, they have been added together or pooled so that they do not break the rule stated on p. 141, and this reduces the number of categories to 5. The fourth column is the individual contributions to $X^2$ for each pair of observed and expected values. Summing the individual contributions yields a value of 2.277 for $X^2$, which is a small value, reflecting the fact that the observed values are close to the expected values. To find the appropriate critical value, we need the degrees of freedom and the significance level.

The degrees of freedom are given by the same formula as before (df = $C - i$) where $C$ is the number of categories and $i$ is the number of pieces of information from the data which are used to calculate the expected values. The number of categories is now 5, because we pooled the last four categories and it is this value which must be used when calculating the degrees of freedom. To get the expected values, we had to use two pieces of information, the total sample size ($n$) and the mean number of colonies per plate ($\bar{Y}$), so $i = 2$. The degrees of freedom are therefore $5 - 2 = 3$.

The critical value of $\chi^2$ at the 5% significance level is 7.815 and our calculated value is less than this, so we have no evidence for the alternative hypothesis. We can conclude that the differences between the two distributions are most likely to be due to chance and that therefore the distribution of colonies is compatible with a Poisson distribution. This is consistent with what we know of the biology of the situation, namely that the spores of micro-organisms have no control over where they land. This should lead to a random pattern of dispersion, because the presence of one spore on a plate will have no effect on whether a subsequent spore lands on that plate.

---

**BOX 6.6  CHI-SQUARED GOODNESS-OF-FIT TEST TO A POISSON DISTRIBUTION: FREQUENCY OF MICRO-ORGANISM COLONIES ON PLATES.**

Null hypothesis:  Distribution of colonies follows a Poisson distribution.

Alternative hypothesis:  Distribution of colonies does not follow a Poisson distribution.

(Mean estimated from sample = 1.5 colonies/plate)

Significance level:  5%.

| Number of colonies per plate | Observed frequency | Expected frequency | $(O - E)^2/E$ |
|---|---|---|---|
| 0 | 19 | 20.08 | 0.058 |
| 1 | 34 | 30.12 | 0.499 |
| 2 | 18 | 22.59 | 0.933 |
| 3 | 14 | 11.30 | 0.647 |
| 4 | 3 ⎫ | ⎫ | ⎫ |
| 5 | 1 ⎬ 5 | ⎬ 5.91 | ⎬ 0.139 |
| 6 | 1 | | |
| 7 | 0 ⎭ | ⎭ | ⎭ |
| Total | 90 | 90 | $X^2 = 2.277$ |

Degrees of freedom (df) = No. of categories $(C)$ – No. of pieces of information required $(i)$

$$= 5 - 2 = 3$$

Critical value of $\chi^2$    = 7.815

Decision rule:  Reject null hypothesis if calculated value is equal to, or greater than, critical value.

Decision:  No evidence for the alternative hypothesis.

Box 6.7 has the data on the number of flowering spikes of the common broomrape (*Orobranche minor*), which we met in Chapter 5. The original number of categories was 22 but, because of small expected numbers, this was reduced to 8 by pooling. The calculated value of $X^2$ is greater than the critical value at the 5% significance level. In fact, the chances of getting a value this large (or larger) if the distribution was a Poisson distribution is remote so we have virtually conclusive evidence for the alternative hypothesis. The fact that the variance/mean ratio is greater than 1, tells us that the pattern of dispersion is contagious, which means that flowering spikes occur together (i.e. in the same quadrat) more often than expected on the basis of a random pattern. There could be several explanations for this, and you should be able to come up with some of them. One attractive explanation arises from the fact that this species is parasitic; it has no green tissue, is incapable of photosynthesis and instead obtains nutrients from the roots of species of clover (*Trifolium*) to which it is

---

**BOX 6.7   CHI-SQUARED GOODNESS-OF-FIT TEST TO A POISSON DISTIBUTION: THE SPATIAL DISPERSION OF FLOWER SPIKES OF COMMON BROOM RAPE (*OROBANCHE MINOR*)**

Null hypothesis:   Frequency distribution of flower spikes in quadrats follows a Poisson distribution.

Alternative hypothesis:   Frequency distribution of flower spikes does not follow a Poisson distribution.

(Mean estimated from sample = 4.33 flower spikes/quadrat)

Significance level:   5%

| No. of flower spikes/quadrat | Observed frequency | Expected frequency |
|---|---|---|
| 0 | 33 } 37 | 0.98 } 5.24 |
| 1 | 4 | 4.27 |
| 2 | 9 | 9.24 |
| 3 | 2 | 13.35 |
| 4 | 3 | 14.46 |
| 5 | 4 | 12.53 |
| 6 | 4 | 9.05 |
| 7 | 1 | 5.60 |
| $\geq 8$ | 15 | 5.51 |

Calculated value of $X^2$   = 239.52

Degrees of freedom  = No. of categories ($C$) – No. of pieces of information required ($i$)

$$= 8 - 2 = 6$$

Critical value of $\chi^2$   = 12.592

Decision rule:   Reject null hypothesis if calculated value is equal to, or greater than, critical value.

Decision:   Very highly significant evidence for the alternative hypothesis ($P < 0.001$).

---

connected. If the distribution of the host species is contagious, then we might expect this to be paralleled by the parasite. Once again we may have several competing hypotheses.

You might be tempted to try the same procedure with the data on the spatial dispersion of stomata from Chapter 5, but you would encounter a problem due to the small sample size. Several of the expected values are small and so the rule referred to earlier is broken. If you try and get round this by pooling categories, you end up with only two categories and your value of $X^2$ will have no degrees of freedom. In the case of the Poisson distribution, there is an alternative test based on the $\chi^2$ test which you can use; it is described in the final section of this chapter.

## Chi-squared goodness-of-fit to a binomial distribution

You can use exactly the same procedure as the one above to test whether an observed frequency distribution fits the distribution expected on the basis of a binomial model. For instance, you could do it on the distribution of kernel colour in maize (Box 5.1) and the distribution of frequency of hits with the point quadrats (Box 5.3). There is only one point to note. In the first case it would be an intrinsic hypothesis because we are assuming that the proportion of yellow kernels is 0.5, while in the second it would be an extrinsic hypothesis because we estimated the proportion of hits ($\hat{p}$) from the data and used this to calculate the expected values. This will have to be taken into account when calculating the degrees of freedom.

## Chi-squared goodness-of-fit with continuous variables

The preceding examples have all been situations in which the variable naturally fell into categories. Goodness-of-fit tests can also be applied to variables which are basically continuous, but which are placed in categories for convenience. This categorization may take place at the time that the observations are made, because in some situations it may be difficult to record a continuous variable accurately or it may be done at a later stage. The next two examples illustrate this.

Box 6.8 is from a study on the foraging behaviour of the bumble-bee (*Bombus leucorum*) on flower spikes of viper's bugloss (*Echium vulgare*). The question of interest was whether bees have any preference as to the height at which they alight on the spike of flowers. Clearly, it would have been quite difficult to measure the exact height at which each bee landed so the observer simply recorded whether the landing position was in the top, middle or bottom third of the spike.

The null hypothesis is that bees have no preference so we would expect equal numbers of visits in the three zones. This will therefore be a test of an extrinsic hypothesis. The expected numbers are simply calculated by dividing up the total number of observations in the ratio 1 : 1 : 1 and $X^2$ calculated as usual. Only one piece of information is needed from the data so the degrees of freedom are 2 .The calculated value is larger than the critical value and so we have significant evidence for the alternative hypothesis. Inspection of the relationship between the observed and the expected results shows that there is an excess of visits to the bottom of the flower spike.

Of course this method of recording the results is subjective and it would be very easy for bias to creep in, because there is no clearly defined line between the categories. Once the observer has realized that bees are tending to land at the bottom of the spike, there may be a tendency to place observations which are really at the bottom of the middle category into the bottom category. This is the same problem as we had with the woodlice and we really need a clear criterion for each category. One way to do this would be simply to mark the borderlines between top, middle and bottom in some way, such as with a piece of string round the stem.

---

**BOX 6.8 CHI-SQUARED GOODNESS-OF-FIT TEST WITH A CONTINUOUS VARIABLE: FORAGING HEIGHT OF THE BUMBLE BEE (*BOMBUS LEUCORUM*) ON VIPER'S BUGLOSS (*ECHIUM VULGARE*)**

Null hypothesis:   Frequency of bee visits the same at different heights on the flower spike.

Alternative hypothesis:   Frequency of bee visits different at different heights on the flower spike.

Significance level:   5%.

|  | Top | Middle | Bottom | Total |
|---|---|---|---|---|
|  | Height on flower spike | | | |
| Observed frequency of visits | 8 | 12 | 22 | 42 |
| Expected frequency of visits | 14 | 14 | 14 | 42 |

Calculated value,

$$\chi^2 = \frac{(8-14)^2}{14} + \frac{(12-14)^2}{14} + \frac{(22-14)^2}{14} = 7.429$$

Degrees of freedom:   Number of categories (C) – Number of pieces of information (i)

$$= 3 - 1 = 2$$

Critical value of $\chi^2 = 5.991$

Decision rule:   Reject null hypothesis if calculated value is equal to, or greater than, critical value.

Decision:   Significant evidence for the alternative hypothesis ($P < 0.05$).

---

### Chi-squared goodness-of-fit to a normal distribution

As we saw in Chapter 3, when we measure any continuous variables we assign variates to classes because of the limitations on the precision of our measurements. We may then combine classes for convenience in constructing frequency distributions. These classes can be used in a goodness-of-fit tests provided that we have some way of calculating what the frequencies should be in every class.

One way is of doing this is to assume that the variable follows a normal distribution as we did for the body lengths of isopods (in Chapter 5). We can then compare the observed and the expected frequencies in each class using $\chi^2$. Box 6.9 shows this procedure (the

## BOX 6.9 CHI-SQUARED GOODNESS-OF-FIT TEST TO A NORMAL DISTRIBUTION: BODY LENGTHS OF THE ISOPOD (*SPHAEROMA RUGICAUDA*)

Null hypothesis:  Body length follows a normal distribution.

Alternative hypothesis:  Body length does not follow a normal distribution.

Mean = 4.951 mm and standard deviation = 0.5379 mm; estimated from sample.

Significance level:  5%.

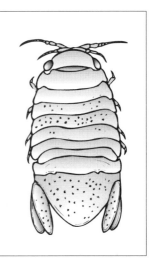

| Body length (mm): upper limit of class | Observed frequency | Expected frequency | $(O - E)^2/E$ |
|---|---|---|---|
| 3.55 | 0 | 1.84 | 1.840 |
| 3.75 | 5 | 3.40 | 0.753 |
| 3.95 | 9 | 7.28 | 0.406 |
| 4.15 | 20 | 14.76 | 1.860 |
| 4.35 | 30 | 25.44 | 0.817 |
| 4.55 | 29 | 38.40 | 2.301 |
| 4.75 | 44 | 50.56 | 0.851 |
| 4.95 | 44 | 58.00 | 3.379 |
| 5.15 | 69 | 57.88 | 2.136 |
| 5.35 | 50 | 50.68 | 0.009 |
| 5.55 | 50 | 38.60 | 3.367 |
| 5.75 | 29 | 25.68 | 0.429 |
| 5.45 | 14 | 14.84 | 0.048 |
| 6.15 | 4 | 7.44 | 1.591 |
| 6.35 | 1 ⎫ | 3.44 ⎫ | |
| 6.55 | 2 ⎬ 3 | 1.1 ⎬ 5.2 | 0.931 |
| > 6.55 | 0 ⎭ | 0.60 ⎭ | |
| | $n = 400$ | $n = 400$ | $X^2 = 20.718$ |

Degrees of freedom = No. of categories ($C$) – No. of pieces of information required ($i$)

$$= 15 - 3 = 12$$

Critical value of $\chi^2$ = 21.026

Decision rule:  Reject null hypothesis if calculated value is equal to, or greater than, critical value.

Decision:  No evidence for alternative hypothesis ($P < 0.10$).

expected frequencies have been calculated by multiplying the relative frequencies in Box 5.8 by the sample size). The calculated value of $X^2$ is 20.178. There are 17 categories initially, but pooling the last three categories because they have small expected numbers, reduces this to 15. This is another test of an intrinsic hypothesis, because the expected frequencies are not derived from some external theory. To calculate them, we had to use the mean and the standard deviation and the sample size. Altogether, we had to use three pieces of information from the data so $i = 3$ and the degrees of freedom are $15 - 3 = 12$. The critical value of $\chi^2$ at the 5% significance level is 21.026 and our calculated value falls just below this. We have no significant evidence for the alternative hypothesis. The differences between the observed and the expected frequencies are sufficiently small to be likely to be due to sampling variation.

This procedure, like many statistical tests, is not ideal. The classes we use are arbitrary and if we had divided the data up differently, we would get a different value of $X^2$ and possibly come to a different conclusion. Although it is not very likely, we could even choose to use classes which suit our preconceptions, that is, which minimize the differences between the observed and the expected distributions. In addition, it will only work satisfactorily if the sample size is large. The reason is the same as the one we considered in the case of the Poisson distribution; any pooling of classes with small expected values could mask important differences, especially in the tails of the distribution. There is an alternative test, the Kolmogorov–Smirnoff test for normality, which is described in Neave & Worthington (1988). It is based on a completely different idea and is particularly useful for small samples.

## The dispersion test: an alternative test for a Poisson distribution

In Box 6.10 there are four sets of data on spatial dispersion. One is the data on stomata from Chapter 5, while the second concerns the aquatic larvae of a small insect the blackfly (*Simulium*). The numbers of larvae were counted in 12 small, equally sized randomly chosen areas on a single rock. The third set relate to counts, made under the microscope, of the numbers of the protozoan *Tetrahymena* in 30 random squares of a haemocytometer while the last set are the numbers of dog whelks (*Nucella lapillus*) in randomly placed quadrats (from Box 5.7). They all illustrate a common problem – small sample size. For example, with 15 observations and several categories, there will not be many observations in each category. If we calculate the expected frequencies on the basis of a Poisson distribution as before, some will be smaller than the rules (p. 141) allow. Many of the categories will need to be pooled and this may begin to affect the validity of the test.

Fortunately there is an alternative test available, based on $\chi^2$, although the formula given in Box 6.10 is quite different from the one we have been using. (An explanation of the formula is given towards the end of this section.) This test also uses the critical values in a different way. To calculate the value of $X^2$ all we need to do is multiply the index of dispersion by $(n - 1)$, that is by the sample size minus one. We can see where the test gets its name from, when we remember that the index of dispersion is the other name for the variance/mean ratio. When we come to look this value up in the tables we have to use both tails of the distribution of $\chi^2$ rather than just the right-hand tail which we used before. We can see the reason for this by concentrating on one of the examples.

Take the case of the *Simulium* larvae. The null hypothesis is that the dispersion of the larvae is random, which means that the frequency distribution of numbers per quadrat should follow a Poisson distribution. If this were true then we know that, in theory, the variance/mean ratio should be 1. In practice it may be more or less than one because of sampling

**BOX 6.10 THE CHI-SQUARED TEST OF DISPERSION**

| | Stomata | Simulium larvae | Tetrahymena | Nucella lapillus |
|---|---|---|---|---|
| Index of dispersion $I = s^2/\bar{Y}$ | 0.16 | 0.29 | 1.37 | 6.14 |
| $n$ | 15 | 12 | 30 | 15 |
| $X^2 = I \times (n-1)$ | 2.291 | 3.182 | 39.654 | 85.951 |
| df | 14 | 11 | 29 | 14 |
| $\alpha_2 = 5\%$ | | | | |
| Critical values of $\chi^2$ | | | | |
| Lower | 5.629 | 3.816 | 16.047 | 5.629 |
| Upper | 26.119 | 21.920 | 45.722 | 26.119 |
| Pattern of dispersion | Regular | Regular | Random | Contagious |

variation (which will be affected by sample size), but it is unlikely that it will deviate a great deal from this value. On this basis a calculated $X^2$ value, using the formula above, is likely to have a value around $(n-1)$, which in this case would be around 14.

The alternative hypothesis is that the pattern of dispersion is not random, which means that it is either regular or contagious. As we saw in Chapter 5, if it is regular then the numbers of occurrences in each sampling unit are likely to be more similar to one another than they would be if the distribution were random, so the variance is likely to be low and less than the mean. The variance/mean ratio is likely to be much less than 1 and, as a result the $X^2$ value is likely to be much less than $(n-1)$, that is, much less than 14. Small values of $X^2$ are likely if the dispersion pattern is regular. (Note that we have to keep saying that this is what is likely, because of course we are dealing with samples. A sample from a regularly dispersed population could, by chance, be so variable that it gives rise to a large $X^2$ value – it's possible, but not likely.) If the dispersion pattern is contagious, the argument goes the other way. Now variance:mean ratios are likely to be much larger than one and $X^2$ values much larger than 14. We can now see why we use both tails of the distribution. $X^2$ values falling in the left-hand tail are indicative of a regular pattern of dispersion while values falling in the right-hand tail are indicative of a contagious pattern. A pattern of random dispersion will produce $X^2$ values which tend to fall in the middle of the distribution.

To find the critical values, look in the $\alpha_2$ row at the top of Table A6 and find the two columns with the correct significance level (5%). These give the left-hand and right-hand critical values at the 5% significance level (which are the values which cut off 2.5% of the distribution in each tail). The degrees of freedom in this case are one less than the sample size $(n-1)$, so the critical values are as given for each set of data.

For the stomata and the blackfly larvae, the calculated value is smaller than the lower critical value so we can conclude that we have significant evidence for a regular pattern of

---

**BOX 6.11    THE DISPERSION TEST: WHERE THE FORMULA COMES FROM**

$$\chi^2 = \frac{\Sigma(Y - \mu)^2}{\sigma^2} = \frac{1}{\sigma^2}\Sigma(Y - \mu)^2$$

$$= \frac{1}{\text{Population variance}} \times \text{Sum of squares}$$

$$= \frac{\text{Sample variance}}{\text{Population variance}} \times (n - 1)$$

$$= \frac{s^2(n - 1)}{\bar{Y}}$$

---

dispersion. For the larvae the biological interpretation is as follows. The larvae are found attached to rocks in fast flowing water and they use modified appendages on their heads to filter small particles out of the water. If individuals were in clumps, then those in the middle or at the downstream edge would get rather little food. Aggressive encounters between individuals lead to an even spacing which minimizes interference with feeding. Can you see what the biological significance of the regular dispersion is in the case of the stomata? For the protozoan, there is no evidence for either sort of non-random dispersion, while for the snails the calculated value of $X^2$ is much larger than the upper critical value, so we have highly significant evidence for a contagious pattern.

If you are curious about where this formula comes from, it can be quite simply explained. However, you can skip the next paragraph if you feel that you don't want to go into this! First of all you need to remember the formula for $\chi^2$ which is given again in Box 6.11. This can be rearranged slightly to show that $\chi^2$ is the ratio of the sum of squares, $\Sigma(Y - \mu)^2$, to the population variance. Then we can alter the top line so that it uses the sample variance rather than the sum of squares, because there is a simple relationship between them. We know that the variance is equal to the sum of squares divided by $(n - 1)$ (Box 4.8), so the sample variance multiplied by $(n - 1)$ is the same as the sum of squares and can be substituted in the top line.

The final question is why has the population variance $\sigma^2$ on the bottom line been changed to the sample mean $\bar{Y}$? The null hypothesis in this case is that the distribution is a Poisson distribution and, if this is true, then the mean should equal the variance. We can therefore use the sample mean as an estimate of what the population variance would be if the null hypothesis were true. $\chi^2$ is therefore the same thing as the variance/mean ratio multiplied by $(n - 1)$, but again this is all based on a continuous distribution. Because of this, our calculated value, based on a discontinuous distribution cannot be a value of $\chi^2$, which is why we call it $X^2$. However as before, $X^2$ has a distribution that is close enough to a $\chi^2$ distribution for us to be able to use $\chi^2$ tables.

This test is obviously easier to carry out than the procedure which we adopted with the colonies and the flower spikes because there is much less calculation involved and you may be wondering why we don't always use the simpler procedure even if sample sizes are large. The answer is that the two procedures are testing different things and are therefore not equivalent. The first test examines the extent of the difference between the observed and the expected frequencies in every category. The second test is actually comparing the observed variance of the sample on the top line, with an estimate of what the variance would be if the distribution was of the Poisson type. That is why the sample mean, which should estimate the variance if the distribution is of the Poisson type, appears on the bottom line. The first test has the disadvantage that it needs a large sample size and also that it can lack sensitivity if categories have to be pooled. The second test suffers from one drawback. It is possible to have a distribution whose mean and variance are equal but which is not a Poisson distribution.

# 7

# Single samples: the reliability of estimates

We have already met this problem several times. It was outlined in Chapter 1 using the examples of the numbers of snails per unit area, the concentration of phosphate and the proportion of different types of blood cells. In Chapter 2 we came across it in our discussion of the proportion of damp turns made by woodlice and again in Chapter 6, where we had to use the sample mean $\bar{Y}$ as an estimate of the population mean $\mu$ of a Poisson distribution. The issue should by now be clear – in all these cases we have a sample of observations and there is always an element of chance in sampling. As a result, the sample statistics are unlikely to have the same value as the population parameters.

Let's clarify this by looking again at the frog blood cell example, where our question is what proportion of the cells are neutrophils in healthy frogs. Suppose that we found four neutrophils in a random field of view in which there were 12 cells altogether. The sample fraction would be 4/12 and the sample proportion would be 0.33, yet if we were to have chosen a different field of view at random, we could have got a different value. Even if we used the whole slide this would not give the population proportion; that would be the value we obtained if we examined all the blood of all the healthy frogs of this species. It is, of course, possible that we could have struck lucky with our one sample and, by chance, obtained a sample proportion which was equal to the population proportion. However, this is not very likely and even if this was the case there is no way that we could know that it had happened.

Our sample statistic is only an estimate of the population parameter and, because it is a single value, it is known as a **point estimate**. Now on its own, a point estimate is of little value because we have no idea about how close it is likely to be to the population parameter. What we also need is a measure of the likely closeness, that is, a measure of the **reliability** of this point estimate. This will involve quantifying, in some way, the amount of sampling variation which our point estimate is subject to.

The general argument is as follows. If we can show that the effect of sampling variation is likely to be small, then we can be fairly sure (but not absolutely certain) that our sample statistic which is an estimate will be quite close to the population parameter. Conversely, if the effect of sampling variation is likely to be large, then our estimate could be quite different from the value of the population parameter. To quantify the amount of sampling variation we will need to think about the frequency distribution of the statistic, based on all possible samples. We will start by using the frog blood cell example.

## The reliability of a sample proportion

So what determines the extent of sampling variation with a variable of this type and how do we quantify it? Well, as you might expect, sample size is one factor and the effect of this is illustrated in Figure 7.1. For the sake of this exploration, I have assumed that we are omniscient, that is, we know that the population proportion of neutrophils cells is actually 0.400000.

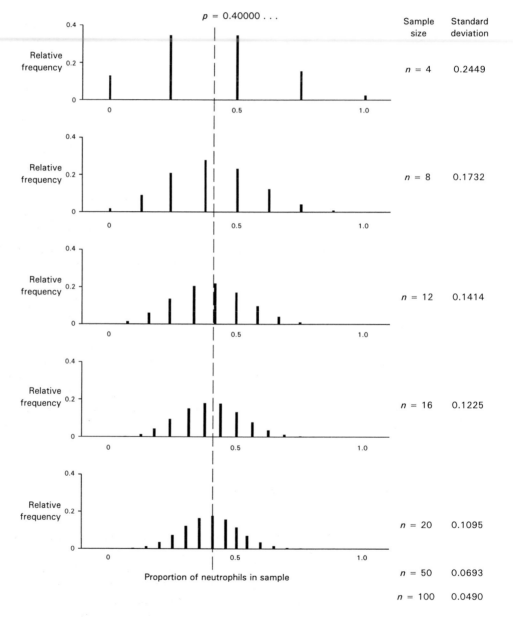

**Figure 7.1** The effect of sample size on the sampling distribution of the proportion of neutrophils when the population proportion, $p$ = 0.40000, is shown by the dotted line.

This enables us to use the tables of binomial distributions to find out the frequency of occurrence of samples with different proportions of neutrophil cells and to see how these distributions change with sample size. These distributions are of course sampling distributions and they show us the extent of chance variation in the sample proportion. It should come as no surprise that as the sample size increases, extreme sample proportions become less frequent and the bulk of the sampling distribution becomes concentrated in the middle. It is also clear that the tighter packing of the sample proportions occurs around the population proportion ($p$), which in this case we know is 0.4. This is not telling us anything new; it simply means that large samples are better (i.e. yield more reliable estimates) than small samples. Large samples are more reliable because the values of the sample proportions are less widely dispersed around the population proportion. We now need to quantify this effect.

To do this, we use the standard measure of how "spread-out" a distribution is, namely the standard deviation. We first met this as a measure of the variation of items in a sample around the sample mean; remember that there it was a sort of average deviation of the items from the mean. For a binomial distribution, we already know (Box 5.2) that the standard deviation of a proportion can be calculated from the formula $\sqrt{pq/n}$. Although $p$ and $q$ come into the formula, the most important part of this is the sample size ($n$) on the bottom line, because it shows us that as the sample size goes up, so the standard deviation goes down. The calculated standard deviation for each of the distributions is included on Figure 7.1, and we can see that these standard deviations decrease as sample size increases. The distributions for $n = 50$ and $n = 100$ have not been drawn and only the standard deviations are given. The smaller the standard deviation, the more tightly the sample proportions are packed around the population proportion. This means that the standard deviation quantifies how close, in general, sample proportions are to the population proportion.

There is a slight complication in the terminology. The term standard deviation is strictly reserved for the measure of variation of the items in a sample around the sample mean (Ch. 4). Here we are looking at the variability of sample proportions (which are sample statistics) around the population proportion. Where a standard deviation is being used to describe the variability in a sample statistic, it is given a different name the **standard error** or **s.e.** for short. In the case of the binomial distribution, the distinction between a standard deviation and a standard error is not very clear – we'll meet a better example later in this chapter.

The standard error is the simplest numerical measure we can get of the variability of a sample statistic and hence of its reliability and it is a term we shall meet frequently in the rest of the book.

## Using the standard error as a measure of reliability

You may be wondering where all this is getting us, because the argument so far has been based on knowing what the population proportion ($p$) is. In practice this is just what we don't know! What we do is to reverse the argument above.

We saw above that when we did know the population proportion, the standard error had a value which reflected the variation or spread of sample proportions around the population proportion. Small values imply that, in general, sample proportions are close to the population proportion, while large values imply that in general they are not. Now suppose, instead, that we have a sample proportion ($\hat{p}$) and we know that it has a small standard error (i.e. that the sample proportion came from a tightly packed distribution). In this situation, we can be fairly confident (but not absolutely certain) that the sample proportion will be quite close to the population proportion. If, however, the sample proportion has a large standard error, then we can be much less confident about its reliability. So how do we know what the standard error is? The answer is that we don't (we would need to know $p$ to calculate it)! However, we can estimate the standard error using the sample proportion ($\hat{p}$). The formula is simply $\sqrt{\hat{p}\hat{q}/n}$ rather than $\sqrt{pq/n}$. Box 7.1 gives an example of this calculation using our blood cell data and you might like to calculate standard errors for some of the other proportions we have come across.

Whenever we report a point estimate of a parameter, it should always be accompanied by a measure of its reliability and the standard error is the simplest such measure. To present this information, we simply attach the standard error to the point estimate written in the form " numerical value of point estimate $\pm$ numerical value of standard error" (Box 7.1). We now have what is called an **interval estimate**. Alternatively we can represent it graphically (Box 7.1b) using error bars which extend from a point one standard error above the sample proportion to a point one standard error below the sample proportion. Either way, the reader is given an indication of the likely extent of variability associated with the estimate and hence its reliability. Incidentally you may see an estimate given in the form "estimate $\pm$ 2 standard errors". This is quite acceptable and there are good reasons for it which we will meet later in this chapter.

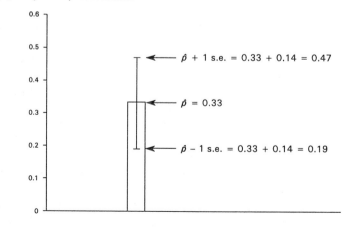

**BOX 7.1   CALCULATING THE STANDARD ERROR OF A SAMPLE PROPORTION**

(a)  Sample proportion of neutrophils $\hat{p} = 4/12 = 0.33$

Estimated standard error $= \sqrt{\hat{p}\hat{q}/n} = \sqrt{0.33 \times 0.67/12}$

$= 0.14$

Sample proportion $\pm$ standard error $= 0.33 \pm 0.14$

(b) Graphical presentation

$\hat{p} + 1$ s.e. $= 0.33 + 0.14 = 0.47$

$\hat{p} = 0.33$

$\hat{p} - 1$ s.e. $= 0.33 + 0.14 = 0.19$

Now, although the standard error is the simplest measure of reliability available, it has a number of limitations. We know that a small standard error is better, than a large one because we can be more confident that the estimate is close to the parameter, but there is not an exact and useful interpretation of the actual value. The interval represented by our error bars is certainly not a range which definitely includes the population proportion, nor can we say anything about the chances of this interval including the population proportion. In particular, it does not allow us to quantify the meaning of terms such as more (or less) confident. To do this, we have to calculate a more refined measure of reliability called a confidence interval, but this idea is best developed and explored in the context of the reliability of a sample mean.

# How reliable is our sample mean?

Exactly the same general problem arises with estimates of other sorts of variables such as mean body length or mean number of colonies per plate. The sample mean ($\bar{Y}$) is only an estimate of the population mean ($\mu$) and again we need a measure of its reliability. As before, we can begin our quest by calculating a standard error and the easiest way to see what this involves is to do another thought experiment.

## The distribution of sample means

Suppose we have a variable with a frequency distribution in the population like the one shown in Figure 7.2a, with a mean $\mu$ and a standard deviation s and that we take a random sample of $n$ observations from this population (Fig. 7.2b). I have made $n = 10$ in my example and each dot represents an individual value of $Y$. We calculate the mean of this sample, $\bar{Y}_1$, which is marked with the arrow. We then replace these ten individuals and take another sample of 10 observations and again calculate the mean $\bar{Y}_2$. Some of the observations in this second sample may by chance have the same values as observations in the first sample, others will be different; but the second mean is not likely to be the same as the first.

Now imagine repeating this procedure until we have taken every possible sample of size 10 (i.e. every different combination of 10 observations). For each sample, we calculate the mean and then we draw a frequency distribution of these means of samples of size 10 (Fig. 7.2c). This would show us all the possible values of the mean of samples of size 10 and their frequencies of occurrence, in other words, it would be a sampling distribution. What could we say about this distribution of sample means in terms of its shape and location?

The first point is that not all mean values are going to be equally likely to occur. A sample with a very high or a very low mean can only arise if all the observations in that sample have very high (or very low) values. These extreme observations are rare in the population, so the chances of getting a sample containing only 10 high (or 10 low) observations is very small. On the other hand, there are many different combinations of 10 observations that will give a sample with an intermediate mean. As a result the distribution of sample means will have a hump in the middle and two tails.

Secondly, the distribution will be less widely spread than the original distribution. The reason for this has already been touched on and has to do with the fact that this is a

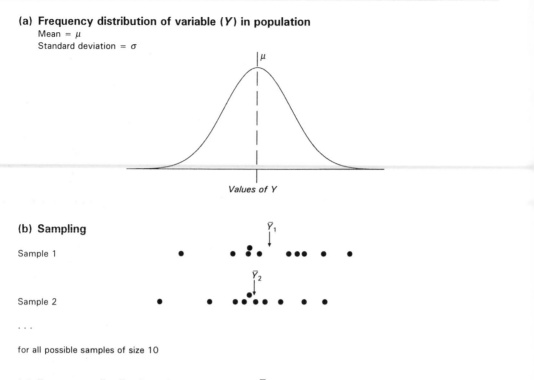

**(a) Frequency distribution of variable (*Y*) in population**

Mean = $\mu$

Standard deviation = $\sigma$

*Values of Y*

**(b) Sampling**

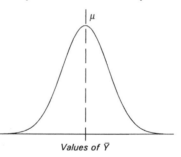

Sample 1

Sample 2

. . .

for all possible samples of size 10

**(c) Frequency distribution of sample means (*Ȳ*) for all possible samples of size 10**

*Values of Ȳ*

**Figure 7.2** The distribution of sample means.

distribution of sample means. The effect of using the mean of a sample, rather than the observations themselves, is to even out the variability due to extreme individual observations. Because of this, the means of samples are always going to be less variable than the individual observations.

The third point cannot be arrived at by this common-sense route but it can be shown mathematically. The mean of the distribution of sample means will be equal to the population mean $\mu$. With a bit more information we shall be able to use this distribution of sample means in the same sort of way as we used the distribution of sample proportions in the last section.

160

## The standard error of a sample mean

So far we have described the shape of this distribution of sample means in general terms, but to be of any real use, we need to make our description quantitative. In particular, we need to know how tightly packed the sample means are around the population mean. We can measure this by using the standard deviation of the distribution. Again, because we are dealing with the distribution of a statistic ($\bar{Y}$), we shall call the standard deviation of this distribution of sample means the standard error. If we can calculate the standard error, then we can apply the same logic as we used in the section on sample proportions.

If the standard error is small, then the observations (which in this case are means of samples of size 10) are very tightly packed around the mean of the distribution, which in this case is also the population mean ($\mu$). If the sample means are tightly packed, then most of the possible mean values in samples of this size are near to $\mu$. It follows that if we only have one mean, but we know that it comes from a distribution which has a small standard error, then the chances are that this one mean will be fairly close to the true mean. Remember that a small standard error means that the distribution is tightly packed. The converse holds true if the standard error is large. So a knowledge of the standard error (= standard deviation of the distribution of sample means) will give us a measure of how close a sample mean is likely to be to the population mean.

So what determines the size of the standard error, i.e. the spread of a distribution of sample means? The easiest way to explore this is to carry on with our thought experiment. First, imagine repeating the procedure, but with samples of 100 observations instead of 10 (Fig. 7.3a). In a small sample, one extreme observation can easily lead to a distortion in the mean, so the distribution of sample means might be as shown by the solid line. Large samples are more likely to have a balance of high and low observations, leading to less variation between their means as shown by the dashed line. Both distributions will be centred around the population mean $\mu$.

It follows from this that the distribution of sample means for large samples will be more closely packed around the population mean than the distribution of sample means of small samples (Fig. 7.3a). Indeed it can be shown mathematically that the standard error is simply proportional to $1/\sqrt{n}$, where $n$ is the sample size. You can see from this formula that as the size of the sample increases, so the standard error gets smaller, which means that the distribution becomes increasingly tightly packed around the population mean ($\mu$). Again this simply confirms what common sense tells us, namely that the mean of a large sample is more reliable than the mean of a small sample. Incidentally the logical conclusion of this is that as the sample gets larger and larger, so the standard error gets smaller and smaller until it reaches zero. At this point, our sample comprises the whole population and there is no problem about the reliability of our estimate. The sample mean will actually be the population mean.

Something else affects the standard error. Let's go back to our original sample size of 10 and think about what the effect would be if the variable had a larger standard deviation initially. This is shown by the dotted curve at the bottom (Fig. 7.3b), where you can see that the frequency distribution of $Y$ is more spread out than before. Again, there is a common-sense answer: the more variable the original observations, the more variable will be the means of the samples. Under these conditions, when we take samples of size 10, the distribution of their sample means should be more widely spread than that produced by samples of the same size from a less variable population. In fact it can be shown mathematically that

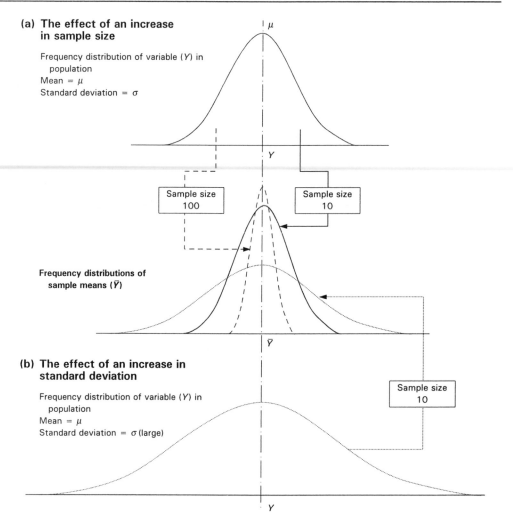

**(a) The effect of an increase in sample size**

Frequency distribution of variable ($Y$) in population
Mean = $\mu$
Standard deviation = $\sigma$

Sample size 100

Sample size 10

**Frequency distributions of sample means ($\bar{Y}$)**

**(b) The effect of an increase in standard deviation**

Frequency distribution of variable ($Y$) in population
Mean = $\mu$
Standard deviation = $\sigma$ (large)

Sample size 10

**Figure 7.3** The standard error of the sample mean: what determines its size?

the standard error is directly proportional to the standard deviation of the original distribution $\sigma$.

These are the only two things which affect the spread of the distribution of sample means and hence the standard error; in summary this decreases with increasing sample size and decreasing $\sigma$. The standard error is simply equal to $\sigma/\sqrt{n}$.

Let's just recap on what we have shown. If we take repeated samples of size $n$ from a population with mean $\mu$ and standard deviation $\sigma$, then the means of those samples will have a distribution that is centred around $\mu$. The spread of this distribution is measured by its standard deviation, which is given a special name, the standard error, because it is a distribution of sample means. The standard error is given by $\sigma/\sqrt{n}$. This relationship is quite general and does not depend on the shape of the population from which the samples are taken. It does not have to be a normal distribution nor does it have to be a continuous distribution.

## Calculating a standard error

What we have just covered is all very well in theory, but how in practice do we obtain a value for the standard error. After all, we cannot go through the procedure of actually taking all the possible samples of a given size; in fact we are only likely to have one sample at our disposal. This means of course that we know the sample size ($n$), but what about the population standard deviation ($\sigma$)? This is another parameter and so we can never know it exactly; however, the spread of the observations in our sample should broadly reflect the spread of the variable in the population. So what we do is to use the sample standard deviation ($s$) as an estimate of the population standard deviation ($\sigma$). Our one sample therefore carries enough information for us to calculate an estimate of the standard error using the formulae in Box 7.2. Again, whenever we report a sample mean, we ought also, as a minimum, to give the standard error of the mean, in one of the forms described earlier (Box 7.1).

Incidentally, the formulae in Box 7.2 illustrate a major difference between the standard deviation of the items in a sample and the standard error (= the standard deviation of the distribution of sample means around the true mean). As we have just seen, increasing the

---

**BOX 7.2  CALCULATING A STANDARD ERROR OF THE MEAN (DATA FROM BOXES 4.3 & 4.6)**

(a) **Formula for calculating standard error of the mean:**

$$\text{Estimated standard error} = \frac{s}{\sqrt{n}}$$

$$\text{where } s = \text{sample standard deviation} = \sqrt{\frac{\Sigma\left(Y - \overline{Y}\right)^2}{n - 1}}$$

$$n = \text{sample size}$$

(b) **Standard error of mean length (mm) of female *Sphaeroma rugicauda*:**

$\overline{Y} = 4.951$ mm, $s = 0.538$ mm, $n = 400$

$$\text{Standard error} = \frac{0.538}{\sqrt{400}} = 0.027$$

Estimated mean length $= 4.951 \pm 0.027$ mm

(c) **Standard error of mean activity ($\mu$M/min) of acid phosphatase at pH 3:**

$\overline{Y} = 11.24$, $s = 1.26$, $n = 5$

$$\text{Standard error} = \frac{1.26}{\sqrt{5}} = 0.56$$

Estimated mean activity $= 11.24 \pm 0.56 \ \mu$M/min

---

sample size will reduce the standard error, but it will not cause any systematic reduction in the sample standard deviation ($s$). The formula for $s$ has the term $(n - 1)$ on the bottom line (Box 7.2). When we add another observation to our sample, we not only increase the value of $(n - 1)$ on the bottom line of the formula for $s$ but we also increase the values of the sum of squares on the top line. The value of $s$ will fluctuate around the value of $\sigma$ as each additional item is added, the direction of the fluctuation depending on whether we happen to add an observation which is large or one which is small. As the sample size gets larger, the effect of each additional observation becomes smaller and the value of s gradually approaches $\sigma$. If the sample is large enough, say $n$ equal to or larger than 100, then $s$ the sample standard deviation, is really a very good estimate of $\sigma$, so that the standard error is a very good approximation.

As we saw above, a small standard error is better than a large one, because it implies a more reliable estimate, so how can we reduce the standard error? Clearly one way would be to increase the sample size, but the relationship between size of sample and size of standard error is not totally straightforward, because it is the square root of the sample size which is important. This means that to halve the standard error we would have to quadruple the sample size and, as a result, the sample size might have to be impracticably large. The other way to reduce the standard error would be to reduce $\sigma$, the standard deviation of the population. This can often be easily achieved, especially in laboratory experiments. Variability can be reduced by the use of experimental material of the same genotype and raised under uniform conditions. Careful practical work, both in the way the material is treated during the course of the experiment and in the way the variable is measured will also reduce variability.

## More on the shape of the distribution of sample means

The standard error is used to give an indication of the closeness of a sample mean to the population mean in exactly the same way that we used the standard error of a proportion and it suffers from the same drawbacks. Small standard errors are better than big ones, but the exact interpretation of the numerical value again is not clear. All we can say is that the smaller the standard error, the closer the sample mean is likely to be to the population mean. To be more informative, we need to quantify the word "likely" in terms of probabilities and to do this we need to know more about the shape of the distribution of sample means.

The easiest place to start is with large samples, because then we can use ideas relating to the normal distribution which we have already come across. We have to go back to our earlier thought experiment in which we took every possible sample from a parent distribution and consider only large random samples (the meaning of large we will have to leave till later). We already know that the means of these samples will have a distribution with a mean of $\mu$ and a standard deviation of $\sigma/\sqrt{n}$ (which we can estimate by $s/\sqrt{n}$), but it will have another most useful property. If the samples are large it will be a normal distribution. This can be proved mathematically and is something we shall have to take on trust.

Actually it is a more complicated than this. It would be more correct to say that, irrespective of the shape of the frequency distribution of the variable ($Y$), the frequency distribution of sample means ($\overline{Y}$) will tend towards a normal distribution as the sample ($n$) size increases. If the distribution of $Y$ itself follows a normal distribution or something very close to a normal distribution, then the means of quite small samples (say, $n = 4$ or 5) will follow a normal distribution. If $Y$ follows a distribution which is far from a normal one (e.g. is badly skewed as in a Poisson distribution), then sample sizes have to be much larger in order that the means follow a normal distribution. Unfortunately, there are no hard and fast rules about how large sample sizes have to be except that with sample sizes of the order of 100 (or more), we will be fairly safe if we assume that means of samples ($\overline{Y}$) will follow a normal distribution. The value of making this assumption is that we can use the properties of normal distributions to make more meaningful statements about the reliability of the sample mean.

## From standard errors to confidence limits for the mean based on a large sample

We already know that in a normal distribution, 95% of all the observations have values between limits which lie 1.96 standard deviations below the mean and 1.96 standard deviations above the mean (Fig. 5.6). Put another way, there is a probability of 0.95 that a randomly chosen observation will have a value within the interval between these two limits.

If we now apply this to a distribution of sample means which has a normal distribution, the same principle applies, except that the observations are now sample means rather than individual values of $Y$. Of the sample means, 95% will lie within the interval which spans from 1.96 standard errors below the mean to 1.96 standard errors above the mean (remembering that standard error = standard deviation when we are talking about a distribution of means). We know that the mean of the distribution of sample means is $\mu$ and the standard error is $\sigma/\sqrt{n}$. It follows that 95% of sample means will lie within the interval from $\mu - 1.96\sigma$ to $\mu + 1.96\sigma/\sqrt{n}$ (written $\mu \pm 1.96\sigma/\sqrt{n}$ for short). This is illustrated in Figure 7.4a.

To make the next step, we need to imagine taking at random some values from this distribution of sample means (which is of course equivalent to taking random samples of size $n$ from the original population of $Y$). Twenty-three such values of are shown in Figure 7.4b. In the long run, 95% of these should lie within the interval $\mu \pm 1.96\sigma/\sqrt{n}$ which is marked by the two dashed lines. (The solid line marks the position of $\mu$.) Another way of looking at this is that 19 out of every 20 sample means should lie between the two limits of this interval, although again it is important to emphasise that this will only be true in the long run. By chance a set of 20 means taken at random could all lie within these limits, alternatively fewer than 19 of the 20 could lie within the limits. I have emphasized this important point by having 2 out of the 23 values outside the limits.

Now in practice, we shall only have taken a single random sample and so have only a single mean at our disposal – what can we say about it? Well, if 95% of all such means lie within the limits described above, then the odds are in favour of the mean of our single (random) sample being one of those that lies within these limits, rather than one of those which lies outside. However, because of the element of chance we cannot be completely sure or confident that our one sample mean will lie within these limits. In fact, our level of confidence could be quantified as 95% to reflect the fact that 95 out of every 100 means will lie within these limits. Once in a while, actually on 5 occasions out of every 100, we shall by chance pick a random sample whose mean lies outside these limits. Where does this get us,

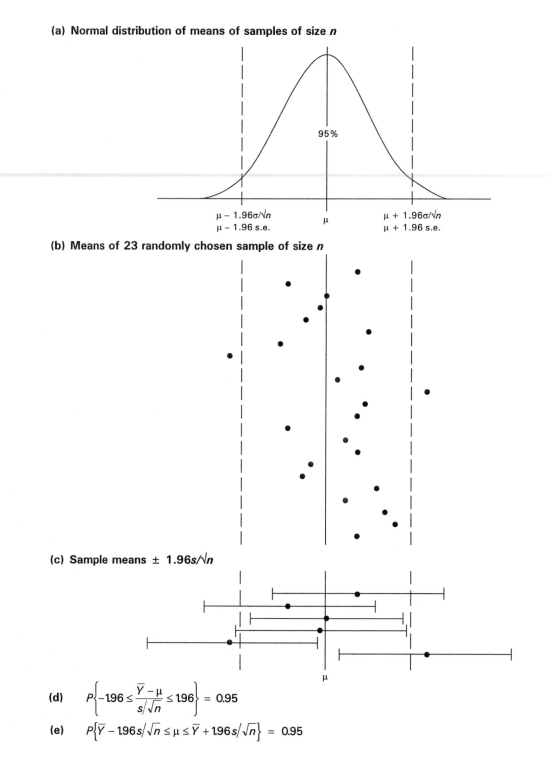

**(a) Normal distribution of means of samples of size *n***

95%

$\mu - 1.96\sigma/\sqrt{n}$
$\mu - 1.96$ s.e.

$\mu$

$\mu + 1.96\sigma/\sqrt{n}$
$\mu + 1.96$ s.e.

**(b) Means of 23 randomly chosen sample of size *n***

**(c) Sample means $\pm$ 1.96*s*/$\sqrt{n}$**

$\mu$

**(d)** $\quad P\left\{-1.96 \leq \dfrac{\overline{Y} - \mu}{s/\sqrt{n}} \leq 1.96\right\} = 0.95$

**(e)** $\quad P\left\{\overline{Y} - 1.96s/\sqrt{n} \leq \mu \leq \overline{Y} + 1.96s/\sqrt{n}\right\} = 0.95$

**Figure 7.4** From standard errors to 95% confidence intervals for the mean based on a large sample.

remembering that we set out to obtain a measure of the precision of our sample mean?

The easiest way to see this is to go back to our thought experiment and the 23 sample means shown in Figure 7.4b. Each of these means is based on an (imaginary) sample of $n$ values of $Y$, so for each of these samples we could calculate the standard deviation ($s$). Each value of s could be used as an estimate of $\sigma$ and, because the sample size is large, $s$ will be a very good estimate of $\sigma$, so good in fact that we will assume for the moment that $s = \sigma$. For each sample we could then calculate a value of $s/\sqrt{n}$ and a value of $\pm 1.96s/\sqrt{n}$. Because we are assuming that $s = \sigma$, the values calculated for each sample will have the same value as $\pm 1.96\sigma/\sqrt{n}$ the interval around $\mu$ which we have just been using.

Now, of course we don't know $\mu$, but what happens if we attach this interval to the value of $\overline{Y}$ instead? We can do this diagrammatically (Fig. 7.4c) using lines on each side of the sample mean in the same way as we did when representing the standard error. I have done this for the first four means and also for the two extreme ones. It should be clear just from the geometry of the diagram that if we were to calculate intervals in this way for a large number of sample means, then in 95 out of every 100 cases the interval would include or cover the true but unknown mean $\mu$, represented by the solid vertical line. Basically the sample means which are drawn from the central 95% portion of the distribution are the ones whose intervals have this property; sample means from the 2.5% of the distribution in each tail would yield intervals which fail to include the true mean. Intervals like these are called **95% confidence intervals**. The value of $\overline{Y} - 1.96\sigma/\sqrt{n}$ is the **lower 95% confidence limit** and the value of $\overline{Y} + 1.96\sigma/\sqrt{n}$ is the **upper 95% confidence limit**. The value of 95% is chosen as a matter of convention and is called the **confidence level**, denoted by $\gamma$. In the next section we'll see how to calculate such a confidence interval (which is very easy) and also learn more about what it tells us.

A mathematical explanation of a confidence interval is given by the two expressions at the bottom of the Figure 7.4 and, although we don't need to understand fully all the notation, they do give us an alternative perspective which will be useful later. The first expression (d) is a statement involving a familiar procedure namely the standardisation of a normal distribution. You should recognize the bit in the middle of that expression which is simply calculating how far away a randomly chosen value of $\overline{Y}$ is from the mean of the distribution in terms of the standard deviation (here of course equal to the standard error). We know that the answer will be a value of $z$. The rest of the equation tells us what we already know, namely that the probability ($P$) that this value of $z$ lies between –1.96 and +1.96 is 0.95.

The second expression (e) is merely a rearrangement of the first; it tells us that the probability of $\overline{Y} - 1.96\sigma/\sqrt{n}$ being less than or equal to $\mu$ and $\overline{Y} + 1.96\sigma/\sqrt{n}$ being greater than or equal to $\mu$ is 0.95. If you think about it, on those occasions when $\overline{Y} - 1.96\sigma/\sqrt{n}$, the lower limit of the interval, is equal to or less than $\mu$, then $\mu$ must be in the interval. Likewise, when the upper limit of the interval, $\overline{Y} + 1.96\sigma/\sqrt{n}$, is equal to or greater than $\mu$, then $\mu$ must also be in the interval. It follows that the population mean must lie between these two limits. Putting these two ideas together, the second expression is telling us that the probability that such an interval includes $\mu$ is 0.95. Put another way, in the long run 95 out of every 100 such intervals would contain the true but unknown population mean $\mu$.

It follows that if we have one random sample with its accompanying 95% confidence interval, we can be fairly confident (actually 95% confident) that this interval will be one of those that does include the population mean.

## Calculating the confidence limits for the mean based on a large sample

In practice we have a single large sample at our disposal so that we have one value of $\bar{Y}$ and one value of $s$, which we can use as our very good estimate of $\sigma$. Numerical values for the upper and lower 95% confidence limits are calculated as shown in Box 7.3; the formula is the same as the one which we derived above with the exception that we write it with $s$ instead of $\sigma$, because in practice $s$ is what we use. (Although we used $\sigma$ when developing the theory, it is of course not known exactly, even with large samples. Actually the confidence intervals drawn in Figure 7.4 have all been made slightly different lengths to represent this.) You can see that the formula amounts to multiplying the standard error of the mean by 1.96 and then obtaining the upper and lower 95% confidence limits by adding or subtracting this quantity from the mean. The confidence interval is the distance between the two limits. These limits can be presented graphically or in writing in exactly the same way as the standard error; the abbreviation **CL** stands for confidence limits. It was pointed out earlier that a common way of reporting the reliability of a sample mean is to present it as the "mean $\pm$ two standard errors" rather than the "mean $\pm$ one standard error". We can now see why. Multiplying the standard error by 2, rather than by 1.96, will give us an interval which is nearly a 95% confidence interval, in fact it will be slightly too wide.

What does this interval tell us? We know that 95% of all such intervals include the population mean and therefore we can be reasonably confident (but not completely sure) that our one interval will be one of those that does include the population mean, rather than one of those that does not. In fact we can be 95% confident. It follows that we can be 95% confident that the true mean has a value between 4.898 and 5.004 mm. In this case, the confidence interval is very small and as a result our sample mean of 4.951 mm is a very reliable estimate of the true mean. The wider the confidence interval the less reliable the estimate.

Note that, although we use the sample mean when calculating and reporting confidence limits, they are not confidence limits for the sample mean. The sample mean is known so it does not require confidence limits. The confidence limits are for the unknown population mean; they delimit a band of values which we can be fairly certain include the unknown population parameter.

Incidentally, although we can say that we are 95% confident that our particular interval includes the population mean, it would be incorrect to say that there is a 95% probability that this particular interval includes the population mean. Although these look like the same thing they are not. The latter would imply that if we calculated this particular interval on 100 separate occasions, then on 95 of these occasions the true mean would lie in the interval and on 5 it would not. Since we are talking about a particular interval its value would always be the same so this would imply that the population mean had different values on different occasions, which it cannot have.

It should be clear from the preceding account that, by chance, our single sample could have come from the tails of the distribution and that had it done so, our calculated confidence interval would not include the true mean. In our example we used a 95% confidence interval, that is, we set our confidence level at 95%. Under these conditions there is a 5% chance that we will obtain one of these extreme samples, so there is a small chance that we will be wrong in asserting that $\mu$ lies within our calculated interval.

The choice of a 95% confidence level is merely a matter of convention, in the same way that significance levels are set at conventional values. We could attempt to improve things by working to a higher confidence level, e.g. $\gamma = 99\%$. Using the same arguments as be-

---

**BOX 7.3 CALCULATING 95% AND 99% CONFIDENCE LIMITS FOR THE MEAN BASED ON A LARGE SAMPLE**

**(a) Formula for calculating confidence limits:**

Confidence limits = Sample mean $\pm$ ($z$ $\times$ Standard error)

where value of $z$ depends on confidence level ($\gamma$).

**(b) 95% confidence limits for mean of body lengths of *Sphaeroma rugicauda*** (Data from Box 7.2):

$\bar{Y} = 4.951$ mm

s.e. = 0.0267

Mean $\pm$ 95% CL = $\bar{Y} \pm$ (1.9600 $\times$ s.e.)

$\qquad\qquad\qquad = 4.951 \pm$ (1.9600 $\times$ 0.027)

$\qquad\qquad\qquad = 4.951 \pm 0.053$ mm

Lower limit $\qquad = 4.951 - 0.053 = 4.898$ mm

Upper limit $\qquad = 4.951 + 0.053 = 5.004$ mm

**(c) 99% confidence limits for mean of body lengths of *Sphaeroma rugicauda*:**

Mean $\pm$ 99% CL = $\bar{Y} \pm$ (2.5758 $\times$ s.e.)

$\qquad\qquad\qquad = 4.951 \pm 0.069$ mm

Lower limit $\qquad = 4.882$ mm

Upper limit $\qquad = 5.020$ mm

**(d) Confidence interval:**

95% confidence interval = 5.004 − 4.898 = 0.106 mm

99% confidence interval = 5.020 − 4.882 = 0.138 mm

fore, we would then be able to say that 99% of samples would produce intervals which include the true mean. The odds would be more strongly in favour of the one sample which we have got being one of those 99. If we want to do this we shall have to change the value of $z$. To obtain the correct value, you need to turn to Table A5 and find the row at the top labelled $\gamma$ (for confidence level). Look along this row until you find 99% and then locate the value of $z = 2.5758$. The corresponding confidence limits are given by the sample mean $\pm 2.5758$ standard errors and the effect of this is to increase the size of the confidence interval (see Box 7.3). We are more confident that this interval includes the population mean (after all the interval is wider). But there is a price to pay. Because the interval is wider we are less sure about what the value of the population mean is.

So far so good, but we must not lose sight of the fact that all of the preceding argument has been based on several assumptions. A confidence interval will only be a valid interval estimate of the population mean if we have taken our sample at random. If we have consciously or unconsciously chosen atypical individuals, then this will be reflected in the sample mean, the standard error and ultimately the confidence interval. We also had to assume that the sample means have a normal distribution. As was mentioned earlier this will always be true if the original (parent) distribution is normal (irrespective of sample size) and it will also be true for non-normal parent distributions if sample sizes are large enough. But what happens if this is not the case? Finally, the argument depended on our being able to use $s$ as an estimate of $\sigma$. I have alluded (without justification) to the fact that this is reasonable if the sample is large, but what happens if it is small? These latter two issues will now be examined.

## Confidence limits for the mean based on a small sample from a normal distribution

This is probably the easier of the two situations to deal with and the problem hinges on the use of $s$, the sample standard deviation, as an estimate of $\sigma$, the population standard deviation. When sample sizes are large (say, 100 or more), $s$ is a very good estimate of $\sigma$. That is why up till now, we have been able to substitute the value of $s$ for the value of $\sigma$ in our calculations.

However, when sample sizes are smaller than this, the effects of sampling variation are more marked and values of $s$ can differ quite considerably from $\sigma$. As a result, using $s$ to calculate a confidence interval (when we should be using $\sigma$) can lead to serious errors. We have, therefore, to make a modification to the procedure to take account of the uncertainty associated with the value of $s$. In outline, this new procedure involves replacing the value of $z$ ($= 1.96$) as the multiplier of the standard error with a different multiplier, the value of which is adjusted to take into account the uncertainty in $s$. To find the multiplier we need to investigate another distribution. This distribution is used not only to calculate confidence limits based on small samples, but also in other statistical procedures involving small samples. We will meet these in Chapter 8.

Figure 7.5 outlines the nature of the problem. We start with our parent population as before, showing us the distribution of $Y$, although now we have to be more restrictive than in the large sample case. We have to assume that it is a normal distribution. The left-hand pathway does not tell us anything new, because it deals with large samples (it is included to highlight the contrast with small samples). Just to recap on the large sample case, we know that large samples taken from a normal distribution (Fig. 7.5a) produce a distribution of sample means which is itself normal (Fig 7.5b). We can standardize this distribution because with a large sample we know that $s$ gives us a good estimate of $\sigma$. Standardization,

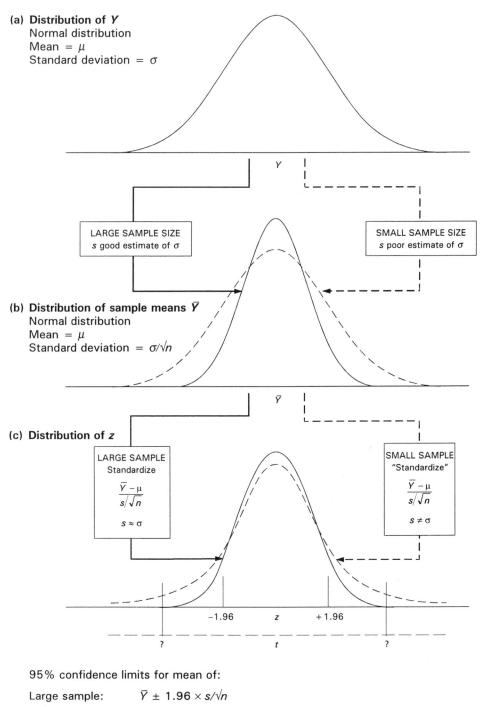

**(a) Distribution of _Y_**
Normal distribution
Mean $= \mu$
Standard deviation $= \sigma$

Y

LARGE SAMPLE SIZE
_s_ good estimate of $\sigma$

SMALL SAMPLE SIZE
_s_ poor estimate of $\sigma$

**(b) Distribution of sample means $\bar{Y}$**
Normal distribution
Mean $= \mu$
Standard deviation $= \sigma/\sqrt{n}$

$\bar{Y}$

**(c) Distribution of _z_**

LARGE SAMPLE
Standardize

$$\frac{\bar{Y} - \mu}{s/\sqrt{n}}$$

$s \approx \sigma$

SMALL SAMPLE
"Standardize"

$$\frac{\bar{Y} - \mu}{s/\sqrt{n}}$$

$s \neq \sigma$

−1.96   _z_   + 1.96

?   _t_   ?

95% confidence limits for mean of:

Large sample:      $\bar{Y} \pm 1.96 \times s/\sqrt{n}$

Small sample:      $\bar{Y} \pm t \times s/\sqrt{n}$

**Figure 7.5**  Confidence limits for mean based on a small sample from a normal distribution.

171

that is calculating values of $(\bar{Y} - \mu) \div s/\sqrt{n}$, converts the distribution into a distribution of $z$ (Fig. 7.5c). As we saw at the bottom of Figure 7.4, it is this which enable us to say that 95% of means will yield values of $z$ in the range $-1.96$ to $+1.96$. From this, we know that the 95% confidence limits are given by multiplying the standard error by 1.96.

Now let's examine the right-hand pathway which deals with the small sample case. Small samples from a normal distribution produce a normal distribution of sample means (Fig. 7.5b) as do large samples. The only difference will be that the standard error of the mean (= standard deviation of this distribution) will be larger, because n is smaller. It is when we try to standardize this distribution that the problem arises. Because we are using small samples, the values of $s$ will vary markedly from sample to sample. The effect of this will be to lead to more variation in the values of $(\bar{Y} - \mu) \div s/\sqrt{n}$ than we would get if $s$ was always equal to (or very close to) $\sigma$. As a result, the frequency distribution of the values of $(\bar{Y} - \mu) \div s/\sqrt{n}$ will have a different shape from the standard normal curve (Fig. 7.5c). It will still be symmetrical but, compared with the standard normal distribution, it will have a greater proportion of the observations in the tails and a smaller proportion in the centre. In fact, it will no longer be the standard normal distribution and as such we cannot call it a distribution of $z$. It is instead a distribution of something called $t$.

Because of the altered shape of this distribution the interval from $-1.96$ to $+1.96$ no longer includes 95% of the observations. In particular, because this new distribution has a greater proportion of the observations in the tails, the interval between $-1.96$ and $+1.96$ will include less than 95% of the distribution. To compensate for this we shall need to set the limits further apart, that is, to use values smaller than $-1.96$ and bigger than $+1.96$. Exactly where we have to set these limits will depend on the exact shape of the distribution of $t$.

## Properties of $t$ distributions and tables of $t$

A distribution of $t$, which, remember is merely a frequency distribution of all possible values of $(\bar{Y} - \mu) \div s/\sqrt{n}$, can be described mathematically by a formula, the details of which need not concern us. It turns out that the shape of the curve is related to the size of the sample and that the smaller the sample, the more of the distribution lies in the tails. Mathematically, what is important is not the sample size ($n$) but the **degrees of freedom**, or **df** for short. We have met these elsewhere; in this case the degrees of freedom are ($n - 1$), that is, one less than the sample size. The $t$ distribution for a sample of size 6 is, therefore, described by the curve with five degrees of freedom (written shorthand as $t_{(5)}$). You don't need to worry about why this is so, the important thing is to realize the effect of decreasing the sample size (Fig. 7.6). Note how the curves become increasingly depressed in the centre and spread out in both tails as the sample size and hence the number of degrees of freedom gets smaller.

The fact that these curves can be mathematically described means that it is possible to work out the two values of $t$ for each curve which cut off known percentages ($\alpha$) in each tail of the distribution, e.g. 2.5%. It follows that a known percentage of the distribution ($\gamma$), e.g. 95%, must lie between these two values of $t$. It is these values which we use as the multiplier in calculating a confidence interval. The actual values appropriate to different curves have been worked out by mathematicians which is how I obtained the values on each of the two curves in Figure 7.6. They illustrate how the two values of $t$ have to be set further apart as the sample size decreases.

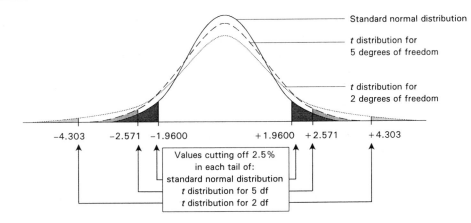

**Figure 7.6** The shapes of *t* distributions.

Table 7.1 contains part of a more extensive table of values of *t* to be found at the end of the book (Table A7). The columns in this table are set out in exactly the same form as those in the table of critical values of *z* (Table A4) and the only difference is the additional column at the left of the table headed df. The three rows at the very top of the table tell us which column we need and you will probably recognize some of the information, e.g. $\alpha_2$, $\alpha_1$, 5% and 1%, but for the moment we need to concentrate on the last of these rows, the one labelled $\gamma$. This ($\gamma$) is simply the proportion of the distribution which lies *between* specified values of *t*, rather than the proportion outside the limits; this means it is the proportion which is not in the tails. As we saw earlier this proportion is the confidence level and two of the values of $\gamma$ (95% and 99%) are highlighted, because they are the two most commonly used confidence levels. Down the left-hand side of the table is a column headed df which is for the degrees of freedom, which we use to find out which row we should be using. In the body of the table are seven columns of figures which are the values of *t* appropriate to different circumstances: for 95% confidence limits we need the value of *t* in the $\gamma = 95\%$ column and the row corresponding to the correct value of the degrees of freedom (df).

**Table 7.1 Critical values of *t* (see Table A7).**

| $\alpha_1^R$ | 10% | 5% | 2.5% | 1% | 0.5% | 0.1% | 0.05% |
|---|---|---|---|---|---|---|---|
| $\alpha_2$ | 20% | 10% | 5% | 2% | 1% | 0.2% | 0.1% |
| $\gamma$ | 80% | 90% | 95% | 98% | 99% | 99.8% | 99.9% |
| df = 1 | 3.078 | 6.314 | 12.706 | 31.821 | 63.657 | 318.309 | 636.619 |
| 2 | 1.886 | 2.920 | 4.303 | 6.965 | 9.925 | 22.327 | 31.599 |
| 3 | 1.638 | 2.353 | 3.182 | 4.541 | 5.841 | 10.215 | 12.924 |
| 4 | 1.533 | 2.132 | 2.776 | 3.747 | 4.604 | 7.173 | 8.610 |
| 5 | 1.476 | 2.015 | 2.571 | 3.365 | 4.032 | 5.893 | 6.869 |
| 6 | 1.440 | 1.943 | 2.447 | 3.143 | 3.707 | 5.208 | 5.959 |
| 7 | 1.415 | 1.895 | 2.365 | 2.998 | 3.499 | 4.785 | 5.408 |
| 8 | 1.397 | 1.860 | 2.306 | 2.896 | 3.355 | 4.501 | 5.041 |
| 9 | 1.383 | 1.833 | 2.262 | 2.821 | 3.250 | 4.297 | 4.781 |
| 10 | 1.372 | 1.812 | 2.228 | 2.764 | 3.169 | 4.144 | 4.587 |

## Calculating confidence limits for the mean based on a small sample

Box 7.4 gives an example of the use of this table to calculate the 95% confidence limits for the mean activity of the enzyme acid phosphatase at pH 3. We have already calculated the standard error (Box 7.2). The sample size was 5, so the degrees of freedom will be 5 – 1 = 4 and the value of $t$ for the 95% confidence level will be 2.776. Note that the variable is likely to follow a normal distribution which is an assumption underlying the use of the

---

**BOX 7.4   CALCULATING CONFIDENCE LIMITS FOR THE MEAN BASED ON A SMALL SAMPLE FROM A NORMALLY DISTRIBUTED POPULATION.**

(a) **Formula for calculating confidence limits:**

Confidence limits = Sample mean ± ($t$ × Standard error)

where value of $t$ depends on confidence level ($\gamma$) and degrees of freedom (df).

(b) **95% confidence limits for mean activity (µM/min) of acid phosphatase at pH 3:**

$\bar{Y} = 11.24$, s.e. = 0.56, $n = 5$

Where value of $t$ is for $\gamma = 95\%$ and degrees of freedom = $n - 1 = 5 - 1 = 4$

$t = 2.776$

Mean ± 95% CL = 11.24 ± (2.776 × 0.56)
= 11.24 ± 1.55 µM/min

Lower 95% CL = 11.24 – (2.776 x 0.56) = 9.68 µM/min

Upper 95% CL = 11.24 + (2.776 x 0.56) = 12.80 µM/min

(c) **95% confidence limits for mean activity (µM/min) of acid phosphatase at pH 5:**

12.0
15.3
15.1
15.0
13.2

$\bar{Y} = 14.12$

s.e. = 0.65

Mean ± 95% CL = 14.12 ± (2.776 × 0.65)
= 14.12 ± 1.80 µM/min

Lower 95% CL = 14.12 – 1.81 = 12.31 µM/min

Upper 95% CL = 14.12 + 1.81 = 15.93 µM/min

---

$t$ distribution. There is another example in Box 7.4 and you could try checking that you get the same answer. Should we have wanted the 99% confidence interval for the mean of same data then the $t$ value would have been 4.604.

We are now in a position to examine an issue that was raised earlier when we were constructing confidence intervals for large samples using $z$, namely the (unjustified) assertion that the sample standard deviation $s$ is a good estimate of $\sigma$ if the sample size is large enough. If we look down the table of $t$ values (in the $\gamma = 95\%$ column) we find (as we already know) that $t$ gets smaller as the degrees of freedom get larger. Eventually, when the degrees of freedom are very large (actually infinite) $t = 1.9600$, which is our familiar value of $z$. The standard normal curve therefore belongs to this family of $t$ curves.

However, if we look at this more closely, we see that $t$ does not decrease steadily with increasing $n$. Rather, it declines very rapidly to begin with and then more slowly, but by the time we have got to df $= 100$ the value of $t$ ($= 1.9840$) is very close to 1.9600. As a consequence using 1.96 as the multiplier when constructing confidence intervals for means of samples of size 100 or more would be acceptable; it will result in the confidence interval being slightly narrower than it should be. This approximation was a convenient one to use because it enabled us to develop the reasoning behind a confidence interval using the familiar normal distribution, without having to simultaneously deal with the unfamiliar and more complicated t distribution. It was also convenient in the past, when extensive tables of values of $t$ were not available. These days there is little excuse for not using the appropriate value of $t$ for large samples.

Because the value of $t$ changes slowly once the degrees of freedom become large, there is little point in listing values for all degrees of freedom. Should you need the value of $t$ for a number of degrees of freedom which are not listed, then the simplest thing to do is to take the value of $t$ corresponding to the next smallest number of degrees of freedom. This will mean that your value of $t$ is slightly larger than it should be, so your confidence limits will be a touch generous.

## Confidence limits for the mean based on a small sample from a non-normal distribution

Unfortunately, we cannot leave the subject of confidence limits yet, because of a further complication which in practice is frequently encountered. We often find ourselves having to use small samples of a variable which we know does not follow a normal distribution. We have already met a number of situations in Chapters 5 and 6 where this is the case, for example variables which follow a binomial or a Poisson distribution. As described previously, the means of large samples from any distribution tend towards being normally distributed, allowing the calculation of confidence limits using $z$. The means of small samples will only be normally distributed if the variable itself has a normal distribution. What this means is that we cannot use $t$ to construct confidence limits for variables which follow a binomial or a Poisson distribution. The same will be true for variables with other distributions, such as those resulting from a contagious pattern of spatial dispersion.

There is no one simple solution for these three cases. In fact, there are often several possible solutions available for each situation, which vary in their ease of use and in the extent to which they involve approximations. One case which can often be dealt with by the methods already described is where a non-normal distribution has arisen because we have inadvertently sampled two different normal distributions together, for instance, where our sample includes males and females, or offspring and adults. If we can separate the measurements for males and females into two separate samples, then each sample will have been drawn from a normal distribution. We proceed by calculating means and confidence limits separately for the two samples.

## Confidence intervals for a binomial proportion

We saw earlier how to calculate the standard error of a sample proportion where the underlying distribution is a binomial. Unfortunately there is no accurate way in which we can convert the standard error to confidence limits, by using a convenient multiplier, particularly if $n$ is small and $\hat{p}$ is extreme. Under these conditions, the binomial distribution is skewed and very different in shape from a normal distribution. The quickest way to deal with this is to use the chart shown in Box 7.5, which enables us to read off directly the 95% confidence limits for a given sample proportion for a given sample size.

To use a chart like this, we simply locate the sample proportion on the horizontal axis

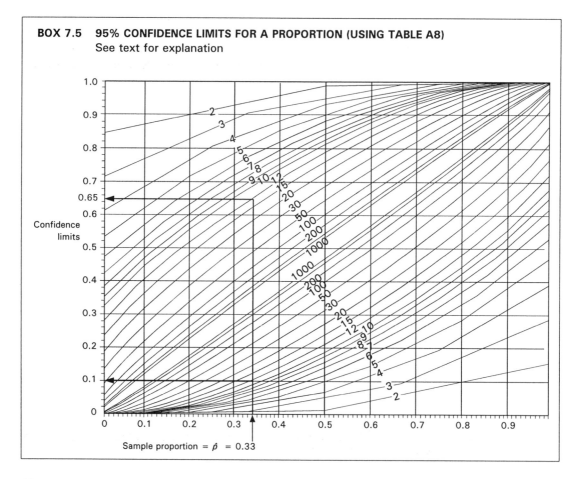

**BOX 7.5 95% CONFIDENCE LIMITS FOR A PROPORTION (USING TABLE A8)**
See text for explanation

and follow the vertical line upwards from there until it intersects the two curving lines for the sample size involved. We then read off these two points of intersection on the vertical scale. The two values obtained are the lower and upper confidence limits. In our blood cell example, where $\hat{p}$ is 0.33 and $n = 12$, the upper and lower limits are read off as 0.10 and 0.65. While this is a very quick and simple method, you can see that it has several drawbacks. If we want to find confidence limits for $10/17 = 0.588$, we first have difficulty locating this value exactly on the horizontal scale. Second, there is no curve for $n = 17$, because there is not enough room for a line for every sample size. We can always imagine a curve about half way between those for $n = 15$ and $n = 20$, but even then we would have difficulty in reading off exact values on the vertical scale. The nearest we could get would be something like 0.33 and 0.81. Although this is probably sufficiently accurate for elementary work, you should indicate that these limits are only approximate.

The chart provides some additional insights into the nature of confidence limits. First, it dramatically shows the effect of increasing the sample size and gives us a good idea about what is a suitable sample size. For example, if our sample proportion of 0.33 had been based on a sample size of 100 cells, then the confidence limits would have been 0.24 and 0.43, a considerable narrowing of the interval compared with a sample of size 12. If $n = 1000$, the confidence limits would be even closer (0.30–0.36) and the interval even narrower.

Figure 7.7 shows how the size of the confidence interval decreases as $n$ increases. Remembering that we can be 95% confident that such an interval includes the population proportion ($p$), we can see that a sample size of 12 is not very informative. The figure also tells us something about the design of an experiment or a sampling programme, because we

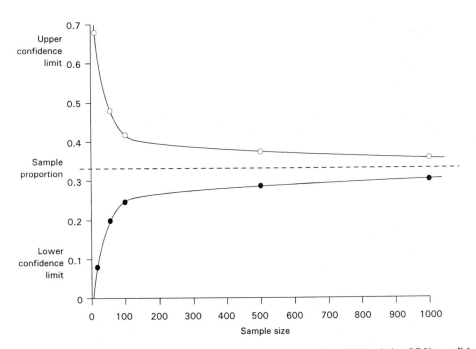

**Figure 7.7** The effect of increasing the sample size on the width of the 95% confidence interval for a sample proportion of 0.33.

177

can see from the graph that the size of the confidence interval decreases rapidly with increasing sample size to begin with but then decreases more slowly. Samples of less than 10 are best avoided, while the extra work involved in taking a sample of size 100 is well worth it because it brings a large increase in reliability. Whether it is worth counting 500 cells to get a very reliable estimate of $p$ depends on the importance which we attach to reliability and the cost of getting the additional information. Counting 1000 cells is probably not worth the extra effort.

One further feature, which we might note in passing, is that confidence limits for a proportion may not be symmetrical about the sample proportion, something which you can see from the examples above and which shows up clearly on the chart. It is most marked when $p$ has a value far from 0.5 and when sample sizes are small and it merely reflects the fact that for all proportions, there are fixed upper and lower limits. A proportion can only take values between 0 and 1. If a sample proportion is 0.9 the upper confidence limit cannot be larger than 1.0, but the lower limit can be smaller than 0.8. In contrast a normal distribution theoretically ranges from plus infinity to minus infinity.

You may be wondering how a chart like this is produced and what to do if you wanted to find 95% confidence limits really accurately. The answer is given in Figure 7.8, using the frog blood cell example. What you have to do is start with your observed sample fraction of four neutrophils out of 12 cells and ask the following two questions. (a) At what upper value of the population proportion (call it $p_u$) does the chance of getting four or fewer neutrophils equal 2.5%? (b) At what lower value of the population proportion (call it $p_l$) does the chance of getting four or more neutrophils equal 2.5%? These upper and lower values of $p$ correspond to the upper and lower 95% confidence limits.

To find the answers is very tedious, because you have to look at the lower and upper tails of a series of pairs of binomial distributions corresponding to a range of values of $p_u$ and $p_l$. The idea is to gradually make $p_u$ larger and $p_l$ smaller until the tails are just the right size. I have done this using the binomial tables (Table A2) but, because the tables do not give every value of $p$, the nearest I could get was a lower limit of 0.10 and an upper limit of 0.65. These are the same answers as we got from the chart. You can see that both the probability of getting four or less and the probability of getting four or more is still slightly bigger than 2.5%. To make them exactly 2.5% would require a slight decrease in the value of $p_l$ and a slight increase in the value of $p_u$. More accurate values obtained by using a computer program are 0.0994 and 0.6516. Exact confidence limits have been calculated in this way and tabulated, but the tables are far too extensive to reproduce here. A shortened version is available in Sokal & Rohlf (1981).

Under some conditions, we can use another approach which, like the chart, also involves approximation. It depends on the fact that as we increase the sample size, so the shape of a binomial distribution approaches that of a normal distribution, something we came across in Chapter 5 and which is also apparent in Figure 7.1. The extent to which this approximation is a good one also depends the value of $p$. If $p$ is in the range 0.4–0.6, then the binomial distribution is symmetrical and more or less normal in shape even when $n$ is small. At more extreme values of $p$ the binomial distribution is markedly skewed at these small sample sizes and much larger samples are needed to produce something approaching a normal distribution. The gist of all this is that if we think that our sample has been taken from a binomial which is sufficiently close to a normal distribution, then we can simply calculate a 95% confidence interval by multiplying the standard error by 1.96. Various rules of thumb exist for deciding under what conditions this approximation is valid and one of these is given in

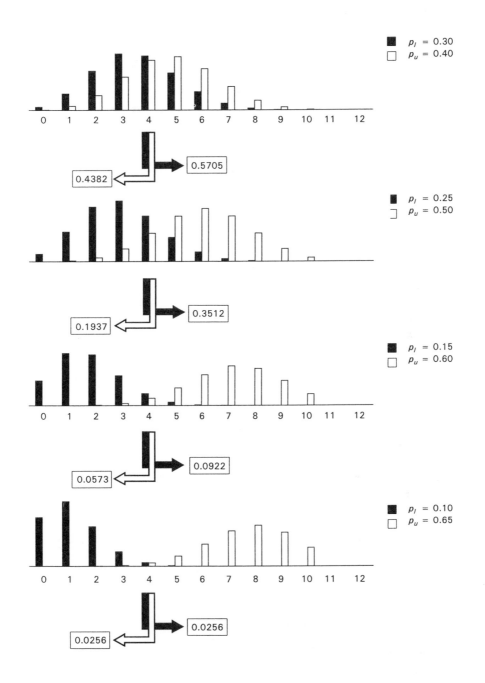

**Figure 7.8** Exact confidence limits for the proportion of neutrophil cells: how are they calculated? The figure at the end of this arrow ⇐▭ is the proportion in the lower tail of the binomial distribution ▭ for the upper value of $p$. It corresponds to the probability of obtaining 4 or fewer neutrophils in a sample of 12 cells. The figure at the end of this arrow ▬▶ is the proportion in the upper tail of the binomial distribution ■ for the lower value of $p$. It corresponds to the probability of obtaining 4 or more neutrophils in a sample of 12 cells.

---

**BOX 7.6   CONFIDENCE LIMITS FOR A BINOMIAL PROPORTION BASED ON A LARGE SAMPLE AND USING THE NORMAL APPROXIMATION.**

(a) **Formula for calculating (approximate) confidence limits:**

Confidence limits = Sample proportion $\pm$ ($z$ $\times$ Standard error)

where value of $z$ depends on confidence level ($\gamma$).

The approximation is valid if $\hat{p}$ is greater than $5/n$ and less than $1-5/n$ where $n$ is the sample size.

(b) **95% confidence limits for sample proportion of yellow kernels (Box 5.2):**

Sample proportion $\hat{p}$ = 0.505

Sample size $n$ = $92 \times 20$ = 1840

$5/n$ = 0.003,   $1 - 5/n$ = 0.997

Observed proportion lies within this range so approximation is valid.

Estimated standard error = $\sqrt{\hat{p}\hat{q}/n}$

$$= \sqrt{0.505 \times 0.495/1840}$$

$$= 0.012$$

Confidence limits = $\hat{p} \pm$ (1.96 $\times$ s.e.)

Sample proportion $\pm$ 95% CL = 0.505 $\pm$ 0.023

---

Box 7.6, but you can see how it requires that sample sizes be larger if $\hat{p}$ is very large or very small. We can certainly use this approximation to find confidence limits for the proportion of yellow kernels.

## Confidence limits for counts from a Poisson distribution

The problems here are very similar to those we encountered with the binomial, which, given the similarity between the shapes of the distributions, should not be very surprising. The solutions are also similar. Supposing that we have a single count of three cells in a haemocytometer square, then this is a point estimate of $\mu$. It is reasonable to assume that the cells are distributed at random and on this basis we can find the confidence limits for this single count, just as we could find limits for a single sample proportion.

The basis of the method is the one outlined in Figure 7.8 except that now we use Poisson

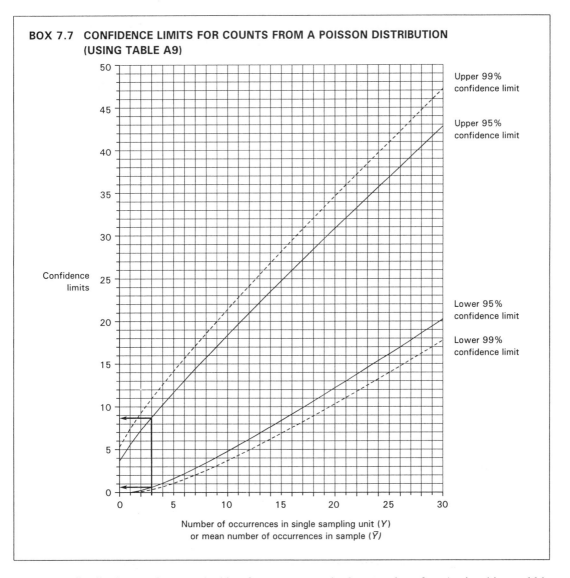

**BOX 7.7  CONFIDENCE LIMITS FOR COUNTS FROM A POISSON DISTRIBUTION (USING TABLE A9)**

distributions and we are looking for an upper and a lower value of $\mu$. Again, this would be very time consuming and so we make use of a chart (Box 7.7) as we did for the binomial distribution. This gives us confidence limits of about 0.6 and 8.6 cells/square; if we had used the exact method, we would have obtained values of 0.619 and 8.76 cells/square. If you wish, you can check this roughly by using the tables of the Poisson distribution. You should be able to find that if $\mu = 0.60$, then the probability of getting three or more cells in a square is 0.0232, while if $\mu = 8.8$ the probability of getting three or fewer cells in a square is 0.0244. These are both slightly smaller than 0.025, implying that we have over-shot a little; the lower limit needs to be a bit higher and the higher limit a bit lower. This chart can be used for a value of $Y$ based on a single sample unit, if $Y$ is 30 or less. Alternatively, we can use it for a mean based on several sampling units, providing that the total ($\Sigma Y$) is 30 or less. In these cases we find the mean along the bottom of the chart. So if we

181

had counted a total of 10 cells in five squares, the sample mean is 2 and confidence limits are about 0.25 and 7.2. Basically it is a method for small samples.

For large samples from a Poisson distribution, we adopt a different strategy. Although we didn't deal with it in Chapter 5, you may have noticed that as $\mu$ gets larger, so the Poisson distribution becomes more symmetrical. This is very obvious in Table A3. By this stage it will probably not surprise you to learn that as this happens, the shape of a Poisson distribution approaches the shape of a normal distribution. It is generally accepted that this approximation becomes close enough for most practical purposes when $\mu$ exceeds 30. We can also use this approximation if the sample size is sufficiently large.

Therefore, if we have a single count of 30 or more, or counts on several sampling units and the total of the counts ($\Sigma Y$) is greater than 30, we can calculate confidence limits based on an approximation to a normal distribution as has been done in Box 7.8. This involves multiplying the standard error by the appropriate value of $t$, although we use a slightly different formula for the calculation of the standard error. The normal formula is $s/\sqrt{n}$, which is equivalent to $\sqrt{(s^2/n)}$. Because we know that the variance of a Poisson distribution always has the same value as the mean we can use $\bar{Y}$ instead of $s^2$. In Chapter 6, we found no evidence that the distribution of *Tetrahymena* was other than a Poisson distribution. Because $\Sigma Y = 104$, which is larger than 30 we can use this method to calculate confidence limits for the mean (Box 7.8). We could also use the method on the micro-organism colony data, because $\Sigma Y = 135$.

---

**BOX 7.8  CALCULATING CONFIDENCE LIMITS FOR THE MEAN BASED ON A SAMPLE FROM A POISSON DISTRIBUTION USING THE NORMAL APPROXIMATION**

(a) **Formula for calculating (approximate) confidence limits:**

Confidence limits = Sample mean ± ($t$ × Standard error)

where value of $t$ depends on confidence level ($\gamma$) and degrees of freedom.

(b) **95% confidence limits for mean number of *Tetrahymena* cells per haemocytometer square (data from Box 5.6):**

$\bar{Y} = 3.47, n = 30$

Total number counted (= 104) is greater than 30, so normal approximation is valid.

Mean ± 95% confidence limits $= \bar{Y} \pm (t \times \text{s.e.})$

$$= \bar{Y} \pm \left( t \times \sqrt{\bar{Y}/n} \right)$$

$$= 3.47 \pm \left( 2.0452 \times \sqrt{3.47/30} \right)$$

Sample mean ± 95% CL = 3.47 ± 0.70 cells/square

---

## Transformation

This still leaves unresolved the question of confidence limits for the mean when the sample is taken from a contagious distribution, which is a particularly common occurrence in ecological studies. The basis of the procedure which we have to adopt is quite simple to grasp

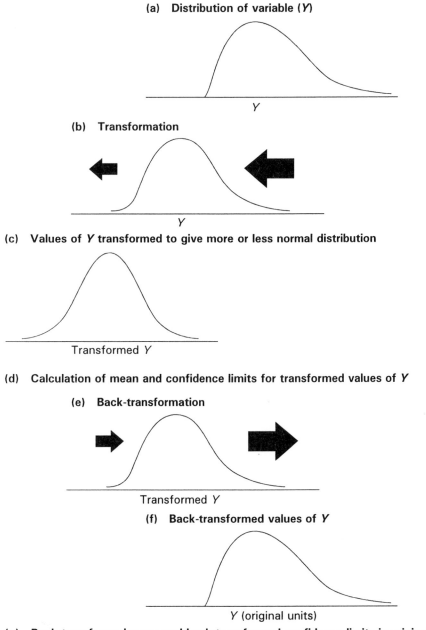

(a)   **Distribution of variable ($Y$)**

$Y$

(b)   **Transformation**

$Y$

(c)   **Values of $Y$ transformed to give more or less normal distribution**

Transformed $Y$

(d)   **Calculation of mean and confidence limits for transformed values of $Y$**

(e)   **Back-transformation**

Transformed $Y$

(f)   **Back-transformed values of $Y$**

$Y$ (original units)

(g)   **Back-transformed mean and back-transformed confidence limits in original units**

**Figure 7.9**   The principle of transformation.

and although the calculations are rather more involved, with planning, they can be easily carried out on a calculator.

The procedure is called **transformation** and the principle is illustrated in Figure 7.9. Basically it involves mathematically altering the values of the variable ($Y$) so as to make the shape of the distribution normal (or approximately normal). In this example, it involves extending one tail and compressing the other, which incidentally moves the whole distribution along the horizontal axis. We can then find the mean and the confidence limits for the transformed variable as before, using the appropriate value of $t$. The final stage is to "de-transform" the variable into the original units, a process known as **back-transformation**. This gives us the mean and confidence limits in the original units.

The type of mathematical alteration required depends on the underlying distribution, five common ones are given in Table 7.2. For example, if the variable comes from a contagious distribution and there are some values of zero, we change each value of $Y$ to $\log(Y+1)$. This is easy on a calculator, you add one to each value of $Y$ and then take the logarithm. The only tricky one is the arcsin transformation for a binomial proportion. Here, the transformed value is the angle whose sine is the square root of the proportion, which must be expressed as a number between 0 and 1 not a percentage. You can do this transformation on a calculator by keying in the proportion (e.g. 0.30), taking the square root (0.5477) and then pressing the $\sin^{-1}$ key, which may involve using the INV button as well. For $p = 0.30$ the answer should be 33.211.

**Table 7.2  Five common transformations.**

| Variable | Transformed values |
|---|---|
| Counts from a contagious distribution with some zero counts | $\log(Y+1)$ |
| Counts from a contagious distribution with no zero counts | $\log Y$ |
| Counts from a Poisson distribution with some zero counts | $\sqrt{Y+0.5}$ |
| Counts from a Poisson distribution with no zero counts | $\sqrt{Y}$ |
| Binomial proportions | $\arcsin\sqrt{p}$ |

The calculation of the mean, standard deviation, etc., proceeds as before. You should find that this can all be done in a single operation on a calculator with a standard deviation function without having to write down any intermediate answers. You key in a value for the variable (but don't enter it), then press the appropriate key(s) to transform the value and then enter the transformed value. When you have entered all the data in this way, pressing the appropriate keys will produce the mean and the standard deviation of the transformed data. Dividing this standard deviation by $\sqrt{n}$ will give you the standard error which can then be multiplied by the relevant value of $t$. You will probably find that your calculator allows you to store this answer in the normal memory, even though it is in standard deviation mode. This quantity can then be added to/subtracted from the (transformed) mean to give the upper/lower (transformed) confidence limits, which can then be back-transformed straight away. To back-transform you simply carry out in reverse the operations used in the original transformation. There is a worked example of the $\log(Y+1)$ transformation in Box 7.9.

As you can see, the confidence interval for the transformed data is smaller than that for the original data, so transformation is worthwhile. One disadvantage is that the back-transformed mean, properly called a derived mean, is now lower than the original mean.

---

**BOX 7.9  CALCULATING 95% CONFIDENCE LIMITS FOR THE MEAN BASED ON A SMALL SAMPLE FROM A CONTAGIOUS DISTRIBUTION USING THE LOG ($Y + 1$) TRANSFORMATION: 95% CONFIDENCE LIMITS FOR MEAN NUMBER OF DOG WHELKS (*NUCELLA LAPILLUS*) IN 0.5m × 0.5m QUADRATS**

| Frequency of quadrats | Number of snails/quadrat | | | | | |
|---|---|---|---|---|---|---|
| | $\longrightarrow$ TRANSFORMATION $\longrightarrow$ | | | | | |
| | $Y$ | $Y + 1$ | $\log (Y + 1)$ | | | |
| 5 | 0 | 1 | 0.0000 | | | |
| 4 | 1 | 2 | 0.3010 | | | |
| 2 | 2 | 3 | 0.4771 | | | |
| 1 | 5 | 6 | 0.7782 | | | |
| 1 | 6 | 7 | 0.8451 | | | |
| 1 | 7 | 8 | 0.9031 | | | |
| 1 | 15 | 16 | 1.2041 | | | |
| | | | $\longrightarrow$ BACK-TRANSFORMATION $\longrightarrow$ | | | |
| | | | | antilog | $-1$ | |
| $\bar{Y}$ | 2.7 | | 0.3926 | 2.5 | 1.5 | Mean |
| $s$ | 4.1 | | 0.3865 | | | |
| s.e. = $s/\sqrt{n}$ | 1.1 | | 0.0998 | | | |
| $t$ | 2.145 | | 2.145 | | | |
| Lower 95% CL $\bar{Y} - 2.145 \times$ s.e. | 0.5 | | 0.1786 | 1.5 | 0.5 | Lower 95% CL |
| Upper 95% CL $\bar{Y} + 2.145 \times$ s.e. | 5.0 | | 0.6066 | 4.0 | 3.0 | Upper 95% CL |
| | ORIGINAL | | TRANSFORMED | | | BACK-TRANSFORMED |

Report as untransformed mean ± back-transformed confidence limits

Mean = 2.7 snails/quadrat

Lower 95% CL = 0.5 snails/quadrat

Upper 95% CL = 3.0 snails/quadrat

---

This is a mathematical consequence of the manipulations. Although there are sophisticated techniques for correcting this, the simplest solution is as follows. Report the mean based on the original data and the confidence limits based on the transformed data.

Transformation is not restricted to the calculation of confidence limits. It may also be needed in some of the tests we shall meet in subsequent chapters.

## What use are confidence limits?

Confidence limits have practical uses because there are many situations in which we may need to know the limits within which a parameter is likely to be. A medical researcher who is studying the incidence of a disease needs to know the proportion of people infected now, so that a realistic range of values can be put into a mathematical model, which will enable

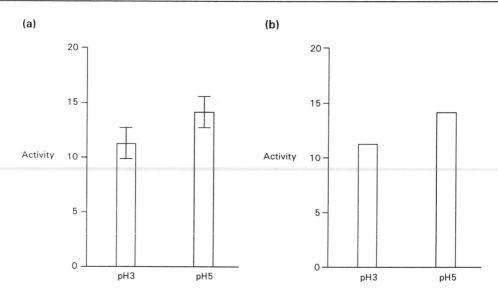

**Figure 7.10**   Activity of acid phosphate (µM/min) at pH3 and pH5.
(a) Correct and (b) incorrect presentation of estimates.

the future spread of the disease to be predicted. Worst and best case scenarios can be examined based on upper and lower estimates of the proportions infected and the social and financial costs of the disease can be assessed.

Similarly, a population biologist estimating the size of a biological population (i.e. the number of individuals) of an endangered species will want to know how reliable the estimate is. Past experience may dictate that a population of only 100 individuals needs careful management, so an estimate of 30 individuals with 95% confidence limits of 20 and 80 provides a fairly sound justification for conservation measures. An estimate of 240 individuals with confidence limits of 3 and 900 would not.

As I have stressed throughout this chapter an estimate should always be reported as an interval estimate, whether in tabular or graphical form. This enables the reader to get an idea of the reliability of the estimate (Fig. 7.10a). A graphical presentation of two point estimates of the mean activity of acid phosphatase (Fig. 7.10b) is incorrect, because it gives no indication of the reliability.

# 8

# Tests on a single sample: association and correlation between two variables

In all of the preceding examples and tests, each sampling unit has been described with respect to one variable. For example, we decided whether the colour of a kernel was yellow or black, counted the number of colonies occur on a plate and measured the concentration of a solution. We have also examined single samples to find out whether they have been taken from populations of variables which have a specified distribution and made estimates of the population parameters and their reliability.

As I mentioned in Chapter 1, we can extend our description by taking a single sample and describing each sampling unit with respect to two variables. The data are then bivariate. This is particularly useful when we are looking for other factors which might affect the variable of interest. If we are using two nominal variables, we could then ask whether particular states of one variable tend to occur with particular states of the other variable more or less often than we would expect. For instance, we might describe individual people with respect to eye colour and hair colour and ask whether people with blue eyes tend to have fair hair. We would be looking for an **association** between the two variables.

If we count the number of snails in a sample of quadrats and also measure the amount of a food plant in the same quadrats, we can ask the same sort of question. Is the number of snails related in any way to the abundance of plants, that is, do they tend to vary together in some way? Once again, we are looking for an association, but with these types of variable, the term **correlation** is usually used. The statistical test used in these situations depends on the type of variable involved.

## The chi-squared test for an association between nominal variables

We can illustrate this type of approach with an example taken from a descriptive study of the vegetation of a blanket bog on Dartmoor. The question is whether certain species tend to occur together, forming recognizable assemblages or communities. The study involved many species, however, the principle can be illustrated using just one pair of species, sundew (*Drosera rotundifolia*) and heather (*Calluna vulgaris*). The statistical analysis will involve the use of $\chi^2$ as the test statistic, although the test procedure is different to that used in Chapter 6.

The fieldwork is simple. We place quadrats at random and simply classify each quadrat on the basis of two criteria: is sundew present or absent and is heather present or absent?

Each of these is obviously a nominal variable with two categories, but you can see that we are describing each sampling unit (quadrat) in terms of two variables (i.e. its state with respect to two species). There are four possible types of quadrat, which are listed in Box 8.1a, along with their observed frequencies.

The null hypothesis is that there is no association between sundew and heather, or if you prefer, that the occurrence of sundew is independent of (i.e. not affected by) the presence of heather. What this means is that the two species occur at random with respect to one another. If this is so then, in the population (of quadrats), the proportion of quadrats with sundew present should be the same in quadrats with and without heather. We can see that these proportions do not look equal. Out of the sample of 124 quadrats with heather present, the proportion with sundew present is 0.49 (61/124). Of the 37 quadrats without heather, the proportion with sundew present is 0.22 (= 8/37). However, the difference in sample proportions could simply be due to the vagaries of sampling so we need a statistical test. Because both variables are nominal, $\chi^2$ will be appropriate.

---

**BOX 8.1   CHI-SQUARED TEST OF ASSOCIATION BETWEEN SUNDEW (*DROSERA ROTUNDIFOLIA*) AND HEATHER (*CALLUNA VULGARIS*)**

**(a)**

| | Number of quadrats |
|---|---|
| Sundew present, heather present | 61 |
| Sundew present, heather absent | 8 |
| Sundew absent, heather present | 63 |
| Sundew absent, heather absent | 29 |

Proportion of quadrats with sundew and heather present = 61/124 = 0.49

Proportion of quadrats with sundew present and heather absent = 8/37 = 0.22

Null hypothesis:   The occurrence of sundew is independent of the occurrence of heather. (i.e. no association).

Alternative hypothesis:   The occurrence of sundew is not independent of that of heather (i.e. there is an association).

Significance level:   5%, two-tailed.

**(b)  2 × 2 Contingency table of observed frequencies**

| | Sundew present | Sundew absent | Row totals | |
|---|---|---|---|---|
| Heather present | 61 | 63 | 124 | |
| Heather absent | 8 | 29 | 37 | |
| Column totals | 69 | 92 | 161 | Grand total |

**(BOX 8.1 continued)**

**(c) Calculating the expected frequencies under the null hypothesis**

Estimate of proportion of quadrats with sundew present $= (61 + 8)/161$ $= 0.429$

Estimate of proportion of quadrats with sundew absent $= (63 + 29)/161$ $= 0.571$

Expected frequency of quadrats with sundew present and heather present
$= 0.429 \times 124$ $= 53.2$

Expected frequency of quadrats with sundew present and heather absent
$= 0.429 \times 37$ $= 15.9$

Expected frequency of quadrats with sundew absent and heather present
$= 0.571 \times 124$ $= 70.8$

Expected frequency of quadrats with sundew absent and heather absent
$= 0.571 \times 37$ $= 21.1$

Total $= 161.0$

**(d) Contingency table with observed and (expected) frequencies:**

|  | Sundew present | Sundew absent |
|---|---|---|
| Heather present | 61 (53.2) | 63 (70.8) |
| Heather absent | 8 (15.9) | 29 (21.1) |

$$X^2 = \frac{\Sigma(|O - E| - 0.5)^2}{E} = 7.756$$

Degrees of freedom $=$ Number of categories $(C)$ – Number of pieces of information $(i)$
$= 4 - 3 = 1$

Handy rule for degrees of freedom $=$ (Number of rows – 1) $\times$ (Number of columns – 1)
$= (r - 1)(c - 1) = (2 - 1)(2 - 1) = 1$

Critical value of $\chi^2 = 3.841$

Decision rule:  Reject null hypothesis if calculated value is equal to, or larger than, critical value.

Decision:  Highly significant evidence for an association between the two species $(P < 0.01)$.

## Setting up the contingency table

Data like these, in which each sampling unit has been classified with respect to two variables, are said to be **cross-classified**. They can be presented in the form of a table called a **contingency table**, which is simply a table with a number of rows and columns (Box 8.1b). In this case we have a square table of two rows and two columns, or in shorthand a $2 \times 2$ contingency table, which as we can see, gives us four boxes or cells, corresponding to the four categories, with an observed frequency in each. Note that for the purposes of the $\chi^2$ test these must be frequencies and not proportions or percentages.

To complete the table we need: (a) the two row totals (which are the total number of those quadrats with heather present and the total of those with heather absent); and (b) the two column totals (which are the total number of quadrats with sundew present and the total of those with sundew absent). These totals are known as the **marginal totals**. Finally, we need the **grand total**, that is the total number of sampling units, which is simply the sum of either the row totals or the column totals – they should both give the same answer. The $\chi^2$ test involves comparing the observed frequencies with a set of expected frequencies using the standard equation – the only problem is what are the expected frequencies?

## Calculating the expected frequencies

It should be clear that we are not interested in testing whether the proportions of the different types of quadrat fit with some particular theoretical ratio such as 1:1 or 3:1, so we cannot work out expected values on this basis. What we want to know is whether, for example, the proportion of quadrats with sundew is the same in quadrats where heather is present as it is in quadrats where heather is absent. We shall have to estimate the expected values using information from the data itself. The only way we can do this is by assuming initially that the null hypothesis is true. If it is, the proportion of quadrats with sundew will be the same in quadrats with and without heather, so these two categories can be added together. The best estimate of the proportion of quadrats with sundew present is therefore $(61+8)/161 = 0.429$ (Box 8.1c). Likewise, the estimate of the proportion of quadrats without sundew is $(63+29)/161 = 0.571$.

These expected proportions are relative frequencies and have to be converted to frequencies based on the sample size so that they can be compared to the observed frequencies. This is easy (Box 8.1c). For example, the expected frequency of quadrats with sundew and heather present is equal to the total number of quadrats with heather present, multiplied by the proportion which are expected to have sundew present. The total number of quadrats with heather present was 124, of which we expect 0.429 to also have sundew present, so the expected frequency is simply $0.429 \times 124 = 53.2$. Similar calculations can be used to obtain the other expected frequencies.

That is the theoretical basis for calculating the expected frequencies but there is a much easier way to do it. For example, the expected frequency in the top left-hand cell of the table (quadrats with both species present) was given by $124 \times 0.429$ which is $124 \times 69/161$. This is simply the two marginal totals which relate to that cell, multiplied together and divided by the grand total. The other combinations of marginal totals can be used in the same way to obtain expected frequencies for the other cells. Do not be tempted to round these figures off to the nearest whole number on the grounds that you could never expect to get 53.2 quadrats in a category; this would be incorrect. Don't forget that the term "expected frequency" actually refers to the mean number you would get if you repeatedly took samples of size 161 from the population. Before we can calculate the value of $X^2$ we shall need to find out how many degrees of freedom there are, so we know whether to incorporate Yates' correction.

## The degrees of freedom

The degrees of freedom can be worked out using the same rules as in Chapter 6, that is: df = number of categories $(C)$ – number of pieces of information required $(i)$. The number of categories $(C)$ is obviously 4, but what is the value of $i$, the number of pieces of information needed to calculate the expected frequencies? Well from the data we took the proportion of quadrats with sundew, the proportion of quadrats without sundew, the number of quadrats with heather, the number of quadrats without heather and the grand total in order to calculate the four expected values. At first sight it looks as though $i = 5$, giving negative degrees of freedom, which would cause serious problems.

In fact, two of these "pieces of information" do not have to be calculated from the data. To calculate the proportion of quadrats with sundew, we use two pieces of information, the total number of quadrats with sundew and the grand total. Once we have calculated this proportion as 0.429 we automatically know what the proportion of quadrats without sundew is; it must be 1 – 0.429. So this requires no further information from the data. To calculate the expected frequency of quadrats with both species present we have to use another piece of information, the number of quadrats with heather (124). Given that we already know and have used information on the total sample size (161) we automatically know what the number quadrats without heather is; it must be 161 – 124. As a result we only use three pieces of information from the data, so $i = 3$ and the df = 4 – 3 = 1.

Although this is the principle involved in working out the degrees of freedom for a contingency table, in practice, we can use a handy formula (Box 8.1d). We simply take the number of columns $(c)$ and subtract one, the number of rows $(r)$ and subtract one and then multiply the results $(c–1) \times (r–1)$. The answer is the number of degrees of freedom.

## Calculating the value of $X^2$

We can now calculate the value of $X^2$ using the standard formula. One common fault is to only use two of the categories, say those with sundew present. This is incorrect, you must use the values of $(O–E)^2/E$ from all four categories. This is a test with one degree of freedom and so our calculation of $X^2$ must incorporate Yates' correction. Each difference between an observed and expected frequency must be reduced by 0.5, giving a calculated value of $X^2$ of 7.756 with one degree of freedom. The standard 5% significance level is adequate for this type of problem, giving us a critical value of 3.841.

Since the expected frequencies are based on the null hypothesis being true, large differences between observed and expected frequencies suggest that it is not true. Large differences will produce large calculated values of $X^2$ so the decision rule is the same as for a goodness-of-fit test. Our result is highly significant. We reject the null hypothesis and conclude that we have very strong evidence for an association between the two species; the distribution of the two species with respect to one another is not random.

However, we should not leave the analysis here, because we can now enquire about the direction of the association, because there are two alternatives to consider. Sundew and heather could tend to occur together. This would mean that when we find sundew in a quadrat, the chances are that heather will also be present, and if sundew is absent from a quadrat, then heather is likely to be absent as well. This would be an example of a **positive association**. In its most extreme form it would result in all the quadrats having either both species present or both species absent and no quadrats having only one of the species. A **negative association** is one in which the two species tend not to occur together. Quadrats with both species present or absent would be rare, while quadrats with only one of the species would be common. Which sort of association have we got here?

There are two ways in which we can tell. We can look at the relationship between the observed and the expected values where you can see that there are more quadrats than expected with both species present and with both species absent. Conversely, for quadrats in which only one species occurs, there are fewer quadrats than expected. In other words, the two species tend to occur together – if one is present, there is a high chance that the other will also be there. Alternatively, we can examine the percentages. Overall 124 quadrats out of 161 (= 77%) have heather present. If there were no association, the percentage of quadrats with heather and with sundew should be about 77% and the proportion of quadrats with heather but without sundew should also be about 77%. In fact, 61 out of the 69 quadrats (88%) with heather have sundew present, while 63 of the 92 quadrats (69%) with heather do not have sundew. Heather is more likely to be found if sundew is present so in this case we have a positive association.

## A quick formula for a 2 × 2 table

Working out the value of $X^2$, as described above, is laborious using a simple calculator, but there is a short-cut formula available (Box 8.2). The example is another set of data from the same survey, involving a third species, the bog moss *Sphagnum cuspidatum*.

As you can see, the formula does not involve calculating the expected frequencies and individual values of $(O-E)^2/E$, but simply uses the observed frequencies (labelled as $a$, $b$, $c$ and $d$), the marginal totals ($a+b$, $c+d$, $a+c$, and $b+d$) and the sample size ($n$). The easiest way to use this is as follows. First calculate $ad$ and put it in the memory and then calculate $bc$. Subtract this from $ad$ using the M- key and then recall the answer from the memory. The vertical lines mean that you need the absolute value of this quantity, so if it has a minus sign use the change sign key to remove it. Subtracting $n/2$ is equivalent to incorporating Yates' correction. You can then square the answer, multiply it by $n$ and divide the answer by the four marginal totals one after the other to give the value of $X^2$.

Although much quicker, this method does have a drawback, particularly when the result is statistically significant. In the preceding example, we could easily see what sort of association there was by examining the direction of the differences between the observed and expected values. Using the quick formula, we do not have the expected values so they may have to be calculated separately. However, if the two sample sizes are similar or the proportions very unequal, you will probably be able to discern the type of association by direct inspection of the observed frequencies. Alternatively, you could convert the observed frequencies to relative frequencies, which will also enable you to see what sort of association is involved; indeed, you may have done this before carrying out the statistical test as part of a

**BOX 8.2   QUICK FORMULA FOR CALCULATING $X^2$ IN A 2 × 2 CONTINGENCY TABLE: TEST OF ASSOCIATION BETWEEN HEATHER (*CALLUNA VULGARIS*) AND BOG MOSS (*SPHAGNUM CUSPIDATUM*)**

Null hypothesis:   The occurence of heather is independent of the occurrence of bog moss (i.e. there is no association).

Alternative hypothesis:   The occurence of heather is not independent of the occurrence of bog moss (i.e. there is an association).

Significance level:   5%, two tailed.

|  | Bog moss present | Bog moss absent |  |
|---|---|---|---|
| Heather present | 18 = $a$ | 35 = $b$ | 53 = $(a+b)$ |
| Heather absent | 61 = $c$ | 47 = $d$ | 108 = $(c+d)$ |
|  | 79 = $(a+c)$ | 82 = $(b+d)$ | 161 = $n$ |

$$X^2 = \frac{\left(|ad - bc| - N/2\right)^2 \times n}{(a+b)(c+d)(a+c)(b+d)}$$

$$= \frac{\left[|(18 \times 47) - (61 \times 35)| - 161/2\right]^2}{53 \times 108 \times 79 \times 82}$$

$$= \frac{\left(|846 - 2135| - 80.5\right)^2 \times 161}{53 \times 108 \times 79 \times 82}$$

$$= \frac{(1289 - 80.5)^2 \times 161}{53 \times 108 \times 79 \times 82}$$

$$= 6.341$$

Degrees of freedom = (Number of rows − 1) × ( Number of columns − 1) = 1

Critical value of $\chi^2$ = 3.841

Decision rule:   Reject null hypothesis if calculated value equal to, or greater than, critical value.

Decision:   Significant evidence for an association between the two species ($P < 0.05$).

preliminary inspection of the data. Remember though, that you **must not use percentages or proportions** in the calculation of $X^2$.

In this example there is also a highly significant association, but if we examine this further we find that it is a negative association. Now, the number of quadrats with both species present is smaller than expected, as is the number with both species absent. The fact that the two species tend not to occur together is also shown by the proportions. Only 23% of quadrats with bog moss have heather, while the figure is 43% for quadrats without bog moss.

## How do we interpret an association?

If we find a positive correlation like the one above, what can we conclude about the reasons for the two species tending to occur together? You can probably make some suggestions. Sundew might in some way depend on heather. This could take the form of parasitism or semi-parasitism, as was suggested in the case of the broomrape in Chapter 6. Alternatively, perhaps heather modifies the environment in some way to provide a micro-environment which favours sundew. Or, it could be the other way round, perhaps heather depends on sundew. Perhaps neither species depends on the other but they both have common environmental requirements and as a result they tend to be found together. The statistical analysis which demonstrated the association cannot tell us which of these explanatory hypotheses is correct. Just because two variables are associated, we cannot conclude that variation in one causes variation in the other; a correlation does not imply a causal link between the two variables.

The value of the approach lies in the fact that once associations have been identified, explanations can be advanced and the way is open to designing experiments to test hypotheses. (You might like to try designing some experiments to investigate some of these explanations.) In this case, the positive association is due to common environmental needs. Both heather and sundew are also positively associated with slightly raised areas of the bog, the so called hummocks, and it can be shown that they only thrive under the somewhat drier conditions found there. The negative association of heather with this species of bog moss is because they have different environmental needs; the bog moss is adapted to the wetter areas called hollows.

There is one further point which must be made here. The calculated value of $X^2$ does not bear any relation to the strength of the association. The strongest positive association that one could get would be a situation in which the two species always occurred together something that could happen if they were obligate symbionts. In this situation, if one of the species was present, the other one would also have to be there and if one species was absent then the other would absent too. There would be no quadrats with only one species present.

In our example the positive association is obviously weaker than this, because the two species can occur on their own in a quadrat. However the $X^2$ value cannot be used as a measure of this weakness as the following example shows. Imagine that the proportions in each category in the above example are correct (i.e. they are equal to the proportions in the population) but imagine that we had taken ten times the number of quadrats. There would now be 610 quadrats with both species present, 290 with neither species, etc. The value of $X^2$ will now be 77.56, 10 times as large as in our example, but we could not conclude that the association was 10 times as strong. The extent to which the two species occur together is exactly the same as it was before.

## Association with more than two categories

This type of analysis can be extended to situations in which there are more than two categories for each variable. We might, for instance, wish to know if there was any association between eye colour and hair colour so the contingency table might have four categories of hair colour (blonde, brown, black and red) and four categories of eye colour (blue, grey, brown and green).

Box 8.3 contains an ecological example based on observations of caterpillars in their natural habitat. The sampling unit is an individual caterpillar. One variable is behavioural

---

**BOX 8.3   CHI-SQUARED TEST OF ASSOCIATION BETWEEN PLANT SPECIES AND BEHAVIOUR OF CATERPILLARS OF THE SIX-SPOT BURNET MOTH (*ZYGAENA FILIPENDULAE*)**

Null hypothesis:   There is no association between caterpillar behaviour and plant species.

Alternative hypothesis:   There is an association between caterpillar behaviour and plant species.

Significance level:   5%

| Plant species | Behaviour | | | | Row total |
|---|---|---|---|---|---|
| | Inactive | Feeding | Walking | Pupating | |
| Red fescue (*Festuca rubra*) | 80 (51.81) | 0 (13.85) | 10 (33.94) | 23 (13.40) | 113 |
| Bird's-foot trefoil (*Trifolium corniculatus*) | 13 (29.80) | 27 (7.96) | 18 (19.53) | 7 (7.71) | 65 |
| Kidney vetch (*Anthyllis vulneraria*) | 15 (24.30) | 4 (6.49) | 34 (15.92) | 0 (6.28) | 53 |
| Yarrow (*Achillea millefolium*) | 8 (10.09) | 0 (2.70) | 14 (6.61) | 0 (2.61) | 22 |
| Column total | 116 | 31 | 76 | 30 | 253 |

Figures in brackets are expected frequencies.

$$X^2 = \sum \frac{O^2}{E} - N = 153.4$$

Degrees of freedom:   (Number of columns − 1)(Number of rows − 1) = (4−1)(4−1) = 9

Critical value of $\chi^2$ = 16.919

Decision rule:   Reject the null hypothesis if calculated value is equal to, or larger than, critical value.

Decision:   Very highly significant evidence for an association between caterpillar behaviour and plant species ($P < 0.001$).

---

(what the caterpillar was doing), the other is the species of plant on which the caterpillar was found. Because each variable has four categories, the result is a $4 \times 4$ contingency table. The formula in Box 8.2, which does not require the calculation of expected frequencies, can only be used on a $2 \times 2$ table so you will have to calculate them using the longhand method described in Box 8.1. Expected frequencies are given in brackets in the table. Once you have obtained these you can use the alternative formula for calculating $X^2$. (As you calculate each value of $E$ it is possible, with some additional button pushing, to then calculate each value of $O^2/E$ and accumulate these in the memory. This avoids having to write down and then re-enter the expected values and so minimizes the chances of errors in rounding off.)

There is a highly significant association, although it is not possible to describe it as either positive or negative, because of the complex nature of the variables. When we have several categories and the result is significant we must be careful about the biological interpretation because the analysis only tells us that, overall, the differences are significant. We cannot then single out specific differences and label them as significant. If we inspect the difference between each observed and expected value and find a large difference, then it suggests a biologically interesting relationship. So the fact that we observed 80 stationary caterpillars on red fescue, when we expected 51.8 is very suggestive, but we cannot say that this one difference is statistically significant. What we have done is to identify patterns which are worth further investigation, using experiments designed to test more specific hypotheses.

## Correlation between ordinal variables: Spearman's rank correlation coefficient

If we can quantify the two variables by ranking, measuring or counting, then we can test for a correlation, that is, a tendency for the two variables to vary together. The simplest of these tests, Spearman's rank correlation coefficient, is designed for ordinal (ranked) variables. Because it has few underlying assumptions, it can be applied in a wide range of situations.

We can illustrate what is meant by a correlation and also why we need a statistical test, with some data on the pigmentation of orchid leaves and flowers which were illustrated in Chapter 1. A reasonable question to ask would be: "Is the degree of spotting on the leaves related in any way to the degree of spotting on the flowers?". What we want to know is whether the two variables vary together in a systematic way, that is, whether they are correlated (= co-related). If there is a relationship, it could be in one of two directions. Plants with a high degree of leaf spotting could tend to have more spotting on the flowers, and vice versa. This would be a positive correlation. Alternatively, it would be a negative correlation if plants with very spotted leaves tend to have unspotted flowers.

It would be relatively simple to obtain the data. We could collect (or at least examine) a random sample of plants and assess the degree of spotting on the leaves and flowers of each one. This variable could be measured in some way but this would be difficult, so instead, we shall treat it as an ordinal variable. To do this we first order the leaves with respect to their degree of spotting (Fig. 8.1a) and describe the position of each one in the order by a number. This is its rank. We then repeat this for the flowers. This ordering or ranking must be done separately for the leaves and the flowers so that we know the position of each leaf in the order for leaf spotting and the position of each flower in the order for flower spotting.

Although it doesn't actually matter whether we put the items in ascending or descending order, it will make the subsequent interpretation much easier if we rank both variables in ascending order. This means giving the lowest rank to the least spotted leaf or flower.

Due to our preconceptions, a bias could easily creep in when making a somewhat subjective assessment like this, so it is most important that the ranking of leaves and flowers is done independently. We must ensure that a knowledge of the rank we give to the leaf of an individual plant does not influence the rank we give to the flower of the same plant. If, for instance, we are anticipating that there will be a positive correlation, then having given a leaf a particular rank, we may be tempted to give the flower from the same plant a matching rank, even if this is inappropriate. This is an issue of sampling design which can be dealt with in one of two ways. Can you see what they are?

One way would be to take a leaf and a flower from each plant, label them in a way which disguises their origin and give them to someone else to rank. Once they have been ranked, you can use the labels to put the leaf and flower from each plant in their correct pairs. This would be an example of what is called a **blind design**, because the origin of the material is concealed. Alternatively, get two different people to do the ranking with the intact plants, but with one doing the leaves and the other doing the flowers. Obviously neither person should know the purpose of the ranking.

An initial step with any correlation is to present the data on a graph in which one of the variables is plotted on the vertical axis, and the other on the horizontal axis (Fig. 8.1b). The two ranks for each individual plant can be represented by a single point. You can think of these as the co-ordinates of that individual with respect to leaf and flower spotting. A graph like this is called a **scatter diagram** and we will see later that it is a wise preliminary step with bivariate data. If you are wondering which variable to put on which axis, the answer is that it does not matter. Conventionally, the **independent variable** with values denoted by $X$ goes on the $x$ axis and the **dependent variable** denoted by $Y$ goes on the $y$ axis. An independent variable is one which is manipulated or has values chosen by the experimenter and which is thought to causes change in a dependent variable. In our example, neither variable can be considered as independent and so we shall label one as $Y_1$ and the other as $Y_2$. (Strictly this notation is not ideal because we have already used $Y_1$ and $Y_2$ in Chapter 4 as symbols for two different values of the same variable. However, we shall use this simple notation because this overlap is probably less confusing than the use of a more complex notation.)

The diagram looks promising. The points are somewhat concentrated in a band running from the bottom left to the top right of the diagram, which suggests that there is a tendency for plants with lightly spotted leaves to have lightly spotted flowers, and vice versa. In other words, the two variables look as though they are correlated. But there is a problem, with which we should by now be fully familiar and which is illustrated in the two lower diagrams (Fig. 8.1c). Our seven values are, of course, only a sample taken from a population of such values.

The open circles in the left-hand diagram represent the population of values and illustrate the situation in which there is no correlation between the two variables in the population. If there is no correlation between leaf and flower spotting, then in an individual plant any degree of leaf spotting could occur with any degree of flower spotting. The points representing the population would be scattered as shown. Our sample (filled circles) could have been taken from a population like this and, by chance, the individuals with the less pigmented leaves that we happen to have chosen could be ones with less pigmented flowers, and vice versa. The sample shows a correlation which is not present in the population. The alterna-

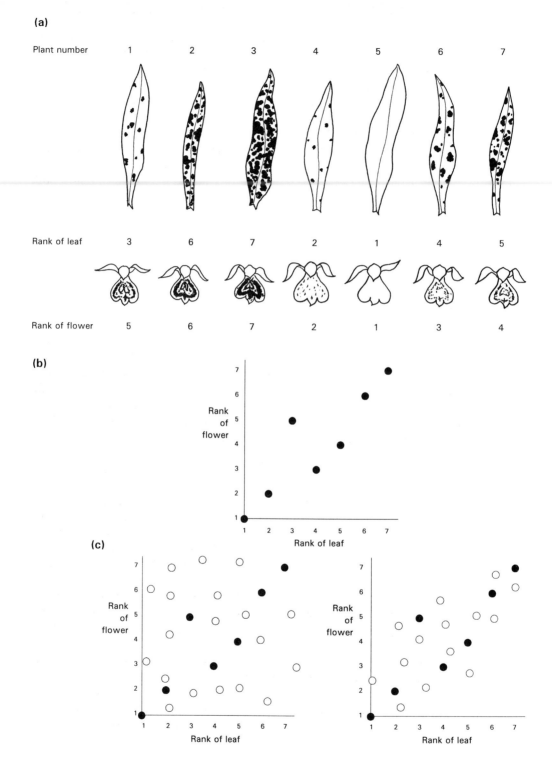

**Figure 8.1** Correlation between two ordinal variables; the amount of spotting on leaves and flowers of the southern marsh orchid.

tive is shown in the right-hand diagram. Now there is a correlation between leaf and flower spotting in the population and it is a positive correlation because plants with lightly spotted leaves tend to have lightly spotted flowers. In this case the correlation in our sample would be reflecting the correlation in the population.

To sort out which of these two explanations can most reasonably account for what we have found will, as usual, need a statistical test. The null hypothesis will be that there is no correlation between the two variables and the test statistic will have to be something which has a known sampling distribution under the null hypothesis. It will also be useful if the test statistic describes the relationship between the two variables in a convenient way. Hopefully, it will also be easy to calculate! An appropriate test statistic is **Spearman's rank correlation coefficient** $(r_s)$. We shall start by seeing how it works and then use it on our example.

## What is $r_s$?

The formula for calculating $r_s$ is given in Figure 8.2a. The most instructive part of the formula is $d$, which is the difference between the ranks of the two variables for any sampling unit. If there is a positive correlation, then for each plant, the rank given to the leaf will be similar to the rank given to its flower and the difference between the ranks $(d)$ will be small. On the other hand, if there is a negative correlation, then there will be little agreement between the rank given to the leaf and the rank given to the flower and $d$ will be large. This suggests that the sum of the values of $d$ could be used as a test statistic; a small total would mean a positive correlation and a large total a negative correlation. In practice, we have to square the values of $d$ before we add them to get round the fact that some of them are positive and some negative. Without this, the positive and negative values always cancel one another out when added together and give zero.

The sum of the squared differences $(\Sigma d^2)$ on its own is not a convenient test statistic because its always positive, whereas a correlation can be either positive or negative. It is more useful if what we calculate has a sign appropriate to the direction of the correlation. In addition, the size of $\Sigma d^2$ will obviously depend on both the degree to which the two variables are correlated and the number of observations involved. The rest of the equation is simply aimed at converting the information carried by $\Sigma d^2$ to a more useful form and you don't need to know what this entails. If you are interested, an explanation follows; if not, then skip the next paragraph!

The maximum value that $\Sigma d^2$ could have occurs when there is a perfect negative correlation and this value increases with increasing sample size. It can be shown that this maximum value would be $(n^3-n)/3$. We first express $\Sigma d^2$ as a fraction of this maximum value so that our measure now varies between 0 for a perfect positive correlation and 1 for a perfect negative correlation; you can think of this as a type of standardization. Mathematically it is the same as calculating $3\Sigma d^2/(n^3-n)$. If we multiply this by 2 (giving $6\Sigma d^2$ on the top line), we now have a measure which can vary between 0 and 2 and subtracting this from one gives us a statistic with the required properties. It now varies conveniently between $+1$ for a perfect positive correlation and $-1$ for a perfect negative correlation. This measure also has a known sampling distribution.

**(a) Formula for Spearman's rank correlation coefficient ($r_s$)**

$$r_s = 1 - \frac{6\Sigma d^2}{n^3 - n}$$

**(b) Producing a sampling distribution for $r_s$ with $n = 3$**

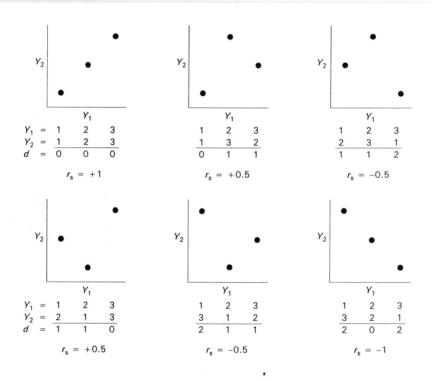

**(c) Sampling distribution of $r_s$ for $n = 3$**

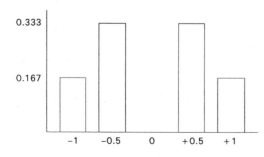

**Figure 8.2** The basis of Spearman's rank correlation coefficient.

## The sampling distribution of $r_s$

The sampling distribution of $r_s$ can be obtained by using the same sort of logic that we used in the woodlouse experiment, although the going gets fairly tough if we use a reasonable sample size. We can, however, illustrate the principle with a sample size of 3 (Fig. 8.2b). For convenience we shall label one variable $Y_1$ and the other $Y_2$. If the null hypothesis is true (i.e. there is no correlation between the two variables) then all possible values of $Y_2$ are equally likely to occur with each value of $Y_1$. This means that the three values of $Y_2$ could occur in any order with respect to the values of $Y_1$. With $n = 3$ there are six different orders, shown in the small scatter diagrams, each giving rise to a value of $r$ between $+1$ and $-1$. Incidentally, you can see from this that the maximum value of $\Sigma d^2$ does occur when there is a perfect negative correlation and that its value is $(n^3-n)/3 = (27-3)/3 = 8$.

Some of these values of $r_s$ occur more than once and we can present them in the form of a frequency distribution (Fig. 8.2c). This is the sampling distribution of $r_s$ for a sample of size three and you can see that it has a rudimentary hump in the middle and two tails. Remember that a sampling distribution is a relative frequency distribution of all possible values of the test statistic, calculated on the basis of the null hypothesis. We use it to find out the probability associated with a particular value so that we can decide whether that value is likely or unlikely. In this case, because of the small sample size, even the most extreme value of $r_s$ has quite a high probability of occurrence (actually 0.167). Clearly there is no possibility of rejecting the null hypothesis with such a small number of observations.

With $n = 5$, there are 21 different values of $\Sigma d^2$ and the sampling distribution has much better defined tails. Values of $r_s$ of $+1$ and $-1$ have a probability of occurrence of 0.0083 which is sufficiently rare to be judged improbable if the null hypothesis is true, making this the smallest sample size for a two-tailed test. With further increases in sample size the sampling distribution of $r_s$ becomes less widely spread and large positive and negative values of $r_s$ become even rarer. As a result, the values of $r_s$ which cut off known proportions in the tails of the distribution, that is the critical values of $r_s$, become less extreme. These critical values have been calculated and tabulated; as usual they depend on the sample size, the significance level and whether we are doing a one-tailed or a two-tailed test.

## Calculating $r_s$ and using tables of critical values

We can now return to our orchid example (Box 8.4). The null hypothesis is that there is no correlation between the two variables and the alternative hypothesis is that there is a correlation. The direction of any correlation is not specified by the alternative hypothesis, so this will be a two-tailed test. Both extreme positive and extreme negative values of $r_s$ will be evidence against the null hypothesis. A significance level of 5% will be adequate.

We can use the ranks which we have already assigned. For each individual we calculate $d$, the difference between the ranks of the two variables. For example, the first individual has rank 3 for leaf spotting and rank 5 for flower spotting, so $d = 3 - 5 = -2$. These values of $d$ are then squared ($d^2$), added together ($\Sigma d^2$) and the answer put into the formula to find $r_s$. The calculated value is $+0.8929$. If you are calculating $r_s$ "by hand", then it is wise to check on your ranking procedure, especially if there are a lot of observations. The largest rank should have a value equal to the sample size and the sum of the differences ($\Sigma d$) should be zero.

---

**BOX 8.4   SPEARMAN'S RANK CORRELATION COEFFICIENT: TESTING FOR A CORRELATION BETWEEN THE DEGREE OF LEAF SPOTTING AND FLOWER SPOTTING IN THE SOUTHERN MARSH ORCHID (*DACTYLORHIZA PRÆTERMISSA*)**

Null hypothesis:  There is no correlation between the degree of leaf spotting and the degree of flower spotting in *Dactylorhiza prætermissa*.

Alternative hypothesis:  There is a correlation between the degree of leaf spotting and the degree of flower spotting in *Dactylorhiza prætermissa*.

Significance level:   5%, two tailed.

| Plant number | Rank of leaf | Rank of flower | Difference in rank ($d$) | Difference squared ($d^2$) |
|:---:|:---:|:---:|:---:|:---:|
| 1 | 3 | 5 | −2 | 4 |
| 2 | 6 | 6 | 0 | 0 |
| 3 | 7 | 7 | 0 | 0 |
| 4 | 2 | 2 | 0 | 0 |
| 5 | 1 | 1 | 0 | 0 |
| 6 | 4 | 3 | 1 | 1 |
| 7 | 5 | 4 | 1 | 1 |
| | | | $\Sigma d = 0$ | $\Sigma d^2 = 6$ |

$$r_s = 1 - \frac{6\Sigma d^2}{n^3 - n} = 1 - \frac{6 \times 6}{7^3 - 7}$$

$$= 1 - 0.1071 = +0.893$$

Critical value = 0.786

Decision rule:   Reject null hypothesis if calculated value is equal to, or larger than, critical value.

Decision:   Significant evidence for a positive correlation ($P < 0.05$).

The table of critical values of $r_s$ (Table A10) is set out in the same way as the other tables. You will notice that only positive values of $r_s$ that is values in the right-hand tail are given, though calculated values of $r_s$ can be either positive or negative. This is done because the distribution is symmetrical. In a two-tailed test, both large negative and large positive values of $r_s$ are evidence against the null hypothesis so we can ignore the sign of the calculated value when using the tables (although we have to use it to tell us the direction of the correlation). We find the critical value ($\alpha_2 = 5\%$, $n = 7$) to be 0.7857. Our calculated value is larger than this, so we can reject the null hypothesis. We have significant evidence for a positive correlation; plants with lightly marked leaves do tend to have lightly marked flowers, and vice versa.

## Tied observations and one-tailed tests

In the preceding example, no attempt was made to measure the degree of spottedness; the two variables were collected as ordinal variables. Spearman's correlation can also be used on measurement or count data, by replacing the actual values by ranks. This can be very useful, because one of the advantages of using $r_s$ is that it involves making no assumptions about the shape of the distribution of the variables. It is an example of a **distribution-free test**. The next example (Box 8.5) shows how to carry out Spearman's rank correlation in these circumstances and also illustrates two other points about the test procedure. As we

---

**BOX 8.5  SPEARMAN'S RANK CORRELATION COEFFICIENT WITH TIED VALUES AND A ONE-TAILED TEST: TESTING FOR A CORRELATION BETWEEN INDEX OF PHYSICAL EFFICIENCY AND EXERCISE**

Null hypothesis:   There is no correlation between index of physical efficiency (IPE) and amount of exercise taken.

Alternative hypothesis:   There is a positive correlation between IPE and amount of exercise taken.

Significance level:   5%, one tailed.

| IPE | Rank | Exercise category | Rank | Difference in rank ($d$) | Difference squared ($d^2$) |
|-----|------|-------------------|------|--------------------------|----------------------------|
| 83  | 3    | 1                 | 2    | 1                        | 1                          |
| 104 | 6    | 4                 | 8.5  | −2.5                     | 6.25                       |
| 108 | 8    | 4                 | 8.5  | −0.5                     | 0.25                       |
| 105 | 7    | 3                 | 6.5  | 0.5                      | 0.25                       |
| 124 | 10   | 5                 | 10.5 | −0.5                     | 0.25                       |
| 180 | 11   | 5                 | 10.5 | 0.5                      | 0.25                       |
| 114 | 9    | 2                 | 4.5  | 4.5                      | 20.25                      |
| 101 | 5    | 3                 | 6.5  | −1.5                     | 2.25                       |
| 77  | 2    | 2                 | 4.5  | −2.5                     | 6.25                       |
| 67  | 1    | 1                 | 2    | −1                       | 1                          |
| 91  | 4    | 1                 | 2    | 2                        | 4                          |
|     |      |                   |      | $\Sigma d = 0$           | $\Sigma d^2 = 42$          |

$$r_s = 1 - \frac{6\Sigma d^2}{n^3 - n} = 1 - 0.1909 = +0.890$$

Critical value of $r_s$ (with sign) $= +0.618$

Decision rule:   Reject null hypothesis if calculated value is equal to, or more extreme than, critical value with a sign appropriate to the alternative hypothesis.

Decision:   Significant evidence for a positive correlation ($P < 0.05$).

---

shall see, the other commonly used procedure is a measure of correlation which assumes that both variables have normal distributions.

In this example, we are asking whether there is a correlation between the physical fitness and the amount of exercise taken by human subjects. Physical fitness was measured by an index of physical efficiency (IPE), which is a measure of how quickly the pulse rate drops to its resting value after a standardized bout of physical activity. As such, it is a derived variable, based on counts and the fitter you are, the higher your IPE. It is difficult to obtain a precise, comparative measure of the amount of exercise taken, because this involves both the type of exercise and the number and duration of the periods of exercise. To overcome this problem, the subjects (first year students again!) were asked to put themselves into one of five ordered categories. For example, category 1 was for people who took no exercise, while category 5 was for those who took more than 5 sessions per week.

Both males and females were involved and there was also a wide age range, due to the presence of some mature students. Both these factors could lead to non-normal distributions of one or both variables. Because one of the variables is ordinal and we know nothing about the shape of the distribution of either variable a distribution-free procedure would be most appropriate.

As a result of the way physical activity has been assessed, some individuals have the same rank for this variable, that is, there are observations which are tied. This could easily have occurred in our orchid example – we might have been unable to decide whether two leaves had different degrees of spotting. The procedure for dealing with tied observations is simple; we assign tied observations the average of the ranks they share. For example, there are three students in category one, so they share ranks 1, 2 and 3. The average rank of these three students is $(1 + 2 + 3)/3 = 2$, so they are each given this average.

The null hypothesis is that there is no correlation, but in this case a one-sided alternative hypothesis would be in order. If there is a correlation between these two variables in the population, then it would surely be a positive one; high physical efficiency would be correlated with high levels of exercise. Because of this, only large positive values of $r_s$ will be evidence against the null hypothesis and so we need to take account of the sign of our calculated value of $r_s$ and compare it with an appropriately signed critical value.

In the case of a predicted positive correlation, this means that a calculated value has to be positive and equal to or larger than the appropriate critical value found in the column headed $\alpha_1^R = 5\%$, which will be 0.536. Our calculated value of $+0.809$ is greater than this critical value and is highly significant evidence for a positive correlation. If we had obtained a value of $-0.809$, then with our one-tailed alternative hypothesis, we would have had no evidence against the null hypothesis because the sign of $r_s$ is not in the predicted direction. Remember that in embarking on a one-tailed test, we have decided in advance that a negative correlation, if it occurs, cannot have any biological significance; instead it must have resulted from one of those vagaries of sampling which will occur from time to time.

If we had been testing a one-tailed alternative hypothesis which predicted a negative correlation, then large negative values of $r_s$ would be of interest. We really need to be looking in the left-hand tail of the sampling distribution, where the negative critical values occur, whereas the tabled values refer to the right-hand, positive tail. However, because the distribution is symmetrical, the critical values in this situation are given by finding the critical value in the appropriate $\alpha_1^R$ column and putting a minus sign in front of it. To reject the null hypothesis your calculated value would have to be negative and equal to or more extreme than the critical value.

# Correlation between two normally distributed variables: the product–moment correlation coefficient

Figure 8.3 is a scatter diagram of the forearm length and the lower leg length in a random sample of adult females (first year students yet again!). Is there a correlation between the two variables? We could use Spearman's test, but there is an alternative, the product–moment correlation coefficient, which is specifically designed for variables of this type. This correlation coefficient is based on the assumption that both variables follow a normal distribution and we know that this will typically be the case for variables such as limb length in a population of adults of the same sex. We will also see that the calculations involve the means and standard deviations of the two variables, which are used as estimates of the population parameters. Because of this the test is known as a **parametric test**. Spearman's correlation coefficient does not involve means and standard deviations so it is known as a **non-parametric** test.

This product–moment correlation coefficient is just like $r_s$, in that it can have a value between $-1$ and $+1$ and has a sampling distribution which is known. However, the calculation is very different.

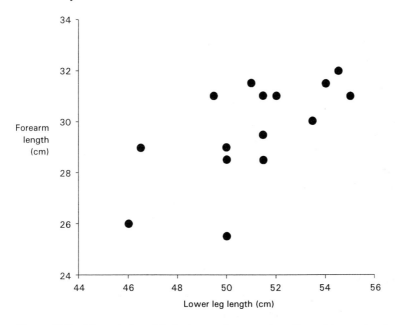

**Figure 8.3** The relationship between forearm length and lower leg length in a sample of 15 female biology students.

## How the product–moment correlation coefficient works

The product–moment correlation coefficient is denoted by $r$ and a formula for calculating $r$ is given in Figure 8.4. Although it looks formidable it has a relatively straightforward interpretation based on a simple principle. It is also fairly easy to calculate, although it is better to use a modified form of the equation for this.

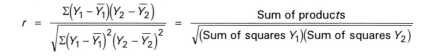

$$r = \frac{\Sigma(Y_1 - \overline{Y_1})(Y_2 - \overline{Y_2})}{\sqrt{\Sigma(Y_1 - \overline{Y_1})^2(Y_2 - \overline{Y_2})^2}} = \frac{\text{Sum of products}}{\sqrt{(\text{Sum of squares }Y_1)(\text{Sum of squares }Y_2)}}$$

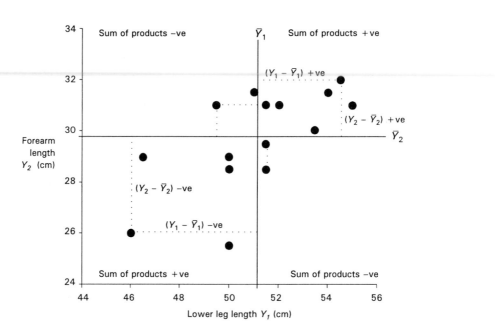

**Figure 8.4** How the product–moment correlation coefficient works.

The general form of some parts the equation probably look rather familiar; indeed the two parts under the line (in the denominator) are the formulae for calculating a sum of squares which we encountered in Chapter 4. The denominator is obviously important, but it is not as interesting as the numerator, which again seems to have some similarities to the formula for a sum of squares. We can quite easily understand what this part of the formula is doing by looking at the scatter diagram for the arm and leg lengths, which has been slightly modified. It now has marked on it a vertical line corresponding to $\overline{Y}_1$ and a horizontal line corresponding to $\overline{Y}_2$.

The top line of the formula tells us to take a point $Y_1Y_2$ on the scatter diagram, find out how far away it is from $\overline{Y}_1$, (i.e. $Y_1 - \overline{Y}_1$) and how far away it is from $\overline{Y}_2$ (i.e. $Y_2 - \overline{Y}_2$) and then multiply these two values together to obtain their product $(Y_1 - \overline{Y}_1)(Y_2 - \overline{Y}_2)$. The $\Sigma$ simply means that we should repeat this for every point and then add together the results to get what is called the **sum of products**. The reason why this is this useful is shown in the diagram. Basically the sign of this sum tells us the direction of the correlation and the size of the sum tells us the strength of the correlation. Let's see why this is the case.

The vertical line running through $Y_1$ divides the points up into those that lie below the mean of $Y_1$ and those that lie above it. The horizontal line does the same thing with respect to the values of $Y_2$ and their mean. As a result, all the points in the top right quadrant are above the mean of $Y_1$ and above the mean of $Y_2$. The deviation from $\overline{Y}_1$ and $\overline{Y}_2$ of any point in this quadrant will be positive and so will the product of the two deviations. Using a similar argument, each point in the bottom left-hand quadrant lies below the mean of $Y_1$ and below the mean of $Y_2$ and so will have a deviation from $\bar{Y}_1$ and a deviation from $\bar{Y}_2$ which is negative. Multiplying the two negative deviations for each point will again yield a positive product. On the other hand, each point in the top left quadrant lies below the mean of $Y_1$, but above the mean of $Y_2$, so its deviation from $\bar{Y}_1$ is negative and its deviation from $\bar{Y}_2$ is positive. As a result the product of the two deviations for each point in this quadrant will be negative. The same will be true for points in the bottom right quadrant.

The sum of products is what we get when we add together the products for every point and you can perhaps now begin to see why it is useful. If, as is shown, the points are concentrated in the bottom right and top left (i.e. there is a positive correlation), then the positive products will more than compensate for the negative products when they are added together and the sum of products will be positive. If, on the other hand, the points are concentrated in the top left and bottom right (a negative correlation), then the sum of products will be negative. So the sign of the sum of products tells us what type of correlation we have.

The size of the sum will depend on the degree of scatter of the points, which depends on the strength of the association. In the example, arm length and leg length show quite a strong correlation. By this we mean that longer leg lengths are very often (but not always) accompanied by longer arm lengths. As a result the points tend lie reasonably close to an imaginary straight line. When this happens most of the points will fall in two diagonally opposite quadrants and the positive products will far outweigh the negative products. As a result the sum of products will be large. If the correlation was weaker then the points would be more scattered around the imaginary straight line. Roughly equal numbers of points would fall in all four quadrants, the negative products would more or less cancel out the positive products and the sum of products would be small. The sign and the size of the sum of products therefore summarize two important facets of the data, in rather the same way that $\Sigma d^2$ does in Spearman's rank correlation coefficient.

However, the sum of products on its own is not much use as a test statistic, because its size will also depend on the range of the values of $Y_1$ and $Y_2$. If $Y_1$ and $Y_2$ both have a range of values between, say, 1 and 100, then the sum of products will be large. If they range between 1 and 10 then the sum of products will be small. To take this into account, we need to convert the sum of products to a standard scale using something which reflects the underlying variability in $Y_1$ and $Y_2$. Variation in $Y_1$ is measured by the sum of squares of $Y_1$, while variation in $Y_2$ is measured by the sums of squares of $Y_2$, which are the two terms which appear on the bottom line of the equation. Multiplying these together and taking the square root gives a measure of the variation in both variables. When we divide the sum of products by this measure, we are looking at the sum of products in relation to the underlying variability, that is we are effectively standardizing the sum of products. It is converted into a number which will lie between $-1$ and $+1$ and this is the product–moment correlation coefficient ($r$).

When $r$ is calculated from a sample it is another sample statistic, which will be an estimate of the population correlation coefficient, denoted by the Greek letter $\rho$ (pronounced "roe"). Under the null hypothesis of no correlation between the two variables $\rho$ will be zero. Under these conditions, we would not expect the value of $r$ calculated from a sample to be zero because of sampling variation, but neither would we expect it to be greatly different from zero. On the other hand, large positive and negative calculated values of $r$ will be unlikely if the null hypothesis is true and so will be evidence for the existence of a correlation. Because the sampling distribution of $\rho$ is known we can compare a calculated value of $r$ with a critical value in the usual way.

---

**BOX 8.6 PRODUCT–MOMENT CORRELATION COEFFICIENT: TEST FOR A CORRELATION BETWEEN LOWER LEG-LENGTH AND FOREARM LENGTH**

Null hypothesis: There is no correlation between lower leg length and forearm length.

Alternative hypothesis: There is a correlation between lower leg length and forearm length.

Significance level: 5%, two tailed.

| Lower leg length $Y_1$ (cm) | Forearm length $Y_2$ (cm) | $Y_1 Y_2$ (cm$^2$) |
|---|---|---|
| 51.0 | 31.5 | 1606.50 |
| 55.0 | 31.0 | 1705.00 |
| 52.0 | 31.0 | 1612.00 |
| 50.0 | 25.5 | 1275.00 |
| 50.0 | 28.5 | 1275.00 |
| 46.0 | 26.0 | 1196.00 |
| 54.5 | 32.0 | 1744.00 |
| 51.5 | 28.5 | 1467.75 |
| 53.5 | 30.0 | 1605.00 |
| 46.5 | 29.0 | 1348.50 |
| 51.5 | 31.0 | 1596.50 |
| 54.0 | 31.5 | 1701.00 |
| 51.5 | 29.5 | 1519.25 |
| 49.5 | 31.0 | 1534.50 |
| 50.0 | 29.0 | 1450.00 |
| $\Sigma Y_1 = 766.5$ | $\Sigma Y_2 = 445.0$ | $\Sigma Y_1 Y_2 = 22786.00$ |
| $s_1 = 2.613$ | $s_2 = 1.961$ | |

$$r = \frac{\text{Sum of products}}{\sqrt{(\text{Sum of squares } Y_1) \times (\text{Sum of squares } Y_2)}} = \frac{\Sigma Y_1 Y_2 - (\Sigma Y_1 \Sigma Y_2 / n)}{(n-1)s_1 s_2}$$

$$= \frac{22786 - (766.5 \times 445/15)}{14 \times 2.613 \times 1.961} = 0.648$$

Degrees of freedom = Sample size $-2$ = $15 - 2$ = $13$

Critical value = 0.514

Decision rule: Reject null hypothesis if calculated value is equal to, or greater than, critical value.

Decision: Highly significant evidence for a positive correlation ($P < 0.01$).

## Calculating *r* and using tables of critical values

Box 8.6 shows the steps involved in calculating $r$ from a sample using a modified formula which makes things very easy. As you can see, this modified formula only involves calculating the values of $\Sigma Y_1$, $s_1$, $\Sigma Y_2$, $s_2$ and $\Sigma Y_1 Y_2$. This can be done easily on a calculator with a standard deviation function and if you plan the calculation you only need to write down two intermediate values. The steps are set out in Box 8.7. The sum of products can be calculated by using an expression which is analogous to the easy formula for calculating a sum of squares which we met in Box 4.8. The two sums of squares can be obtained from an expression which uses the two standard deviations, and these can be obtained directly from a calculator. The latter modification is based on a relationship that we also met in Box 4.8. If you are interested the following section is an explanation, if not, skip to the next paragraph. In Box 4.8 we saw how a sum of squares is the same thing as the variance multiplied

---

**BOX 8.7    CALCULATING THE PRODUCT–MOMENT CORRELATION COEFFICIENT ON A CALCULATOR**

(1)   Put calculator into standard deviation mode.

(2)   Enter values of $Y_1$.

(3)   Recover the value of $\Sigma Y_1$ and write it down. (If your calculator will not give you $\Sigma Y$ directly, then obtain it by multiplying $\bar{Y}_1$ by the sample size.)

(4)   Recover the value of $s_1$, multiply it by $(n - 1)$ and store it in the memory (most calculators do not use the normal memory in standard deviation mode, but check yours!)

(5)   Clear the standard deviation function and enter the values of $Y_2$.

(6)   Retrieve $\Sigma Y_2$, using the same system as in (3) if necessary, and immediately divide by n and multiply by the value of $\Sigma Y_1$, which you wrote down earlier. Note the answer, which is $\Sigma Y_1 \Sigma Y_2 / n$.

(7)   Now retrieve $s_2$ and straight away multiply it by $s_1(n - 1)$, which is in the memory to give you the denominator of the equation, which you can then store.

(8)   Now calculate $\Sigma Y_1 Y_2$. You should be able to do this by keying in each value of $Y_1$ and its matching value of $Y_2$ in the following sequence:

$Y_1 \times Y_2 + Y_1 \times Y_2 + \ldots$ etc.

The answer is $\Sigma Y_1 Y_2$.

(9)   Subtract from $\Sigma Y_1 Y_2$ the answer from step (6) which you wrote down to get the numerator.

(10)   Divide the answer from step (9) by what is in the memory to obtain $r$.

---

by $(n - 1)$. In the formula given in Figure 8.4 you can see that the denominator involves the square root of the two sum of squares, so we need the square root of the variances multiplied by $(n - 1)$. The square root of a variance is a standard deviation.

Our calculated value of $r$ in Box 8.6 is only an estimate of the population value and is subject to chance variation because of sampling. As a result, even if there was no correlation in the population ($\rho = 0$) we could by chance get a positive or negative value of $r$. To test whether the calculated value is significantly different from zero we need to know the sampling distribution of $r$. I hope by now that you are familiar with the idea of a sampling distribution so we will not go into this one in any detail. Suffice it to say that it can be described and that critical values can be obtained for one and two-tailed tests and for various significance levels. These critical values have been tabulated and are used in almost the same way as the tables of $r_s$. There is only one difference. With $r_s$ we entered the table using the sample size; with $r$ we have to use the degrees of freedom which are two less than the sample size. The reason is that this is a parametric test and we have estimated two parameters: $\bar{Y}_1$ was an estimate of $\mu_1$ and $\bar{Y}_2$ was an estimate of $\mu_2$.

So, in our example, the sample size is 15 and there are 13 degrees of freedom which, with a two-tailed test at the 5% significance level, give a critical value of 0.514. Our calculated value exceeds this, indeed the probability associated with a calculated value this large is just less than 1%. We have highly significant evidence for a positive correlation.

## A comparison of Spearman's rank correlation and product–moment correlation

As we have just seen, product–moment correlation is based on both variables having a normal distribution. Spearman's correlation makes no such assumption and, although designed for ordinal variables, it can also be used on count and measurement variables. So why bother with $r$, why not use $r_s$ in all situations, especially as you may find it easier to calculate? The simple answer is that they both have advantages and disadvantages but under different circumstances, so there is no universal best procedure.

Let's just look briefly at some of issues. The data in Figure 8.5a come from the blanket bog survey we looked at earlier in the chapter. It is a scatter diagram of the number of shoots of bog asphodel (*Narthecium ossifragum*) in a quadrat plotted against the height of the ground in the quadrat. The two variables clearly vary together in a systematic way, that is, they are correlated, but neither $r_s$ nor $r$ are significant. This is because both measures are only sensitive to a relationship in which the direction is consistent, that is, the pattern formed by the points slopes in one direction or the other, but not in both. The proper word for this is a monotonic (i.e. a single slope) relationship.

Figure 8.5b is a scatter diagram of the number of limpets and barnacles in quadrats on a rocky shore. It shows a fairly clear monotonic, negative correlation. If we calculate $r_s$, we get a value of –0.548, which is highly significant. On the other hand, calculating $r$ on the same data we obtain a value of –0.381, which is not significant. The reason is that $r$ is only designed to detect a linear correlation, that is, one in which the points are scattered around a straight line. The coefficient $r_s$ is sensitive to any type of monotonic relationship in which one variable tends to increase (or decrease) as the other increases, irrespective of the exact form of that relationship. These two examples show why you ought always to put your data

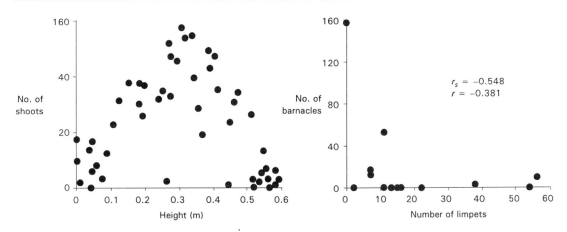

**Figure 8.5** (a) Number of shoots of bog asphodel (*Narthecium ossifragum*) and ground height in quadrat; (b) numbers of barnacles and limpets in quadrats on rocky shore.

in the form of a scatter diagram before doing one of these analyses so that you can check that the analysis is appropriate. If you fail to do this and simply rely on the value of $r$ or $r_s$ you may end up discarding interesting relationships because they are shown not to be statistically significant.

The example in (Fig. 8.5b) might confirm your suspicions that $r_s$ is better than $r$ because it is sensitive to a wider range of relationships. While this is true, there is another dimension to consider. Spearman's correlation is a more general-purpose test, that is, it is more widely applicable and this is because it relies on fewer assumptions. However, it has a disadvantage compared with a specific test such as the product–moment correlation which is based on more restrictive assumptions. Spearman's test will be less powerful, when both are applied to a set of data which meet all the assumptions of the specific test. By less powerful we mean less likely to reject the null hypothesis. So, if the data are suitable for analysis by using $r$, then you should use it, because it is a more powerful test; this means that you stand a better chance of detecting a relationship if there is one.

The product–moment coefficient has an additional property which is useful when it comes to interpreting the strength of the correlation. We know from the way it works that the closer the points are to a straight line, the nearer $r$ is to 1. If $r$ is +1 or –1, then the points actually lie on a straight line, which means that a given value of $Y_1$ always occurs with just one particular value of $Y_2$ (and vice versa). We could then say that all of the variation in $Y_1$ is accounted for or determined by variation in $Y_2$. If the points are more scattered, a particular value of $Y_1$ can occur with a range of values of $Y_2$ (and vice versa). The fact that two sampling units can have the same value of $Y_1$, yet different values of $Y_2$, means that not all of the variation in one variable can be accounted for by variation in the other. In the extreme case, if there is no correlation then any value of one variable can be found with any value of the other variable. Then, none of the variation in $Y_1$ is accounted for by variation in $Y_2$.

A useful measure of the proportion of the variation in one variable which is accounted for by variation in the other is obtained by simply squaring the value of $r$; $r^2$ is known as the **coefficient of determination**. This property does not apply to Spearman's rank correlation coefficient.

Figure 8.6 illustrates the meaning of the coefficient of determination. In our example of arm and leg length (Fig. 8.6a) the points are not as scattered as they are in (Fig. 8.6b)

211

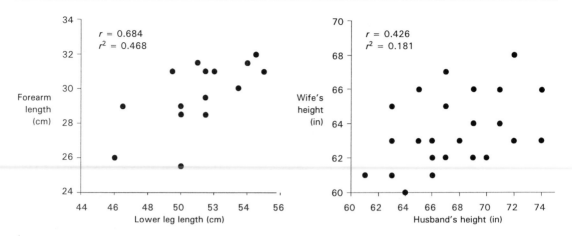

**Figure 8.6** The coefficient of determination: (a) forearm length and lower leg length in female students; (b) heights of husbands and wives.

which shows the relationship between the heights of a random sample of married partners. Both are significant positive correlations but in (a) $r^2 = 0.684^2 = 0.468$, which means that about 47% of the variation in arm length is accounted for by variation in leg length. In (b) the points are more widely scattered and $r^2 = 0.426^2 = 0.181$. Only about 18% of the variation in the height of the male partner is accounted for by variation in the height of the female partner.

## General points about correlation

Many of the general points which we examined in the context of association using $\chi^2$ are also valid when considering the interpretation of $r_s$ and $r$. First of all we have to take samples at random. Obviously if we choose material in such a way that there is a bias then we could end up creating a correlation in the sample when none existed in the population or alternatively obscuring a real correlation to the point where it was no longer detectable.

Second, the demonstration of a correlation does not imply causation. Figure 8.7 shows that the number of nesting storks in Denmark each year (from 1953 to 1977) is significantly positively correlated with the number of human babies born in that year. We would not suggest that this implied a causal connection between the number of storks and the number of births! Correlations can only identify relationships or patterns and suggest possible causal connections; they cannot tell us which explanation is the correct one.

The fact that orchids with a lot of purple spotting on their leaves also tend to have heavily marked flowers does not mean that changes in leaf spotting causes the changes in flower colour. This could be the explanation, but it is much more likely that variation in both characteristics is caused by variation in something else. There could be variation between plants in a gene which controls the amount of spotting and which is expressed in both leaves and flowers. Or it could be that the environment in which the plant has grown affects the amount of spotting of both leaves and flowers.

The same argument applies to the correlation between leg and arm length. Having long

212

legs does not cause long arms. The elongation of both arms and legs during development is likely to be controlled by the same growth hormone and if you have a lot of it then your limbs tend to be long. The fact that $r^2$ is only 0.47 means that long legs are not correlated perfectly with long arms, so that an individual with long legs can have short arms. This presumably happens because there are other genetic and environmental factors at work.

You might be wondering how this applies to the correlation between exercise and physical fitness. Surely it is the amount of exercise taken which causes the level of fitness, rather than the other way round. Although this seems to be the common-sense interpretation of the correlation, there are others. Perhaps people naturally differ in fitness because of genetic factors and those with a low fitness choose not to take much exercise because it is not a pleasant experience!

The only way we can verify that there is a causal connection is with an experiment designed to isolate any effects of the factor of interest. You may find it instructive to try and design some experiments to discriminate between some of the possible explanations for our examples. In many cases this cannot be done, particularly in areas such as medical research, evolution and ecology because we cannot manipulate the factor of interest. We cannot deliberately expose humans to different levels of a potentially toxic factor such as cigarette smoke, radioactivity or sunlight to see if it causes a particular disease but we can see if they are correlated. As a result we have to make use of naturally occurring variation in the factor of interest and tease out the likely causal connections.

Finally, there is another danger in the use of correlations. It is quite easy, particularly in ecological work, to describe each sampling unit with respect to many variables. For example, we might count the numbers of 15 different species in a series of quadrats and also make measurements on five environmental variables. We could then analyse the data by examining all the pairwise combinations of variables to see which were correlated. There

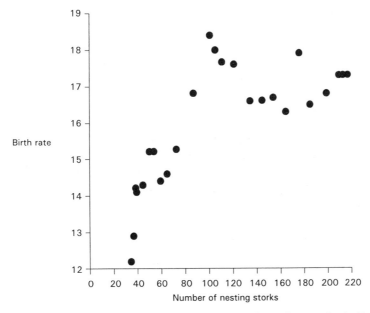

**Figure 8.7** Relationship between number of nesting storks in Denmark (1953–77) and birth rate of humans.

would be 190 different combinations which we could try and therein lies the danger. If we set the significance level at 5%, we know that even when the null hypothesis is true, in five tests out of every hundred, by chance the results will be so extreme that we will reject the null hypothesis. We will make a type I error. If we carry out 190 tests we would expect to find about 10 "significant" correlations even if there were no correlations between the variables in the population. One simple way to avoid this is to set a more stringent significance level such as 1%.

# 9

# Tests using two independent samples: are the two populations different?

We now move on to situations in which we are taking two samples, one from each of two populations. These situations will arise in simple experiments involving just two treatments which could be a test of the effect of two temperatures on enzyme activity, two fertilizer levels on plant growth, or two pollution concentrations on animal survival. Two-sample tests will also be needed in ecological studies. We might wish to compare the frequencies of males and females caught using two different methods, the numbers of a given species in two different habitats or the time spent by pollinators foraging at two different times of day. If the samples have been chosen at random as described in Chapter 4, then they will be independent samples. The meaning of this term will be examined more fully in the next chapter. In all these cases we would want to know whether the two populations from which the samples were taken were different. The basic problem is, of course, that even if the two populations are the same, the two samples are likely to be different because of sampling variation. As a consequence, the fact that the two samples are different does not necessarily mean that the two populations are different.

As you can see from these examples, any type of variable can be involved and we shall also have to be clear about the sort of differences we are interested in. A typical question would be whether the two populations differ in their location (e.g. in their median or mean), but we might also want to know whether two populations differ in their variability. Different variables and different questions will require different statistical tests but the underlying principle is always the same.

All the tests start with a null hypothesis which is that the two samples are drawn from two populations which are the same, which implies that the factor under investigation has no effect on the variable. Any differences between the samples are then merely due to sampling variation, that is, the chance events which occur during sampling. The alternative hypothesis will be that the two samples are drawn from different populations and this implies that the factor under investigation does have an effect on the variable. Our decision to reject the null hypothesis will be taken when the difference between the samples is judged to be too large to be likely to be due to sampling variation.

At first sight, it looks as though a comparison of two sample proportions or means could be achieved using confidence limits. Surely if the means of two samples have confidence limits which do not overlap then we can be fairly confident that the population means are different, while if the confidence limits do overlap then the population means could well be the same? Unfortunately, although the first of these statements is true, as we will see the second is not. The reasons for this are too complex to deal with here, but it means that we cannot tell if there is a significant difference between sample means merely by looking at their associated confidence limits.

# The chi-squared test for a difference between two population proportions

We shall begin our exploration of tests based on two samples with nominal variables which will involve the familiar $\chi^2$ test used in Chapter 8.

## Variables with two categories

Box 9.1 presents some data from an experiment on the possible effects of two different storage media on the viability of mononuclear lymphocytes, one of the types of cell present in human blood. Blood was stored for 24 hours in the two types of medium and centrifuged in such a way as to separate out the cells of interest. They were then stained with ethidium bromide/acridine orange which, under ultraviolet light, stains live cells green and dead cells orange. The number of each type in a sample was counted under the microscope.

As the data show, there is a higher proportion of live cells in the sample from medium A, but on its own, this tells us very little, because we know that proportions in both samples are subject to sampling variation. By extension from the one sample tests in Chapter 6, the question which needs answering is as follows: "Supposing that the proportion of cells surviving in the two treatments is *not* different how likely are we to get two samples of this size with proportions which are this (or more) different?".

The sampling unit is an individual cell and the variable (viability) has two alternative states, because each cell can be classified as either alive or dead. The variable is clearly a nominal variable. Data such as these can be organized in the form of a $2 \times 2$ contingency table and analysed using $X^2$ as the test statistic, in just the same way as we used it in the last chapter. The test goes under an apparently confusing variety of names. It is sometimes called a test of independence, a contingency test or a test for the homogeneity of proportions. We shall see a little later where these names come from and it will become apparent that they all amount to the same thing. (Incidentally, one of the suggestions for the design of the woodlouse experiment in Chapter 2 was to incorporate a control. We would then have had two samples and so the test to be described here would be appropriate.)

The reason that we can use a contingency table is because the data can be cross-classified. Each sampling unit, a cell, can be classified in terms of whether it is dead or alive and whether it is from medium A or medium B. The contingency table which results is set out in Box 9.1. Do not forget that the observed data must be entered as frequencies, not as proportions or percentages. We are going to use the quick formula to calculate $X^2$ and this does not require the expected frequencies to be calculated. If you want to calculate them, then you use the same logic as we used in Chapter 8, which is set out briefly below.

If the null hypothesis is true, then the two samples are drawn from identical populations. The best estimate of the proportion of live cells that we can get is simply to treat the two samples as though they were one, that is, to add them. The frequency of live cells is then 37 + 10 and we have the answer (47) in the row total. Dividing this total by the grand total (89) gives our estimate of the proportion alive as 0.5281. Likewise, the estimate of the

---

**BOX 9.1   CHI-SQUARED TEST FOR DIFFERENCE BETWEEN TWO POPULATION PROPORTIONS: SURVIVAL OF LYMPHOCYTES IN TWO DIFFERENT STORAGE MEDIA**

|                       | Medium A         | Medium B         |
|-----------------------|------------------|------------------|
| Number of cells alive | 37               | 10               |
| Number of cells dead  | 22               | 20               |
| Proportion alive      | 37/59 = 0.627    | 10/30 = 0.333    |

Null hypothesis:   Proportion of live cells is the same in the two media.

Alternative hypothesis:   Proportion of live cells not the same in the two media.

Significance level:   5%.

Contingency table of observed frequencies

|       | Medium A    | Medium B    |               |
|-------|-------------|-------------|---------------|
| Alive | 37 = $a$    | 10 = $b$    | 47 = $(a + b)$ |
| Dead  | 22 = $c$    | 20 = $d$    | 42 = $(c + d)$ |
|       | 59 = $n_1$  | 30 = $n_2$  | 89 = $N$      |

$$X^2 = \frac{\left[|ad - bc| - N/2\right]^2 \times N}{(a + b)(c + d)n_1 n_2}$$

$$= \frac{\left[|37 \times 20 - 10 \times 22| - 40.5\right]^2 \times 89}{47 \times 42 \times 59 \times 30}$$

$$= 6.888$$

Degrees of freedom = (Number of rows − 1)(Number of columns −1)
$$= 1$$

Critical value of $\chi^2$   = 3.841

Decision rule:   Reject null hypothesis if calculated value is equal to, or greater than, critical value.

Decision:   Highly significant evidence ($P < 0.01$) for difference in proportion surviving in the two media.

---

proportion of dead cells is $(22 + 20)/89 = 0.4719$. The total number of cells in sample A was 59, of which we expect 0.5281 to be live, so the expected frequency is simply $0.5821 \times 59 = 31.16$. Do not be tempted to round these figures off to the nearest whole number on the grounds that you could never expect to get 31.16 cells in a category. Don't forget that the term "expected frequency" actually refers to the mean number you would get if you repeatedly took samples of size 59 from the population. If you are going to calculate

217

$X^2$ using the long formula then you have to use all four categories (you cannot use only, say, live cells). There will be four sets of $(O - E)^2/E$ to calculate and you must use Yates' correction, because there is only one degree of freedom.

However you do the calculation the value of $X^2$ is 6.888, which now has to be compared with the appropriate critical value of $\chi^2$. This will depend on the degrees of freedom and the significance level. We already know the quick rule for finding the degrees of freedom (Box 8.1). The standard 5% significance level is adequate for this type of problem giving us a critical value of 3.841 and the decision rule is the same as before. Our calculated value exceeds the critical value and so we can reject the null hypothesis; in fact the probability associated with our calculated value (remember this is the probability of getting a value as big or bigger than 6.888) is less than 0.01.

We therefore have highly significant evidence that the storage medium does affect viability; the chance of getting a difference between two proportions which is as big or bigger than that observed is very small if the null hypothesis is true. The direction of the effect can be obtained by inspection of the observed and expected values. We can see that more than expected of the live cells are found in sample A while more than expected of the dead cells are found in sample B. Viability is higher in solution A.

Finally, why does this type of test have such variety of names? Contingency simply means chance, so it is a test of whether chance, i.e. sampling variation can account for the differences. It is a test of independence because the null hypothesis is that viability is independent of (i.e. is not affected by) the solution used. It is a test of the homogeneity of proportions because the null hypothesis is that the proportions of live cells are the same (i.e. homogeneous) in the two media.

## Variables with more than two categories

This $\chi^2$ test can be extended to situations in which a nominal variable falls into more than two categories. For example, the ABO blood group is a nominal variable in which there are four categories, the phenotypes A, B, AB and O. We might want to test whether two geographically separated groups of people had different frequencies of these phenotypes, by comparing the frequencies in random samples taken from two different localities. Our observed values would form a contingency table of four rows (the four phenotypes) and two columns (the two localities), giving a $4 \times 2$ table. The null hypothesis would be that the genotype frequencies were not affected by (i.e. were independent of) the locality from which the samples were taken.

I have chosen to illustrate this application with a behavioural example (Box 9.2) to emphasize the point that you can categorize a variable such as behaviour. It also illustrates the use of the alternative formula which is easy to do on a calculator. Use the appropriate marginal totals and the grand total to calculate a value of $E$ and then obtain $1/E$, which can be multiplied immediately by the corresponding value of $O^2$. Add the answer into the memory and repeat for all the cells. Subtracting $N$ from what you have accumulated in the memory gives the calculated value of $X^2$.

BOX 9.2   CHI-SQUARED TEST FOR DIFFERENCE BETWEEN PROPORTIONS IN TWO
POPULATIONS WHEN THERE ARE MORE THAN TWO CATEGORIES: ACTIVITY OF
FEMALE COMMON BLUE DAMSELFLIES (*ENALLAGMA CYATHIGERUM*) AT TWO
TIMES OF DAY

Null hypothesis:   Proportions of females in each behavioural category is the same at different
times of day.

Alternative hypothesis:   Proportions of females in each behavioural category is not the same
at each time of day.

Significance level:   5%.

Observed and (expected) frequencies

| Activity | Time of day | | Total |
| | Early morning | Midday | |
| --- | --- | --- | --- |
| Resting | 48 (30.3) | 39 (56.7) | 87 |
| Flying alone | 36 (41.7) | 84 (78.3) | 120 |
| In Tandem | 8 (12.2) | 27 (22.8) | 35 |
| Mating | 4 (7.7) | 18 (14.3) | 22 |
| Egg laying | 0 (4.2) | 12 (7.8) | 12 |
| Total | 96 (96.1) | 180 (179.9) | 276 |

$$x^2 = \sum \frac{O^2}{E} - N$$

$$= 304.422 - 276 = 28.422$$

Degrees of freedom = (Number of rows − 1)(Number of columns − 1)

$$= (5 - 1)(2 - 1)$$
$$= 4$$

Critical value of $\chi^2$ = 9.488

Decision rule:   Reject null hypothesis if calculated value is equal to, or greater than,
calculated value.

Decision:   Very highly significant evidence ($P < 0.001$) for differences in proportions in each
category at different times of day.

## A quick formula for a contingency table with two columns and a large number of rows

When the variable falls into many categories, the contingency table has a large number of rows and calculating the expected values becomes tedious. Fortunately, there is another quick formula we can use (Box 9.3).

We have already met these data on two samples of invertebrates collected from two adjacent but contrasting freshwater habitats, fast flowing shallow riffles and slower, deeper pools. Each organism caught has been identified and assigned to a taxonomic category and so again we have a nominal variable, this time with 14 categories, giving a $14 \times 2$ contingency table. The number in each cell is simply the observed frequency of organisms in that group in that habitat.

The null hypothesis is that the proportion of the individuals in each taxonomic group in the two habitats is the same, i.e. that the proportions are independent of the habitat. This means, for example, that oligochaetes would comprise the same proportion of the total organisms in riffles as they do in pools and that the same is true for the other families.

The meaning of the symbols in the quick formula are as follows: $a_1$, $b_1$, $a_2$, $b_2$, etc., are the observed frequencies in the two cells in rows 1, 2, etc. The totals for each row are $r_1$, $r_2$, etc.; $n_1$ and $n_2$ are the two column totals; and $N$ is the grand total. The easiest way to use this on a pocket calculator is to take the column with the smallest observed values as the column of values of $a_1$, $a_2$, etc. Clear the memory, square the first value of $a$ and divide it by $r_1$ and put it in the memory. Repeat this for the next value of $a$, dividing by $r_2$, and so on for all the values of $a$, adding the answers into the memory as you go. Then calculate $n_1^2/N$ and use the $M-$ key to subtract this from what is in the memory. Recall the result and multiply it by $N$, by $N$ again and then divide by $n_1$ and by $n_2$. The answer is $X^2$. The calculated value of $X^2$ is 176.8 and the degrees of freedom using the formula $(r-1)(c-1)$ are 13, giving a critical value of $\chi^2$ at the 5% significance level of 22.362.

This tells us that overall there is a very highly statistically significant difference in the composition of the fauna in the two habitats but, as in the other applications of $\chi^2$, we cannot say which specific categories of organisms have significantly different frequencies in the two habitats. As with the other short-cut formula you will not have the expected values to help you in the interpretation of a significant result but you can always use the percentages for this purpose. You may well have calculated these already (Box 4.2b). For example, oligochaetes make up a much smaller proportion of the animals found in the riffles than they do of those in the pools and the same is true for the Ephemeridae. The converse is the case for Elminthidae. These observations should form the basis of new, more specific, research hypotheses which will have to be tested by further sampling.

We also need to be careful about the biological interpretation of this analysis. The sampling method could be biased. Perhaps certain types of animal tend to escape capture in the pools because they can swim away from the net, whereas this does not happen in the riffles because the current is too strong. Such animals will seem to be less frequent in the pool samples. Remember too that the analysis is based on proportions. It does not tell us anything about the absolute population densities, that is, it does not show that oligochaetes, for example, are more numerous in absolute terms in pools. All it shows is that they make up a greater proportion of the animals in pools than of the animals in riffles. Overall, there could be many more animals (of all types) in riffles, so that oligochaetes, despite comprising a small proportion, could in absolute terms be more numerous than in pools. The fact that

**BOX 9.3    QUICK FORMULA FOR $\chi^2$ TEST FOR DIFFERENCE BETWEEN PROPORTIONS IN TWO POPULATIONS: COMPARISON OF INVERTEBRATE FAUNA OF FRESHWATER RIFFLES AND POOLS.**

Null hypothesis:    Proportion of organisms in each taxonomic group is the same in the two habitats.

Alternative hypothesis:    Proportion of organisms in each taxonomic group is not the same in the two habitats.

Significance level:    5%.

|  | Riffles | Pools |  |
|---|---|---|---|
| Oligochaeta | $38 = a_1$ | $141 = b_1$ | $179 = r_1$ |
| Hydrobiidae | $17 = a_2$ | $28 = b_2$ | $45 = r_2$ |
| Gammaridae | $11 = a_3$ | $31 = b_3$ | $42 = r_3$ |
| Chloroperlidae | $37 = a_4$ | $27 = b_4$ | $64 = r_4$ |
| Nemouridae | $3 = a_5$ | $24 = b_5$ | $27 = r_5$ |
| Baetidae | $63 = a_6$ | $15 = b_6$ | $78 = r_6$ |
| Ephemerellidae | $56 = a_7$ | $63 = b_7$ | $119 = r_7$ |
| Ephemeridae | $11 = a_8$ | $55 = b_8$ | $66 = r_8$ |
| Ecdyonuridae | $104 = a_9$ | $100 = b_9$ | $204 = r_9$ |
| Caenidae | $3 = a_{10}$ | $18 = b_{10}$ | $21 = r_{10}$ |
| Leptophlebidae | $32 = a_{11}$ | $76 = b_{11}$ | $108 = r_{11}$ |
| Leptoceridae | $33 = a_{12}$ | $72 = b_{12}$ | $105 = r_{12}$ |
| Rhyacophilidae | $63 = a_{13}$ | $52 = b_{13}$ | $115 = r_{13}$ |
| Elminthidae | $25 = a_{14}$ | $4 = b_{14}$ | $29 = r_{14}$ |
|  | $496 = n_1$ | $706 = n_2$ | $1202 = N$ |

Quick formula $X^2 = \dfrac{\left(\sum \dfrac{a^2}{r} - \dfrac{n_1^2}{N}\right)}{n_1 \times n_2} \times N^2 = \dfrac{\left(247.5 - \dfrac{496^2}{1202}\right)}{496 \times 706} \times 1202^2 = 176.8$

Calculated value of $X^2$ = 176.8

Degrees of freedom = (Number of rows − 1)(Number of columns − 1)
                              = (14 − 1)(2 − 1) = 13

Critical value of $\chi^2$  = 22.362

Decision rule:    Reject null hypothesis if calculated value is equal to, or greater than, critical value.

Decision:    Very highly significant evidence for alternative hypothesis ($P < 0.001$). Proportion of organisms in each taxonomic group is different in the two habitats.

more animals overall were caught in pools does not of course mean that animals are more numerous in pools! This would only be the case if sampling effort and efficiency were the same in the two habitats.

# The Mann–Whitney $U$ test for a difference in the median of two populations

This test is used on ordinal variables, the type of variable which results when we put the sampling units into ascending (or descending) order with respect to some property. This enables us to position each sampling unit within the order and to assign it a rank, which is a numerical value corresponding to its position. There are two reasons why we might need to do this. Some variables are difficult to measure or count in the first place, but the sampling units can be ranked. Alternatively, we may make counts or measurements initially and then convert the values to ranks. To test whether the two samples have been taken from populations which differ in location we shall have to use a completely new procedure, the Mann–Whitney $U$ test, which is specially designed for ordinal (ranked) variables.

## How does the test work?

We shall explore the basis of this test using some data on seed germination times under two different treatments. The purpose of the experiment was to see whether exposure of seeds of yellow rattle (*Rhinanthus major*) to low temperature prior to sowing had any effect on the time they took to germinate. A control group of seeds (chosen at random) was kept at room temperature, while the experimental group was kept at 5°C for 2 weeks. Both sets of seeds were then sown in identical conditions, in a fully randomized layout, and the order in which they germinated was noted.

To see how this test works, we are going to have to use unrealistically small sample sizes, so the results in Figure 9.1a are for three control and two experimental seeds. Once we have established the principle of the test, we shall go on to use it with more realistic sample sizes. We have the usual problem. Even if the treatment has no effect, so that the germination time is the same under both conditions, we know that by chance the two samples are likely to be different. The fact that the germination times in the two samples are different cannot, on its own, tell us whether or not they have been drawn from two different populations.

Although we are interested in whether the two samples were from populations which differ in their location, we start as usual by assuming that the null hypothesis is true. This means that we assume that the two populations have the same location. This is illustrated in Figure 9.1b by the two frequency distributions of germination time. So that we have a consistent and general terminology we shall label the experimental one A and the control one B. The variable, germination time, is a continuous variable and the two distributions are located in the same place. Note that I have drawn these as non-normal distributions in order to emphasize an important point about the test – it does not rely on the variable having any particular distribution. The only assumption is that distribution, whatever it is, has the same shape in both populations.

**(a) Results**

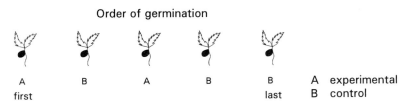

Order of germination

| A | B | A | B | B | A experimental |
|---|---|---|---|---|---|
| first | | | | last | B control |

**(b) Distribution of germination times under the null hypothesis**

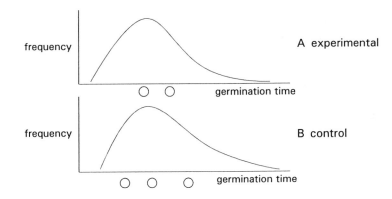

**(c) Two samples combined**

Hypothetical ⒝ ⒜⒝ ⒜ ⒝

Actual ⒜ ⒝ ⒜ ⒝ ⒝

**(d) The test statistic**

For sample A:
Number of B before each A = 0 + 1 = 1 = $U_A$

⒜ ⒝ ⒜ ⒝ ⒝

For sample B:
Number of A before each B = 1 + 2 + 2 = 5 = $U_B$

**Figure 9.1** The Mann-Whitney *U* test for a difference between the medians of two populations: germination times of treated (experimental) and untreated (control) seeds of yellow rattle (*Rhinanthus major*).

The dots under each distribution represent the germination times which we might have observed in a (random) sample of two observations from the experimental population and a (random) sample of three taken from the control population. Each sample should broadly mirror the distribution of the variable in the population and because the two populations are the same, that, is the frequency distribution of the variable is the same, the chances are that the two samples should be similar. The similarity between the two samples could be made to show up by combining the two samples as has been done in Figure 9.1c, while retaining the origin of each observation, which I have done by labelling the observations from both samples.

If the null hypothesis is true, then the observations from the two samples are likely to be well interspersed. They will show considerable overlap as in the first hypothetical sequence. On the other hand, a sequence such as the one we actually obtained in which there is little overlap is less likely if the null hypothesis is true. This line of reasoning suggests a possible test. We need a test statistic, which is a convenient numerical measure of the extent of the overlap of the values in the two samples and which has a known sampling distribution under the null hypothesis. We could then calculate the value of this statistic from our two samples and use the sampling distribution to find the associated probability (i.e. the probability of getting a value as or more extreme than the one which we have actually obtained).

It turns out that an appropriate measure of the overlap between the two samples is given by examining in turn each observation in one of the samples, and counting how many observations in the other sample come before it. Summing these values gives us a test statistic $U$ the sampling distribution of which is known.

We shall use our actual result to show how to obtain the value of $U$ (Fig. 9.4d). We first focus on each observation in sample A in turn and ask how many of the observations in sample B come before it. The first A observation has no observations from B which are before it in the order. The second A observation is preceded by one B observation. The total number of times that an observation from sample A is preceded by an observation from sample B is $0 + 1 = 1$. This quantity is called $U$. We could equally well have focused on sample B where we would have found that one of the observations is preceded by a single observation in A, the other two observations are both preceded by two observations in sample A. This would give us another value of $U$ ($= 5$). We need to distinguish between these two values, so we'll call the first one $U_A$ and the second one $U_B$.

## The sampling distribution of $U$

To make use of a calculated value of $U$, we need to know what all the other possible values of $U$ are and also their probabilities of occurring under the null hypothesis. As with all the other tests we have used, this knowledge of the sampling distribution is essential if we want to know whether our calculated value is a likely or unlikely event.

To understand the basis of this sampling distribution we shall need to represent the order of the germination times when the two samples are combined and we shall do this with our labelled circles. If the null hypothesis is true, then there are a number of possible results, which we can show as a series of sequences (Fig. 9.2a). At one extreme is the sequence in which both seeds in sample A germinate before those in sample B. At the other extreme is the reverse situation in which the two A seeds germinate after the three B seeds and in between are a number of other possible sequences. These comprise all the possible orders of germination that the five seeds could have.

Now if we calculate the value of $U_A$ (or $U_B$) for each sequence, we shall get all the possible values of $U_A$ (or $U_B$) which could occur with two samples of these sizes. You can see that $U_A$ and $U_B$ are complementary in that they just run in opposite directions. Note also that the size of $U$ relates to the direction of the difference between the two samples. If sample A has the longer germination times then $U_A$ is large, whereas if sample B has the larger obser-

**(a)** **Sequences possible under the null hypothesis**

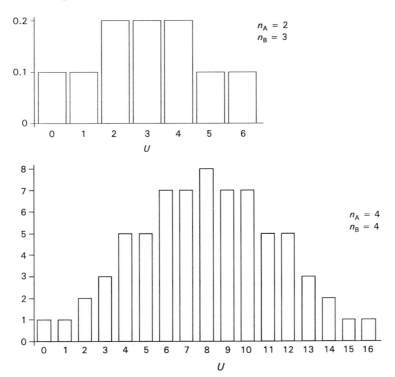

**(b)** **Sampling distributions of $U$**

**Figure 9.2** The Mann-Whitney $U$ test for a difference in medians of two populations: sampling distributions of $U$.

vations then $U_B$ is large. This becomes important later. For our present purpose we can use either so we shall use the values of $U_A$. They show two things of interest. First, very large or very small values occur when there is little overlap of the two samples, which is of course something that is more likely to occur when there is little overlap of the populations. This means that extreme values are evidence against the null hypothesis. Second, some of the values occur more frequently because some of the different sequences can yield the same value. Other values, actually the extreme values, are less frequent because they only arise from a single sequence.

To produce the sampling distribution of $U_A$, we need to know not only what values $U_A$ can take but also the probabilities of getting these different values, which is what we shall now determine. The key point here is that the 10 sequences of results are all equally likely to occur if the null hypothesis is true. The reason is as follows. If there is no difference in germination time in the two treatments then any of the three germination times in sample B could just as well have occurred in sample A; the two observations in sample A could therefore occupy any two positions in the sequence. Conversely, any of the germination times in sample A are just as likely to occur in sample B. This means that each of the sequences shown is equally likely to occur. However, this is not true of different values of $U_A$, because different sequences can give the same value of $U_A$.

The resulting sampling distribution of $U$ values is shown graphically (Fig. 9.2b) and we can see that it has a hump in the middle and two tails, although these are not well developed. It does, however, show the important point that extreme values of $U_A$, reflecting low overlap between samples, are less likely to occur than more moderate values. In this case, even the most extreme values (0 and 6) will occur by chance on one occasion out of every ten, which means they cannot be classed as unlikely events. As such, even the most marked difference between our two samples will not give a value of $U_A$ which would allow us to reject the null hypothesis at the 5% significance level. The reason is that we have used two very small sample sizes to illustrate the principle of the sampling distribution of $U$. To have used larger samples would have been tedious because it would have lead to a much larger number of sequences and you could well have got lost in the welter of possibilities and probabilities!

If we take larger sample sizes, such as 4 and 4, then the sampling distribution of $U$ has a more pronounced hump and tails (Fig. 9.2b). Now there are some extreme values of $U$ which have an associated probability of occurrence under the null hypothesis of 5% or less and so we can set critical values as we have done before. Needless to say, you will not have to go through this procedure to find critical values because, once again, the mathematicians have done the hard work! They have described these distributions for different sample sizes and determined the critical values of $U$ for different significance levels for one-tailed and two-tailed tests. Critical values of $U$ are given in Table A12.

Some final points before we put this test into action. Note that, although the test does not require that the variable has any specific distribution, it is still based on an assumption about the shape of the distribution. The argument which we have just used to obtain the sampling distribution of $U$ depends on the distribution of the variable having the same shape in the two populations. If the two distributions had the same location but different shapes, as is shown in Figure 9.3, then not all sequences will be equally likely. For example, the order AABBBBAA would be more likely than the order BBAAAABB and under these circumstances we would not have been able to produce the sampling distribution of $U$.

The test is a test for differences in location or average. However, it uses information on the order of the observations, not the actual numerical values of the observations. As we saw in Chapter 4, the appropriate measure of location for an ordinal variable is the median, so this test is a test for a difference in median, not mean.

There is also an important point about the relationship between experimental design and statistical testing which emerges from this, namely that a test may require samples of a minimum size. The nature of the variable and its distribution determines the type of statisti-

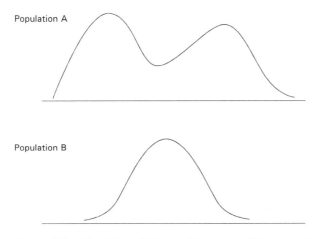

**Figure 9.3** The Mann-Whitney $U$ test for differences in medians of two populations: the assumption that both distributions have the same shape.

cal test which is applicable and this in turn may determine minimum permissible sample sizes. We could do an experiment with samples of sizes 2 and 3, indeed in terms of cost and time it might be quite a reasonable course of action. However it would not be worth doing because we could not possibly detect a difference between the two populations if we had to use the Mann–Whitney $U$ test. The moral here is that you ought to decide what test you are going to do before you carry out the practical work so as to ensure that your sample sizes will meet any such requirement.

I have spent some time explaining what $U$ is and the way in which it is calculated, because this shows us how and why the test works and what its limitations are. In practice there is a quicker and easier way to calculate $U$ which is particularly useful when sample sizes are large, so let's move on and carry out a Mann–Whitney $U$ test. We shall also see how to use the tables of critical values.

## An easier way of calculating $U$ and an actual test

Calculating $U$ in the way described above is not very convenient and it is easy to make mistakes, especially when the sample sizes are large. If you are carrying out this test by hand, I suggest you use a much quicker and easier way to calculate $U$. It also makes it easier to deal with another situation that may occur, namely the presence of two or more observations with the same value.

In Box 9.4 there is a larger set of data from the germination experiment. The procedure is the same to begin with, in that we combine the observations from the two samples and arrange them in order, whilst at the same time retaining the identification of each observation so that we can tell which sample it came from. Then we allocate a number to each observation that specifies its position in the ordered sequence, that is, we assign it a rank. It makes sense to give the lowest rank to the smallest observation and, although this is not essential, it can make things easier at a later stage. The first observation is given the rank of 1, the second observation a rank of 2, etc. The rank of the last observation must be equal to $n_A + n_B$, so use this as a check that you have ranked the items correctly.

**BOX 9.4** **MANN–WHITNEY $U$ TEST FOR DIFFERENCES IN MEDIAN: GERMINATION TIME IN COLD-TREATED AND UNTREATED SEEDS OF YELLOW RATTLE (*RHINANTHUS MAJOR*).**

Null hypothesis:   Treated (A) and untreated (B) seeds have same median germination time.

Alternative hypothesis:   Treated (A) and untreated (B) seeds do not have same median germination time.

Significance level:   5%, two tailed.

Order of germination:

Sample A                                 $n_A = 6$

Sample B                                 $n_B = 6$

Combined   A   A   A   B   A   B   A   B   B   A   B   B     $N = 12$

Ranks:  A   1   2   3       5       7           10           $\Sigma R_A = 28$
        B               4       6       8   9       11   12   $\Sigma R_B = 50$

Check:   $$\Sigma R_A + \Sigma R_B = \frac{N(N+1)}{2}$$

$$28 + 50 = 78 = \frac{12 \times 13}{2}$$

Calculate: $U_A = \Sigma R_A - n_A (n_A + 1)/2 = 28 - 21 = 7$

$U_B = \Sigma R_B - n_B (n_B + 1)/2 = 50 - 21 = 29$

Check:   $U_A = n_A n_B - U_B$

$7 = 36 - 29$

Test statistic:   Larger calculated value of $U$, i.e. $U_B = 29$

Critical value:   31

Decision rule:   Reject null hypothesis if (larger) calculated value is equal to, or greater than, critical value.

Decision:   No evidence for alternative hypothesis ($P < 0.1$).

We then add the ranks for sample A to get a value $\Sigma R_A$ the sum of the ranks in sample A. Alternatively, we could obtain $\Sigma R_B$, the sum of the ranks for sample B. In fact, although we only need one of these values it is a good idea to calculate both because we can then do a quick check on the ranking. The sum of the two values should equal $N(N+1)/2$. We then calculate a value of $U$, by putting one of the values of $\Sigma R$ into the equation. Remember that there are always two values of $U$ ($U_A$ and $U_B$) which we could calculate from any given set of data and these two values are related by the formula $U_A = n_A n_B - U_B$. Which of these two values is used as the test statistic depends on whether we are doing a one-tailed or a two-tailed test, so you will need to calculate both.

For a two-tailed test where the direction of the difference has not been predicted, we use whichever of these two values is the larger. The decision rule for this test is as follows; if the (larger) calculated value of $U$ is equal to, or greater than, the critical value for the chosen significance level, then reject the null hypothesis. The table of critical values of $U$ (Table A12) is set out in much the same way as the other tables, the only difference being that there are a number of subtables. If both sample sizes are equal then either can be used to locate the correct subtable, if not then choose the subtable corresponding to the larger of the two sample sizes. Then locate the other sample size in the left-hand margin. You should be able to find that the critical value at the 5% significance level, for a two-tailed test is 31. The calculated value is 29, so we have no evidence against the null hypothesis. A word of warning: some textbooks, sets of tables and computer programs use the smaller calculated value of $U$ and so will have different critical values and a different decision rule. I have chosen to use the larger value because it means that the decision rule is in line with that of most other tests.

## What to do with tied observations

It is not uncommon to have a situation in which two or more of the sampling units are so similar that we cannot decide what their order should be. We came across this situation in the last chapter. It is particularly likely to happen when the variable is complex and difficult to observe and when our observational method is rather crude as is the case in the example in Box 9.5. This experiment was set up to discover whether the activity of an intertidal gastropod, the purple top-shell (*Gibbula umbilicalis*) was affected by salinity. Individuals were allocated at random to two different salinities (27.5 and 35‰) and their activity after 15 min assessed on a scale of 0 (operculum shut) to 5 (= moving with the body extended).

Using this scale we can put the individuals in order, but because it is a relatively crude measure several individuals have ended up with the tied scores. This makes no difference if the tied observations are in the same sample but it does complicate matters when they are in different samples. The solution to assigning ranks to tied observations is simply to give each one the average of the ranks occupied. For example, there are four snails with the behaviour 1, sharing ranks 3 to 6. The average of these four ranks is $(3+4+5+6)/4 = 4.5$, so this is the rank assigned to each of the four observations. The next two observations share ranks 7 and 8 and so each get rank 7.5. It is easy to lose your way when ranking tied observations, so I suggest you set the data out clearly and use the check mentioned earlier. Once the ranking is done, proceed as before.

## BOX 9.5 THE MANN–WHITNEY $U$ TEST FOR DIFFERENCES IN MEDIAN WITH TIED OBERVATIONS: ACTIVITY OF PURPLE TOP-SHELLS (*GIBBULA UMBILICALIS*) IN TWO SALINITIES OF SEAWATER

Null hypothesis: Median level of activity the same in the two different salinities.

Alternative hypothesis: Median level of activity not the same in the two different salinities.

Significance level: 5%, two tailed.

Activity assessed on a scale of: 0 (operculum shut, shell upside down) to 5 (snail moving with foot extended).

| Activity scores | | | | | | | | | |
|---|---|---|---|---|---|---|---|---|---|
| Salinity A (27.5‰) | 0, | 2, | 1, | 1, | 0, | 1, | 3, | | $n_A = 7$ |
| Salinity B (35‰) | 4, | 1, | 4, | 2, | 5, | 4, | 5, | 5 | $n_B = 8$ |
| | | | | | | | | | $N = 15$ |

| | | | | | | | | | | | | | | | |
|---|---|---|---|---|---|---|---|---|---|---|---|---|---|---|---|
| Scores in order | 0 | 0 | 1 | 1 | 1 | 1 | 2 | 2 | 3 | 4 | 4 | 4 | 5 | 5 | 5 |
| Salinity | A | A | A | A | A | B | A | B | A | B | B | B | B | B | B |
| Ranks | 1 | 2 | 3 | 4 | 5 | 6 | 7 | 8 | 9 | 10 | 11 | 12 | 13 | 14 | 15 |

| Average of tied ranks | | | | 4.5 | | | 7.5 | | |
|---|---|---|---|---|---|---|---|---|---|

| Salinity A | 1 | 2 | 4.5 | 4.5 | 4.5 | | 7.5 | | 9 | | | | | | |
|---|---|---|---|---|---|---|---|---|---|---|---|---|---|---|---|
| Salinity B | | | | | | 4.5 | | 7.5 | | 10 | 11 | 12 | 13 | 14 | 15 |

$\Sigma R_A = 33 \qquad \Sigma R_B = 87$

Check:
$\Sigma R_A + \Sigma R_B = N(N + 1)/2$
$33 + 87 = 120 = (15 \times 16)/2$

Calculate: $U_A = \Sigma R_A - n_A(n_A + 1)/2 = 33 - 28 = 5$

Check: $U_B = n_A n_B - U_A = 56 - 5 = 51$

Test statistic: Larger calculated value of $U = U_B = 51$

Critical value: 41

Decision rule: Reject null hypothesis if calculated value is equal to, or larger than, critical value.

Decision: Significant evidence for alternative hypothesis ($P < 0.01$), level of activity higher in higher salinity.

## Converting counts and measurements to ranks and doing a one-tailed test

The Mann–Whitney *U* test can also be used with count and measurement variables, by simply combining the observations from the two samples into an ordered sequence and then replacing the actual counts or measurements by ranks. We shall see later that there is

---

**BOX 9.6** **THE MANN–WITNEY *U* TEST FOR DIFFERENCES IN MEDIAN USING MEASUREMENTS CONVERTED TO RANKS AND FOR A ONE-TAILED ALTERNATIVE HYPOTHESIS: UREA CONCENTRATION IN URINE OF DRINKERS AND ABSTAINERS**

Null hypothesis:   Median urea concentration the same in urine of drinkers and abstainers.

Alternative hypothesis:   Median urea concentration higher in abstainers.

Significance level:   5%, one tailed.

| Drinkers (A) urea conc. (mg/ml) | Rank, $R_A$ | Abstainers (B) urea conc. (mg/ml) | Rank, $R_B$ | |
|---|---|---|---|---|
| 9.0 | 3 | 6.8 | 1 | |
| 8.6 | 2 | 10.0 | 5 | |
| 9.3 | 4 | 11.5 | 8 | |
| 24.0 | 19 | 31.8 | 20 | |
| 23.0 | 18 | 18.0 | 13 | |
| 10.3 | 6 | 20.2 | 16 | |
| 18.5 | 14 | 38.1 | 22 | |
| 21.5 | 17 | 36.0 | 21 | |
| 18.7 | 15 | 17.2 | 12 | |
| 11.2 | 7 | 15.0 | 11 | |
| 12.7 | 9 | | | $n_A = 12$ |
| 13.0 | 10 | | | $n_B = 10$ |
| $\Sigma R_A = 124$ | | $\Sigma R_B = 129$ | | $N = 22$ |

Check:   $\Sigma R_A + \Sigma R_B = N(N + 1)/2$
$124 + 129 = 253 = 22 \times 23/2$

Calculate:   $U_A = R_A - n_A(n_A + 1)/2 = 124 - 78 = 46$
$U_B = n_A n_B - U_A = 120 - 46 = 74$

Test statistic:   For a one-tailed test the value of *U* which should be the larger under the alternative hypothesis, i.e. $U_B$.

Critical value:   86

Decision rule:   Reject null hypothesis if the value of *U* which should be the larger is equal to, or larger than the critical value.

Decision:   No evidence for the alternative hypothesis.

---

another test which we could use on counts and measurements, but it requires that the variable follows a normal distribution. The advantage of the $U$ test is that it does not make this assumption. As a result, it is widely used to test for differences when there are reasons for thinking the distributions might not be normal or when nothing is known about the distributions.

Box 9.6 illustrates this with some data on urea excretion in humans and also shows you how to deal with one-tailed tests. Urea is a by-product of protein metabolism and we would expect that the total amount excreted in the urine would not be affected by the volume of urine produced. If copious urine is produced because of the need to regulate the body fluid level, then the urea concentration should be reduced. Is this what happens? Students were allocated at random to two groups: abstainers (no fluid in the preceding 3-hr period) and drinkers (of 1 litre of water) in the same period. Urea concentrations in the urine were then measured. This will be a one-tailed test because we have predicted, that, if anything, the concentration should be lower in drinkers. With a one-tailed test, we need to be careful about which value of $U$ we use as our test statistic because we cannot simply choose the one which is the largest. We have to use the value of $U$ *which should be the largest* if the alternative hypothesis is correct.

In our example, the alternative hypothesis is that the concentration will be higher in B than in A. If this were the case then the B observations should tend to come after the A observations so $U_B$ (the number of times As come before Bs) should be large and $U_A$ should be small. So we use $U_B$ as the test statistic.

One-tailed tests are not that common but it is easy to use the wrong calculated value. The following may help. Label the samples A and B. Always rank the observations from small to large which means giving the smallest rank to the first observation. If you do this then the sample which is predicted to have the larger observations should come later in the order. The value of $U$ which is predicted to be the larger is the one with the same label (A or B) as the sample which is predicted to have the larger observations. Note: this only works if you have given the smallest observation the lowest rank.

## The z test for a difference in means of two populations using large samples

Large samples typically arise in ecological investigations, particularly when individual organisms have been measured, so we will use some data on body length for our example. These come from a study of two (biological) populations of the estuarine isopod *Sphaeroma rugicauda*, one from the River Deben the other from the River Colne. A sample of swimming animals was collected from each of the two populations using a standardised netting technique, at the same time every month and 100 of these individuals, chosen at random, were measured. A set of data for the two samples taken in November is given in the legend to Figure 9.4. We can see that the mean sizes of animals in the two samples are different and we know that there are two alternative explanations for this. Either mean sizes in the populations are the same and the differences between the samples are due to sampling variation or mean sizes in the two populations are different. How do we go about distinguishing between these two explanations?

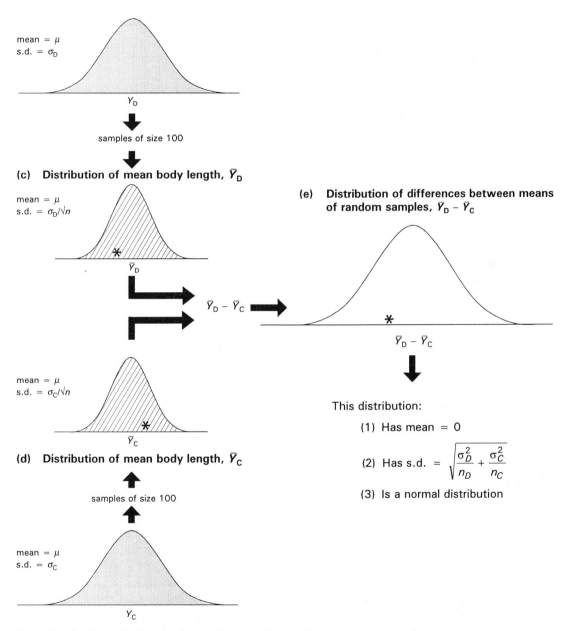

**(a)** Distribution of body length in Deben population, $Y_D$

mean $= \mu$
s.d. $= \sigma_D$

$Y_D$

samples of size 100

**(c)** Distribution of mean body length, $\bar{Y}_D$

mean $= \mu$
s.d. $= \sigma_D/\sqrt{n}$

$\bar{Y}_D$

$\bar{Y}_D - \bar{Y}_C$

**(e)** Distribution of differences between means of random samples, $\bar{Y}_D - \bar{Y}_C$

$\bar{Y}_D - \bar{Y}_C$

mean $= \mu$
s.d. $= \sigma_C/\sqrt{n}$

$\bar{Y}_C$

**(d)** Distribution of mean body length, $\bar{Y}_C$

samples of size 100

mean $= \mu$
s.d. $= \sigma_C$

$Y_C$

This distribution:

(1) Has mean $= 0$

(2) Has s.d. $= \sqrt{\dfrac{\sigma_D^2}{n_D} + \dfrac{\sigma_C^2}{n_C}}$

(3) Is a normal distribution

**(b)** Distribution of body length in Colne population, $Y_C$

**Figure 9.4** The z-test for a difference in means of two populations using large samples. River Deben, mean, $\bar{Y}_D$ = 4.37 mm, standard deviation, $s_D$ = 0.837 mm; River Colne, mean, $\bar{Y}_C$ = 3.37 mm, standard deviation, $s_C$ = 0.739 mm.

233

## How does the test work?

Our first step, as always, is to assume that the null hypothesis is true, in which case the mean of the Deben population will be the same as the mean of the Colne population (i.e. $\mu_{\text{Deben}} = \mu_{\text{Colne}}$). This is illustrated by the shaded curves in Figure 9.4a & b, which represent the distribution of body length in the two populations. These distributions will also have standard deviations $\sigma_{\text{Deben}}$ and $\sigma_{\text{Colne}}$, respectively.

We know from Chapter 7 that if we take samples from a population, the distribution of sample means will have a mean equal to the population mean, although it will be less widely spread than the distribution of the variable itself. The standard deviation of the distribution, which is called the standard error, will be $\sigma/\sqrt{n}$ . The distributions of means of samples of size 100 from the Deben and from the Colne populations can be represented diagrammatically by the hatched curves (Fig. 9.4c & d). They both have the same mean and are less widely spread than the original distributions.

The easiest way to explain the next stage is to use another thought experiment, which is similar to the one we used to get the distribution of sample means in Chapter 7. Imagine taking one mean value $\bar{Y}_D$ (D stands for Deben) at random from the distribution of sample means (Fig. 9.4c) as indicated by the *. This is, of course, equivalent to taking one sample of size 100 at random from the Deben population and calculating its mean. Then repeat this with one mean value $\bar{Y}_C$ taken at random from the distribution (d), also indicated by the *. Now calculate the difference between these two imaginary means $\bar{Y}_D - \bar{Y}_C$. In this case, by chance, $\bar{Y}_C$ is larger than $\bar{Y}_D$, so the difference is negative. This difference is just one of many possible differences between the means of two random samples; obviously if we take a second pair of means at random they are likely to be different from the first pair and to yield a different value of $\bar{Y}_D - \bar{Y}_C$.

Now if we imagine repeating this for all the possible pairs of random samples, we could construct another frequency distribution, which will give us all the possible differences between two sample means and the frequencies with which they occur. This is shown in Figure 9.4e. Note that the horizontal scale is now values of $\bar{Y}_D - \bar{Y}_C$, not values of $\bar{Y}_D$ or $\bar{Y}_C$, and remember that this is all in the mind at this stage! If you don't see where this is all leading or how it can possibly be relevant because in practice you only have one sample from each population, just bear with me.

Having done this, what could we say about the shape, location and spread of this distribution of the difference between the means of the two samples? In short it will be symmetrical, a normal distribution, have a mean of zero and be more widely spread than either of the distributions of sample means. A partial explanation of some of these points follows in the next paragraph, if you are interested.

The symmetry will arise because the two distributions of sample means are symmetrical. This means that for any pair of samples in which $\bar{Y}_D$ is a given amount larger than $\bar{Y}_C$, there will always be a corresponding pair for which $\bar{Y}_D$ is smaller by the same amount. The mean of zero, can basically be explained as follows. If the two population means are the same, then the most likely combination of a sample from each is a pair with the same mean which would give zero as the most likely difference. Because the distribution is symmetrical, zero would also be the mean. This isn't totally convincing and you may simply have to accept that it can be proved mathematically. This is also the only way in which it can be shown that the distribution will be a normal distribution and that it will have wider spread than either of the distributions of sample means. However, we can see the logic of the last point with a simple explanation, using the range rather than the standard deviation. Suppose the maximum and minimum possible values of $\bar{Y}_D$ are 1 mm and 5 mm, and that the same is the case for $\bar{Y}_C$. This gives a range of 4 mm for the distribution of sample means. If we happened to take one sample with the largest possible value of $\bar{Y}_D$ and the other with the smallest possible value of $\bar{Y}_C$ then the difference would be +4 mm, the largest possible. If, by chance, the sample means were the other way round then the difference would be the smallest possible (i.e. −4 mm). In other words the range of the distribution of the difference in sample means would be 8 mm, that is greater than the ranges for each of the distributions of the sample means.

In fact it can be shown that the standard deviation of this distribution is given by the formula in Figure 9.4. The variable which we are dealing with now (the difference between the two sample means) is a statistic, just like a sample mean, so the standard deviation of this distribution is properly called a standard error. It is the **standard error of the difference between $\bar{Y}_D$ and $\bar{Y}_C$.**

This distribution, reproduced in Box 9.7a, is another sampling distribution under the null hypothesis. It tells us what all the possible differences are and one thing is immediately obvious; very large positive and negative differences are unlikely. Because we know exactly

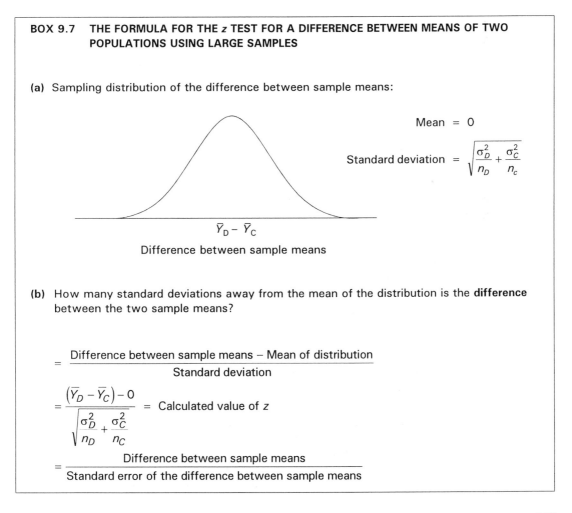

**BOX 9.7    THE FORMULA FOR THE z TEST FOR A DIFFERENCE BETWEEN MEANS OF TWO POPULATIONS USING LARGE SAMPLES**

**(a)** Sampling distribution of the difference between sample means:

Mean = 0

Standard deviation $= \sqrt{\dfrac{\sigma_D^2}{n_D} + \dfrac{\sigma_C^2}{n_c}}$

$\bar{Y}_D - \bar{Y}_C$

Difference between sample means

**(b)** How many standard deviations away from the mean of the distribution is the **difference** between the two sample means?

$$= \frac{\text{Difference between sample means} - \text{Mean of distribution}}{\text{Standard deviation}}$$

$$= \frac{\left(\bar{Y}_D - \bar{Y}_C\right) - 0}{\sqrt{\dfrac{\sigma_D^2}{n_D} + \dfrac{\sigma_C^2}{n_C}}} = \text{Calculated value of } z$$

$$= \frac{\text{Difference between sample means}}{\text{Standard error of the difference between sample means}}$$

what the shape of the distribution is, we can use it to make statements about probabilities and to answer the following question: "What is the probability of obtaining a difference between two sample means which is as extreme or more extreme than the one which we have found?". If the difference between the means is so extreme that the associated probability is lower than our chosen significance level, then this would give us the required evidence for the alternative hypothesis.

To do this we need an answer to a familiar question, which is: "How many standard deviations is our observation away from the mean?". The appropriate formula is given in Box 9.7b. You will see that, despite its apparent complexity, it has a familiar form. The top line is asking the question "How far away is our observation from the mean of the distribution?". Our observation now is a difference between two sample means and the mean of the distribution is zero (under the null hypothesis). On the bottom line we have the standard deviation of the distribution. So dividing one by the other converts the difference on the top line into a difference in units of standard deviation. In other words, the formula is merely standardising the difference between the observation and the mean of the distribution, in exactly the same way as in Chapter 6. The answer is a value of $z$, which we can compare with an appropriate critical value. As the last line shows, the standard deviation in this case is called the standard error for the reason given previously. Also we can simplify the equation by removing the zero on the top line, which will obviously not affect the numerical value.

## Calculating the value of $z$

Now all this might seem very plausible in theory, but how do we get the information to produce this distribution, given that we do not take every possible pair of samples? In fact, we only have two samples at our disposal. This is where we get to the cunning bit, because if you look at the formula you will see that the two samples contain all the information we need. Obviously we know what $\bar{Y}_D$ and $\bar{Y}_C$ are; they are the two sample means and we also know $n_D$ and $n_C$, the sizes of the two samples. Finally, because the samples are large, we know that the sample standard deviations $s_D$ and $s_C$ are very good estimates of the population standard deviations $\sigma_D$ and $\sigma_C$ and can be used in their place. We therefore have all the information to answer the question

So, after all that theory, let us get back to the isopods (Box 9.8). Is there any evidence that mean sizes in the two populations are different? We shall set the significance level at 5% and the test will be two-tailed, because we had no idea, prior to taking the samples, of the direction of any difference. It is important that this decision is made before the data are collected, we would not be justified in deciding to do a one-tailed test on the grounds that one sample mean is larger than the other. We have already met (Table A5) the critical values of $z$ which cut off 2.5% of the distribution in each tail; they are the familiar $-1.96$ and $+1.96$. (The equation in Box 9.7 has been simplified, as suggested, by removing the zero on the top line.)

The calculated value of $z$ can be negative or positive, depending on which sample has the bigger mean and which way round we label them. This introduces a minor complication because the critical values are always given as positive values. There are a number of ways round this. I suggest that if you are doing a two-tailed test (as we are doing here), you put the larger mean first and take the smaller mean away from it so that the value of $z$ will be positive. One-tailed tests are dealt with in the next section.

---

**BOX 9.8    THE z TEST FOR A DIFFERENCE IN THE MEANS OF TWO POPULATIONS USING LARGE SAMPLES: BODY SIZE OF THE ISOPOD *SPHAEROMA RUGICAUDA* FROM TWO LOCATIONS**


Null hypothesis:    Mean size of animals at Deben is the same as mean size of animals at Colne.

Alternative hypothesis:    Mean size of animals at Deben is not the same as mean size of animals at Colne.

Significance level:    5%, two tailed.

|  | Deben | Colne |
|---|---|---|
| Mean (mm) | $\bar{Y}_D = 4.37$ | $\bar{Y}_C = 3.37$ |
| Standard deviation (mm) | $s_D = 0.837$ | $s_C = 0.739$ |
| Variance | $s_D^2 = 0.700$ | $s_C^2 = 0.546$ |
| Sample size | $n_D = 100$ | $n_C = 100$ |

$$z = \frac{\bar{Y}_D - \bar{Y}_C}{\sqrt{\dfrac{s_D^2}{n_D} + \dfrac{s_C^2}{n_C}}} = \frac{4.37 - 3.37}{\sqrt{\dfrac{0.700}{100} + \dfrac{0.546}{100}}} = 8.959$$

Critical value of z:    1.96.

Decision rule:    Reject null hypothesis if calculated value is equal to, or larger than, critical value.

Decision:    Virtually conclusive evidence for alternative hypothesis. Animals larger at Deben ($P < 0.001$).

---

Putting the data into the equation gives $z = 8.959$, a value much larger than the critical value of 1.96. It is larger, too, than the critical value at the 0.1% level, so this would count as a very highly significant result. In fact, the calculated value is so large that the associated probability is vanishingly small and so there is almost no doubt at all that the alternative hypothesis is correct. The mean length of animals from the Deben is almost certainly greater than that of animals from the Colne.

Remember though that we are making an inference about a statistical population, the nature of which is partly determined by the method used. The samples were taken by net and although the technique was identical at the two sites there is still potential for bias. Suppose that the animals are the same size at both sites, but that there is a difference in some environmental factor (e.g. vegetation, temperature and salinity) which influences the activity patterns of animals of different sizes in such a way as to make it less easy to catch small animals at the Deben. On the basis of our samples of swimming animals, those at the Deben will appear to be larger than those at the Colne. One way to check this is to take a sample by another method, which does not depend on the behaviour of the organism. Random cores

237

of turf were taken from the saltmarsh at the same time and the animals washed out in the laboratory and measured. There was no evidence that these animals were different in size from the net-caught animals.

## One-tailed tests

Of course none of this tells us why animals at the Deben are larger than those at the Colne. Is it something to do with a difference in the environment or is it a genetically determined difference? We could discriminate experimentally between these two alternative hypotheses by growing animals in altered environments, either in the laboratory, or better still in the field.

You will find another set of data in Box 9.9 from part of a field experiment in which 20 cages were set up at the Colne site. Juvenile animals of the same size were placed in cages and additional food was added to 10 of the cages chosen at random. Animals were measured three months later. We can analyse these results by using the $z$ test, but this will be a one-tailed test because additional food should if anything increase body size. In this case we arrange the calculation so that the mean which is predicted to be the larger comes first and subtract from this the mean which should be smaller. If there is a difference between the two sample means in the predicted direction then the calculated value of $z$ should be positive and

---

**BOX 9.9 THE $z$ TEST FOR A DIFFERENCE IN THE MEANS OF TWO POPULATIONS USING LARGE SAMPLES WITH A ONE-TAILED ALTERNATIVE HYPOTHESIS: FIELD EXPERIMENT ON THE EFFECT ON MEAN BODY SIZE OF ADDING EXTRA FOOD TO CAGED ISOPODS**

Null hypothesis:   Mean size of animals the same with and without extra food.

Alternative hypothesis:   Mean size of animals with extra food larger than mean size of animals without extra food.

Siginificance level:   5%, one tailed.

|  | Control | Extra food |
|---|---|---|
| $\bar{Y}$ (mm) | 3.02 | 3.42 |
| $s^2$ | 0.72 | 0.84 |
| $n$ | 103 | 97 |

$$z = \frac{3.42 - 3.02}{\sqrt{0.0157}} = \frac{0.40}{0.125} = 3.197$$

Critical value of $z = 1.6449$

Decision rule:   Reject null hypothesis if calculated value is equal to, or greater than, critical value.

Decision:   Very highly significant evidence for alternative hypothesis. Additional food increases mean size ($P < 0.001$).

---

can be compared to the (positive) critical value in the normal way. If the difference is in the opposite direction to that predicted, then the calculated value of $z$ will be negative. It cannot be significant however extreme it is and so there is no need to use the tables of critical values.

## The *F* test for a difference in the variances of two populations

Although all of the two sample tests so far have been tests for differences in location, we now have to make a slight detour to look at a test for differences in variability. One reason for this is that the simplest test for differences in mean when we only have small samples is based on the assumption that the two populations have the same variance. We shall need to be able to check that our data meet this assumption. The second reason is that sometimes we are only interested in whether the two populations are equally variable, as shown in the following example.

When we were talking about experimental variation in Chapter 1, I gave a set of phosphate concentrations measured by a first year student. All the determinations were done on the same solution and there was a lot of variability due to variation in technique. At the same time and using the same solution and the same spectrophotometer, a technician made nine measurements of the concentration. Both sets of results are given in Box 9.10. If we look at the two variances it is clear that the second sample is less variable than the first. Why? It could be that the student worked just as carefully as the technician, in which case the two samples have been taken from populations which are equally variable and the differences in variability are due to sampling variation. Alternatively, there could be a real difference in variability, related to differences in how carefully the practical work was done.

This problem is nothing new. We know that we can get an estimate of the population variance ($\sigma^2$) from the variance of a sample ($s^2$). The value of $s^2$ is of course subject to sampling variation, that's why it is only an estimate. This means that if we have two populations which have the same variance and take one sample from each, the chances are that they will yield different values of $s^2$. Just because two samples have different variances, it does not follow that they have been taken from populations which differ in variance. Clearly though, the bigger the difference between two sample variances, the less likely it is that they have come from populations with the same variance.

In order to test this we need first to be able to describe the extent of chance variation in sample variances which can occur if two populations have the same variances. The way to do this is to find a test statistic which utilises the variances and which has a known sampling distribution. It turns out that the ratio of the two sample variances $s_1^2/s_2^2$, called the **variance ratio**, or ***F*** for short, has the desired properties. The shapes of two typical sampling distributions of $F$ are shown in Figure 9.5.

These distributions are what you would get if the variances were the same in the two populations and you were to take every possible pair of samples (one from each population) and for each pair calculate the ratio of their variances. They are based on the assumption that the variable follows a normal distribution. Needless to say that is not how the curves were produced because the distributions can be mathematically described. Only two distributions are shown here but there is a separate $F$ distribution for each combination of sample sizes. This is because the extent of chance variation in a sample variance is affected by the sample size and we are dealing with two samples which may be of different sizes. In point of fact, the shape of the curve is determined not by the sample sizes themselves ($n_1$ and $n_2$)

**BOX 9.10  THE _F_ TEST FOR COMPARING THE VARIANCES OF TWO POPULATIONS:
PHOSPHATE CONCENTRATIONS MEASURED BY A STUDENT AND A TECHNICIAN**

Null hypothesis:   The two populations have the same variance.

Alternative hypothesis:   Student's measurements more variable than technician's.

Significance level:   5%, one tailed.

| | Phosphate concentration (mg/ml) measured by | |
| | Student | Technician |
| --- | --- | --- |
| | 1.25 | 1.24 |
| | 1.20 | 1.27 |
| | 1.22 | 1.25 |
| | 1.30 | 1.22 |
| | 1.15 | 1.25 |
| | 1.40 | 1.26 |
| | 1.30 | 1.28 |
| | 1.21 | 1.21 |
| | 1.25 | 1.25 |
| Mean | $\bar{Y}_1 = 1.253$ | $\bar{Y}_2 = 1.248$ |
| Standard deviation | $s = 0.073$ | $s = 0.022$ |
| Variance | $s_1^2 = 0.0053$ | $s_2^2 = 0.00049$ |
| Sample size | $n_1 = 9$ | $n_2 = 9$ |
| Degrees of freedom | $df_1 = 8$ | $df_2 = 8$ |

$$F = \frac{\text{Sample variance predicted to be the larger}}{\text{Sample variance predicted to be the smaller}}$$

$$= \frac{0.0053}{0.00049} = 10.719$$

Degrees of freedom:   Numerator df = 8; Denominator df = 8

Critical value:   3.438

Decision rule:   Reject null hypothesis if the calculated value is equal to, or larger than, the critical value.

Decision:   Highly significant evidence for alternative hypothesis. Results of measurements by student more variable than those made by technician ($P < 0.01$).

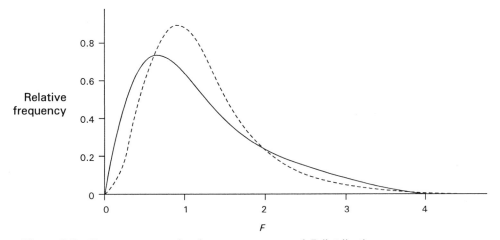

**Figure 9.5** Two representative frequency curves of F distributions.

but by the degrees of freedom $df_1 = n_1 - 1$ and $df_2 = n_2 - 1$. Remember that it is degrees of freedom not sample sizes which are used to calculate the variances in the first place.

The smallest sample variance possible is zero (when all the observations have the same value) which means that the smallest possible value of $F$ is zero. The largest possible value of $F$ is infinity. However, very large and very small values of $F$ will be unlikely if the two populations have the same variance which is why the curves have two tails. The most common outcome will be to pick two samples with more or less equal variances, so the most likely value of $F$ will be around 1. You can see that the curves have a peak in this region.

Each of these curves can be mathematically described and critical values of $F$, cutting off known proportions of the distribution, can be calculated and tabulated (Table A13). We can use them to compare a calculated value of $F$ with a critical value.

## Calculating the value of F

The calculation of the two sample variances is straightforward (Box 9.10), but don't forget that a calculator with a standard deviation function gives you $s$ not $s^2$ and that the values reported here have been rounded. There is, however, one minor complication in the calculation of $F$. Obviously, from two samples like these we can calculate two different values of $F$, depending on which sample is labelled 1 and which is labelled 2. The value we need depends on whether we are doing a one-tailed or a two-tailed test and has to do with the way in which critical values are tabulated.

We saw above how both small and large values of $F$ are unlikely, if the two populations do indeed have equal variances, and so, in theory, critical values will occur in both tails of the sampling distribution. However, because there are so many different sampling distributions of $F$ to be tabulated (one for every combination of two sizes of sample), most tables list critical values in only the right-hand tail in order to save space. The right-hand tail comprises large values of $F$, and to the right of the upper critical value (for a given significance level) lie the values of $F$ which are too large to be considered likely if the populations have equal variances. Because the tables are set up to detect only whether a value of $F$ is unusually large, the value of $F$ that we calculate has to be the one which would be the larger if the alternative hypothesis were true.

In a two-tailed test, the alternative hypothesis is that the population variances are unequal, but which one is the larger is not specified. To calculate $F$, take the sample variance which *is the larger* and divide this by the one which *is the smaller*. In a one-tailed test, the alternative hypothesis does specify which population has the larger variance and this enables us to predict which sample should have the larger variance. To calculate $F$, take the sample variance which *is predicted to be the larger* and divide by the sample variance which *is predicted to be the smaller*. In the example we are dealing with we can probably justify a one-tailed test on the grounds that, if there is going to be a difference in variability, the determinations made by the technician should be less variable. We divide the student variance by the technician variance, giving a calculated value of $F = 10.719$, which we can now compare with the appropriate critical value.

## Using tables of $F$

The tables of critical values of $F$ are quite extensive and are divided into subtables according to the degrees of freedom associated with the variance on the top line (i.e. the numerator) of the equation for $F$. So first find the appropriate subtable. If you can't remember which is the numerator and which the denominator, there is an easy way to make sure you are in the right subtable. The degrees of freedom at the top of the subtable are those relating to the sample on the top line of the calculation. Locate the row corresponding to the value of the degrees of freedom for the other variance in the highlighted margin down the side of the table and find the critical value in the column under the desired significance level. For a one-tailed test use an $\alpha_1^R$ significance level; for a two-tailed test use an $\alpha_2$ significance level. The decision rule has the standard form; reject the null hypothesis if the calculated value is equal to, or greater than, the critical value. In our case both degrees of freedom are 8 and so the critical value is 3.44. We can clearly reject the hypothesis that the two populations have equal variances.

## The $t$ test for a difference in means of two populations using small samples

Although ecological work may yield large samples, in many situations time, money or shortage of experimental material means that we have to use much smaller sample sizes. You may well have guessed that when samples are smaller than about 100, the $z$ test is inappropriate. The reason is exactly the same as the reason why we could not use $z$ when we were looking at confidence limits of the mean of small samples. The standard deviation ($s$) of a small sample is not a good estimate of the population standard deviation ($\sigma$) and this has to be taken into account when we try to assess whether two samples have been taken from populations with the same mean. Box 9.12 gives the data on the activity of the enzyme acid phosphatase at two different pH values which we met when dealing with confidence limits for the mean of a small sample. Does pH affect activity, that is, are the two samples drawn from populations with different means? To answer this question we need a new test, the $t$

test. The principle of this new test is the same as the *z* test, but there are two important differences. Unlike the *z* test it assumes that the variable in both populations follows a normal distribution and it also assumes that the two population variances are the same.

## How does the test work?

The logic of this test is the similar to that of the *z* test, which we worked through in Figure 9.4. The null hypothesis is that the population means are not different and again we need to think of a sampling distribution of the difference between the means of pairs of samples. This distribution will have a mean of zero and a standard deviation which we can estimate. The standard deviation will have to be called a standard error. We then take our observation (the difference between the two sample means) and ask where it falls in this distribution. More specifically (Box 9.11a), we first ask how far away the difference between our two means is from the mean of the distribution, which is known to be zero. We then express this difference in terms of standard deviations by dividing it by the standard deviation (= standard error). Put simply our question is: "How big is the difference in means in terms of the standard error?". The answer is a value of *t*, not a value of *z*. We can then refer this calculated value of t to the table of critical values taking into account the degrees of freedom and the significance level. As with the *z* test it will be extreme positive or negative values of *t* which are unlikely if the null hypothesis is true.

At first sight the formula for calculating *t* looks unfriendly, but it turns out to be similar to the formula for the *z* test. The top line is calculating how far the difference between the two sample means is from the mean of the distribution. In practice, as with the *z* test, we can get rid of the zero and simply use the difference between the two means. The complicated expression on the bottom line calculates the standard deviation of the distribution. The distribution in question is a distribution of differences between sample means, so its standard deviation is called a standard error of the difference between the sample means. You don't actually need to know where this expression comes from, but it isn't so different from the equivalent part of the *z* test formula, so if you are interested read on! If not, skip the next paragraph.

The *t* test is based on the assumption that the two populations have the same variance, which is something that should be checked first using the *F* test. The part of the formula inside the square brackets uses this assumption to obtain an estimate of this variance. It does this by combining the information on the two sample variances, $s_1^2$ and $s_2^2$, to get an estimate of the unknown population variance, called simply $s^2$. It is simply a weighted average of the two sample variances which, quite sensibly, gives more weight to the variance of the larger sample. The weighting is done using the degrees of freedom rather than the sample sizes. Multiplying this by what is in the second bracket is simply equivalent to dividing $s^2$ by $n_1$ and also by $n_2$ and then adding the two answers together, i.e. $s^2/n_1 + s^2/n_2$. This gives us the variance of the distribution of the differences and you can see that this expression has the same form as the one that we had in the denominator in the *z* test. The final step is to take the square root of this, which gives us the standard deviation of the distribution which equals the standard error of the difference.

If the two sample sizes are equal, this formula can be very much simplified (Box 9.11b), which is one good reason for having equal-sized samples. There is another reason which we shall come to shortly.

**BOX 9.11  THE FORMULA FOR THE *t* TEST FOR A DIFFERENCE BETWEEN MEANS OF TWO POPULATIONS USING SMALL SAMPLES**

(a)  General formula:

How many standard deviations away from the mean of the distribution is the difference between the two sample means?

This is given by

$$\frac{\text{Difference between two sample means} - \text{Mean of distribution}}{\text{Standard deviation of distribution of difference between sample means}}$$

$$= \frac{\text{Difference between two sample means} - \text{Mean of distribution}}{\text{Standard error of difference between sample means}}$$

$$= \frac{\left(\overline{Y}_1 - \overline{Y}_2\right) - 0}{\sqrt{\left[\dfrac{(n_1 - 1)s_1^2 + (n_2 - 2)s_2^2}{n_1 + n_2 - 2}\right]\left(\dfrac{1}{n_1} + \dfrac{1}{n_2}\right)}} = \text{Calculated value of } t$$

(b)  For samples of equal size:

$$\text{Calculated value of } t = \frac{\overline{Y}_1 - \overline{Y}_2}{\sqrt{\dfrac{1}{n}\left(s_1^2 + s_2^2\right)}}$$

(c)  Calculating *t* using a calculator based on data in Box 9.12:

 (1)  Standard deviation (SD) mode.
 (2)  Calculate mean ($\overline{Y}_1$) and standard deviation ($s_1$) of sample 1. Note mean. Square $s_1$ to get variance 1 and store.
 (3)  Clear standard deviation mode.
 (4)  Calculate mean ($\overline{Y}_2$) and standard deviation ($s_2$) of sample 2. Square $s_2$ to get variance 2.
 (5)  Divide the larger variance by the smaller to get *F*.
 (6)  Recall variance 1 from memory, multiply by ($n_1-1$), store.
 (7)  Retrieve standard deviation 2 (from SD mode), square it, multiply by ($n_2 - 1$), add to memory to get 14.82. (You may need to get out of SD mode to do this last step.)
 (8)  Divide what is in memory (14.82) by ($n_1 + n_2 - 2$) to give 1.8525 and store.
 (9)  Calculate $1/n_1 + 1/n_2$, multiply by what is in memory and take square root to give denominator (= 0.8608). Store in memory.
 (10)  Retrieve mean of sample 2 (from SD mode), subtract mean of sample 1 (from step (2)).
 (11)  Divide by what is in the memory.

## Calculating the value of $t$

Although our sample sizes are equal so that we could use the simplified formula, we shall use the more complex one. The reason is to show you how the whole calculation can be done with only one intermediate value having to be written down. This avoids any problems with rounding errors.

The steps are set out in Box 9.11c and the actual example is dealt with in Box 9.12. First we need to use the $F$ test (two-tailed) to see whether there is any evidence that the samples have been taken from populations with different variances. You calculate the mean and standard deviation of the first sample and store the variance in the ordinary memory. The only thing you need to note is the mean of the first sample. Then calculate the second mean and standard deviation. If you square the latter you can now carry out the $F$ test, by dividing the larger variance by the smaller. This will be a two-tailed test giving a critical value at the 5% significance level of 9.60 and our calculated value of 1.33 lies comfortably below this. We have no evidence for a difference in variance so we can proceed with the $t$ test. (If the two variances were shown to be significantly different then we would need to think again; this situation is dealt with later in this chapter.)

We have already calculated all the information needed. The variance of one sample is in the ordinary memory and we have a note of the mean. The standard deviation of the other sample is still in the SD mode memory, along with the mean. We first evaluate the denominator, by multiplying each variance by its degrees of freedom, adding the answers and storing the result in the memory. This can then be divided by $n_1 + n_2 - 2$ and the answer stored. Then calculate $1/5 + 1/5$, multiply it by what is in the memory, take the square root of this and store it to obtain the value of the denominator. Then calculate the difference between the two means and divide this by what is stored in the memory.

## Using tables of $t$

We have already looked at the tables of $t$ values in the context of confidence intervals and the same tables can be used to find the critical values for a two sample test. To locate the critical value we need to know three things: whether the test is one-tailed or two-tailed, the significance level and the degrees of freedom (remember that the shape of a $t$ distribution depends on the number of observations).

The rows at the top of the table labelled $\alpha_1^R$ and $\alpha_2$ give various one and two-tailed significance levels and are used to identify the correct column of values. We are doing a two-tailed test with $\alpha_2 = 5\%$ which puts us in the third column. The column to the left of the table is for the degrees of freedom and tells us which row of the table to use. The degrees of freedom are found by adding together the two sample sizes and subtracting two, df = $n_1 + n_2 - 2$, which in our case gives df = 8. The critical value is 2.306 and the decision rule is to reject the null hypothesis if the calculated value of $t$ is equal to, or more extreme than, the critical value. Calculated values can be positive or negative, as for the $z$ test, so I suggest you adopt the same procedure as we did there.

Our calculated value is larger than the critical value, so we can reject the null hypothesis and conclude that mean activity is higher at the higher pH. We can also see that confidence limits cannot be used to test whether two means are different. The 95% confidence limits for these two means overlapped (Box 7.4 and Fig. 7.10) yet the difference between the

**BOX 9.12 THE $t$ TEST FOR DIFFERENCE IN MEANS OF TWO POPULATIONS USING SMALL SAMPLES: MEAN ACTIVITY OF ACID PHOSPHATASE AT pH 3 AND pH 5**

Null hypothesis:   Mean activity of acid phosphatase the same at pH 3 and pH 5

Alternative hypothesis:   Mean activity of acid phosphatase different at pH 3 and pH 5.

Significance level:   5%, two tailed.

| | Activity (μM/min) | |
| --- | --- | --- |
| | pH 3 | pH 5 |
| | 11.1 | 12.0 |
| | 10.0 | 15.3 |
| | 13.3 | 15.1 |
| | 10.5 | 15.0 |
| | 11.3 | 13.2 |
| $\bar{Y}$ | 11.24 | 14.12 |
| $s$ | 1.26 | 1.45 |
| $s^2$ | 1.588 | 2.117 |
| $n$ | 5 | 5 |

$F$ test:   5% significance level, two-tailed; numerator d.f. = 4, denominator d.f. = 4

$F = 2.117/1.588 = 1.333$       Critical value $F = 9.605$

No evidence for alternative hypothesis that population variances are different.

$t$ test:

$$t = \frac{\bar{Y}_1 - \bar{Y}_2}{\sqrt{\left[\frac{(n_1 - 1)s_1^2 + (n_2 - 1)s_2^2}{n_1 + n_2 - 2}\right]\left(\frac{1}{n_1} + \frac{1}{n_2}\right)}}$$

$$= \frac{14.12 - 11.24}{\sqrt{\left[\frac{(4 \times 1.588) + (4 \times 2.117)}{8}\right]\left(\frac{1}{5} + \frac{1}{5}\right)}}$$

$$= \frac{14.12 - 11.24}{\sqrt{1.8525 \times 0.4}} = \frac{2.88}{0.8608} = 3.346$$

Degrees of freedom $= (n_1 - 1) + (n_2 - 1) = n_1 + n_2 - 2 = 5 + 5 - 2 = 8$

Critical value $= 2.3060$

Decision rule:   Reject null hypothesis if calculated value is equal to, or larger than, critical value.

Decision:   Highly significant evidence ($P < 0.01$) for alternative hypothesis. Activity higher at pH 5.

means is highly significant. Overlapping confidence limits do not imply a lack of difference in means.

Finally, we can use the $t$ test when one of the samples has only a single observation ($n_1 = 1$) and this enables us to do a correct test on the beak size of our putative hybrid finch (Box 5.9). Remember that there we assumed that we knew what the population standard deviation was, in fact we only had an estimate. I shall not present the calculations, but you can do them if you want. The standard equation can be used, but there will be only one variance in the denominator and the calculated value of $t$ has $n_2-1$ degrees of freedom. The calculated value of $t = 2.798$, with df = 245. Checking this against the critical value actually leads to the same conclusion as when we used the value of $z$.

## What to do if the assumptions are not met

The t test shares with all the other tests the assumptions of random, independent samples, but it also requires that the two samples are drawn from normally distributed populations with the same variance. What should we do if we think (or indeed know) that one or both of these is not the case? There are three possible solutions.

(a) Carry on with the $t$ test regardless! It has been shown that the results of a $t$ test can still be valid even when there are quite large departures from these assumptions. The $t$ test is said to be robust. This is particularly true if the sample sizes are large and more or less equal and if the test is two-tailed. Unfortunately, there are no handy rules of thumb here which tell you how unequal the variances and how non-normal the distributions can be, before the $t$ test becomes unreliable.

(b) Carry out a Mann–Whitney $U$ test instead, as we did for the urea data. Remember that this test has less restrictive assumptions. In fact with the urea data, the two sample variances are significantly different so the $t$ test would not be strictly appropriate.

(c) Use an appropriate transformation as described in the Chapter 7 and then carry out a $t$ test on the transformed data. Although we looked at these transformations as a way of making the distribution closer to a normal one they also tend to have the property of making the variances equal.

# General points about two-sample tests

Having now dealt with the details of several tests it is worth examining some more general issues.

## What the formulae tell us about sampling and experimental design.

Before we leave the $t$ and the $z$ tests we should just go back to the formulae because it is instructive to view them from a different perspective. The top line is the difference between the two sample means and, as such, is going to reflect the difference *between* the two populations. The bottom line contains a measure of the intrinsic variation *within* each population divided by the sample size. It reflects the variation that occurs *within* each population, the populations in question being populations of mean values. The bottom line is therefore a measure of the extent of sampling variation. Looked at in this way, both $t$ and $z$ are ratios of the difference between the two populations and the variation within them. A large (i.e. significant) value of $t$ or $z$, implying a real difference between the means of the two population, will be found when the difference between samples is large compared to the variation within samples. This tells us something about experimental design.

Let us suppose that a treatment does have an effect on the mean of the variable then, in order to maximize our chances of detecting this effect in an experiment, we need values in the top line to be large relative to the bottom line. The size of the difference between the two sample means will be related to the size of the difference between the two population means, which in turn will be set by the effect of the treatment. It follows that for a given treatment there is nothing that we can do to increase the size of the top line.

Is there anything we could do to decrease the size of the value of the bottom line, that is, to reduce the standard error? This would lead to large values of $t$ or $z$, other things being equal. Increasing the sample sizes would obviously have the desired effect. The logic behind this is that the larger the sample sizes are, the less extreme the effects of sampling variation will be. Unfortunately there may be practical constraints on the maximum sample size that can be used. However, for a given total sample size $(n_1 + n_2)$, the standard error will be a minimum if the two samples have an equal size. So, wherever possible, you should try and arrange this.

Another strategy which would increase the calculated value of $t$ or $z$ would be to decrease the standard deviation $(s)$. Now, you might think that this is fixed in the same way as the mean, but if you think about the origins of variability you will see that this is not necessarily the case. We can make $s$ smaller in a variety of ways. Remember that sampling units within a treatment may differ with respect to a variable for a variety of reasons. They may be genetically different from one another, they may experience different environmental conditions before or during the experiment and our measurement technique will vary. It follows that we can reduce variability by careful attention to the experimental design.

In a laboratory or greenhouse experiment, we could reduce the variability of the biological material by using plants of the same variety or, even better, plants that have been vegetatively propagated. Obtaining genetically identical animals is not so easy. Although some animals (e.g. *Daphnia*) reproduce asexually to give offspring with no genetic differences, this trait is not common. However, we could achieve the same result by using an inbred strain. In all cases, before the actual experiment starts, we should try and raise the experimental material under uniform conditions to minimize environmentally induced variation between individuals. Finally, attention to the experimental technique at all stages (i.e. when the experiment is set up, while it is under way and when the variable is finally measured) will all reduce variability. Clearly, many of these sources of extra variability cannot be reduced in the less "controlled" conditions of an ecological investigation, so that the only option would appear to be to take larger samples. There is, however, a different strategy for reducing variation which we will examine in Chapter 10.

## Parametric versus non-parametric and distribution-free tests

We have already met these terms in Chapter 8. The $z$, $F$ and $t$ tests are parametric tests, because, as their name suggests, they involve the estimation of population parameters such as the mean and the variance and the testing of hypotheses about these parameters. In contrast, the Mann–Whitney $U$ test is a non-parametric test because it does not require estimates of parameters. Statistical tests can also be distinguished on the basis of what assumptions have to be made about the distribution of the variable. For example, the $t$ and $F$ tests assume that the variable has a normal distribution, while the $\chi^2$ and $U$ tests do not rely on the variable having any particular underlying distribution. As a result the latter two are sometimes described as being distribution free.

Now the fact that a distribution-free test has less restrictive assumptions should mean that it can be applied in a wider range of situations. We have already seen that we can use the $U$ test on count or measurement variables by replacing the actual values by rank values. For that matter, we can place count and measurement variables into a few broad categories (e.g. small, medium, large; top, middle, bottom; active, not active) and analyse them using

---

**BOX 9.13 A COMPARISON OF TWO SAMPLE TESTS: HYPOTHETICAL DATA**

**(a) Comparisons of the $t$ test and the Mann–Whitney $U$ test.**

| Activity | |
|---|---|
| 1.7 | 2.6 |
| 1.8 | 2.8 |
| 2.4 | 3.1 |
| 2.9 | 3.5 |
| 3.0 | 3.9 |
| 3.2 | 4.0 |

| Calculated value | Critical value | Decision |
|---|---|---|
| $t = 2.323$ | 2.230 | Siginificant evidence for alternative hypothesis |
| $U = 29$ | 31 | No evidence for alternative hypothesis. |

**(b) Comparison of the chi-squared test with the $t$ test and Mann–Whitney $U$ test.**

| | Continuous variable grouped into categories | | | |
|---|---|---|---|---|
| | 0.0–2.9 Low | 3.0–5.9 Medium | 6.0–8.9 High | |
| Sample 1 | 5 | 13 | 2 | $n_1 = 20$ |
| Sample 2 | 9 | 2 | 9 | $n_2 = 20$ |

| | Calculated value | df | Critical value $\alpha_2 = 5\%$ | Decision |
|---|---|---|---|---|
| $\chi^2$ | 13.66 | 2 | 5.991 | Very highly significant evidence for alternative hypothesis |
| $t$ | 1.057 | 38 | 2.024 | No evidence for alternative hypothesis |
| $U$ | 251.5 | – | 273 | No evidence for alternative hypothesis |

the $\chi^2$ test, which is perhaps the simplest test of all. You may also have decided that these distribution-free tests are easier to understand and to calculate than something like the $t$ test. So why don't we simply use distribution-free tests all the time?

The answer is that the different tests don't necessarily answer identical questions, that is, they are sensitive to different types of patterns in the data. Even when they do a similar job, they may not be equally powerful. To show this, consider the set of hypothetical data in Box 9.13a, which could have come from an experiment similar to the one on the behaviour of top-shells which we analysed earlier. In this imaginary case, rather than merely ranking the individuals, we have actually measured their activity which we could do by measuring how far or how fast they move. A $t$ test on these figures gives a value of $t = 2.323$, which is significant at the 5% level. If we had ranked them instead and used the Mann–Whitney test, we obtain $U = 29$ which is not significant at the 5% level (it has an associated probability between 0.1 and 0.05). If we put the observations into three categories (fast, medium and slow) and try to analyse them using a $\chi^2$ test we find that we cannot, because the expected values are too low.

What this means is that the $t$ test is the most powerful test in that it leads to the rejection of the null hypothesis using a smaller sample size. Against this, we have to weigh the fact that it may take us longer to collect the data using measurements as opposed to ranks or categories, so it might be quicker to rank a larger number of observations and use the $U$ test. We also have to bear in mind that the variable might not have a normal distribution and equal variances which could mean that the $t$ test gives unreliable results.

Different tests also answer different questions. The $t$ and $U$ test are tests for a difference in location. The $\chi^2$ test, on the other hand, is more general purpose because it is sensitive to any type of difference between the two distributions. Consider the example in Box 9.13b in which two samples of a continuous variable have been put into categories. A $\chi^2$ test will pick up the fact that the observations in sample 1 are concentrated in the middle, while in sample 2 they are concentrated at the two extremes. A $t$ or $U$ test on the original data (not given here) would not detect any differences, because both samples have similar locations (i.e. means and medians). In any case, neither of these tests would be appropriate because it is fairly obvious that the data do not meet the required assumptions.

The take-home message of all this is that you need to think carefully about three inter-related questions as part of your experimental or sampling design.

(a) The biological question to be answered; are you interested in differences in location, differences in spread, any differences at all?
(b) What is the most convenient way of describing the variable?
(c) Will the data meet the assumptions of the test which you have decided to employ?

## What a statistical test cannot do

First, it should be obvious, but it is worth repeating, that the answer you get from a test is only as good as the data that are put into the equation. All the tests are based on random samples. If the sampling units in either or both samples are not representative of the populations from which they were drawn, then the statistical test may yield a significant result, even though there is no difference between the populations. Alternatively, the test could fail to detect a difference even when there is one.

This point was explored with the isopod body lengths but you should be able to see how this sort of effect could arise with the other examples. In addition, even if the two samples are random, a significant difference can only be attributed to the difference in treatments if the experiment has been designed properly. If we had looked at the behaviour of top-shells in the high salinity at one time of day and in the low salinity at a different time of day, then we could not say that the difference in activity was caused by the difference in salinity. Activity of many organisms is likely to be different at different times of the day and so we would not be able to distinguish between any effects of salinity and any effects of time of day. We would say that the possible effects of the two factors had been **confounded**. If this is the case there is no way that the statistical analysis can sort out whether salinity affects behaviour. A statistical test is not a magic wand that can put right a badly designed experiment!

A second point is that failure to find a significant difference does not mean that the two populations are the same. All these tests variously look at the difference between the samples in relation to the extent of the differences which could arise due to sampling variation. Small differences between populations are difficult to detect if there is a lot of variability. Under these conditions, if you make the samples small enough, you will not detect a difference between the populations even if there is one. This means that a lack of statistical significance is no guarantee that the treatment has no effect. The urea concentration example shows this clearly, because it is well known that urea concentrations are lower if urine volume is large, yet we found no difference. One reason for this failure to detect a difference between the two samples was that there was a lot of variability within the two samples. This was because the students were variable with respect to factors such as age, sex and diet.

The flip side of this argument is that a statistically significant difference is not necessarily biologically significant. By this we mean that it may not be interesting from a biological point of view or useful from a practical point of view. The reason is that in actual fact the null hypothesis is probably never likely to be true! Animals in two different places are almost certain to be slightly different in size and seeds treated in two different ways are almost certain to have slight differences in germination times. If you took big enough samples, you could probably find statistically significant differences in almost any experiment or investigation, but this would not necessarily mean that the differences were biologically meaningful. For example, if you rework the $z$ test calculations on body length in the isopods but make the two means 3.01 and 3.02, keep the two standard deviations the same as before, but increase both sample sizes to 100,000 you will get a value of $z = 2.14$. This is statistically significant, but demonstrating such a small difference in body size may contribute little to our understanding of the ecology of the organisms.

# 10

# Tests for two related samples

The tests discussed in the previous chapter were concerned with detecting differences between two populations, using simple random sampling or a completely randomized design of experiment. In these cases the samples are said to be independent. What this means is that the value of an observation in one sample is not related to or affected by the value of any observation in the other sample. This stands in contrast to related samples, which form the subject of this chapter.

As we saw with the $z$ and $t$ tests, the statistical analysis of independent samples involves examining the difference between the sample means in relation to the standard error of the difference. The latter is simply a measure of the chance variation which can be expected and it was related to the two things which we know will affect the extent of sampling variation: the variability of the variable in the population and the sample size. The Mann–Whitney $U$ test works on the same principle. The value of $U$, which measures the extent of the overlap between two samples, depends on both the overall difference in location between them and the variation within each sample.

This means that it is always going to be difficult to detect differences between two populations against a background of high variability within each population. In common-sense terms this brings us back to a point raised very early on in the book. If something is highly variable, it is difficult, because of sampling variation, to get a reliable measure of the true value. If our description is unreliable, it follows that it will be difficult to decide whether two populations are different.

As I have emphasized at various points, variation within a population can be reduced by careful attention to the design of the experiment and this is particularly easy in laboratory experimental work, where we have control over how the biological material is produced. Genetic and environmental differences between individuals can be reduced and careful laboratory technique can minimize experimental variation, but even under these ideal circumstances we shall never get rid of the variability completely.

In other situations reducing variability is much more difficult. For example, if we want to examine the effect of a drug on some aspect of human biology, we do not have access to large numbers of genetically identical individuals raised under identical conditions. If we use very variable material allocated at random to control and experimental treatments any effect of the drug may be undetectable against the background variation between subjects. This is what happened in the experiment on urea concentration in Chapter 9.

Similar problems of high levels of variability arise in most ecological work when we are studying organisms in their natural environments. Individual organisms are highly variable in all aspects of their biology, which makes it difficult to detect differences in organisms between sites or times. Similarly, attempts to detect differences between habitats in variables such as nutrient availability or species diversity encounter the problem that within each habitat type there is likely to be a lot of variability. Species diversity in ponds may be affected by pond size, but it will also be affected by many other factors such as age, past management, isolation and surrounding land use. Species diversity within a random sample of ponds of a particular size will be very variable, making it difficult to detect differences in diversity between pond sizes.

## Matched pairs versus independent samples

As we saw in Chapter 3, one way to reduce variability is to use stratified sampling or randomized block experimental designs. It was pointed out that this is more efficient than simple random sampling and completely randomized experiments which means that we can use smaller sample sizes. However, the statistical analysis is generally more complicated except in the special case of two treatments where we can use a matched pairs design. In this design we match the sample units in pairs for any characteristics which we think might contribute to unwanted variation in the variable of interest. The two members of each pair are then allocated at random to the two treatments. Under these circumstances the samples are said to be related. Now the value of, say, the second observation in one sample is not independent of the value of the second observation in the other sample. There are various ways of obtaining matched pairs which depend upon the situation being investigated.

The ideal way to match samples is to organize the experiment so that each sampling unit is both an experiment and a control, either simultaneously or sequentially. For example, suppose we wanted to compare the efficiency of a new treatment for a skin infection with that of an existing treatment, then we could test both formulations on the same individual. We could do this by applying one drug to an infected area on the left side of the body and the other drug to a similarly infected area in the same place on the right side of the body.

We might need to be careful about the details of how this was done, because it could be that there is natural variation between the two sides of the body which influences the course of the infection. In addition there is room for bias on the part of the experimenter, who might tend to apply the new formulation to the side of the body where the infection looked less (or more) severe, or who might apply one formulation more carefully than the other. Finally, the effectiveness of the drug might be difficult to measure and could involve a subjective assessment of the extent to which the infection had been reduced – influenced, perhaps, by preconceptions or the claims of the drug company! Perhaps you can see how these problems can be circumvented.

Similar designs could be used to test for differences in the allergenic effects of two detergents or cosmetics and for differences in virulence between two strains of plant pathogen. In the latter case, the two different strains would be applied to a pair of identical leaves on the same plant. Experiments on tissues and enzymes can be carried out using this design. We could extract enzyme from a number of individuals and then divide each extract into two portions, one being used in the experiment the other in the control.

This principle can be extended to a wide range of situations. If we want to examine the effects of a drug on a variable such as blood pressure we obviously cannot use one side of an individual as the experiment and the other as the control. In these circumstances we can carry out a "before and after" comparison of the blood pressure of the same individuals. Two measurements of blood pressure are made on each subject, one with the drug administered and one in the absence of the drug. Each experimental subject acts as their own control, although again, we would need to be careful about the details, so that other factors, which could affect blood pressure are not different in the two treatments. These factors could include time of day and anxiety levels due to other circumstances. Perhaps even the knowledge that the drug had been administered could affect the subject's blood pressure so we might want to conduct a blind experiment. In a blind experiment the subject does not know what treatment they have received, so that they cannot influence the result. We would also need to decide whether blood pressure in the absence of the drug should be measured before or after the measurement made in the presence of the drug. You might like to think about why this could be an important issue and how you could deal with it.

Matching in pairs can also be achieved in other ways. Each experimental unit can be matched with a control unit which is identical to it in all those factors which we think might affect the variable under investigation. The extent to which this can be achieved depends on both the biological material we are working with and the sort of variable we are investigating and, in general, it is going to be easier to do in laboratory than in the field.

Individuals which are identical twins and which have been raised in the same cage would be good candidates as would a pair of individuals of the same sex from the same litter, especially if the parents were inbred strains. We can also pair for factors relating to the equipment used in an experiment. If we thought that the temperature in an incubator was different on different shelves, we could arrange the experiment so that one experimental and one control culture was on each shelf. Similarly, we could position experimental and control plants in pairs on the greenhouse bench, so that both members of the pair were the same distance from the glass; they would therefore get the same amount of light.

Matching can also be achieved in observational studies when there are a large number of possible subjects with well documented characteristics, from which the investigator can pick pairs who are matched. This is common in medical studies and sociological studies in our own species, where information on diet, lifestyle, medical history, height, weight, socio-economic group, etc., is known. An investigation into the effects of some aspect of diet (e.g. vegetarianism) on the levels of blood cholesterol could compare levels in pairs of individuals matched for all factors apart from the dietary factor of interest. Similarly, we could match small and large ponds in pairs, such that the two members of each pair were the same for all other relevant factors.

In all these cases we could have used independent, that is unpaired samples. We could take a group of patients, randomly allocate them to two subgroups and treat one subgroup with the drug, while the other group acted as the control. The two sub-groups would then constitute two independent samples and we could test for a difference between the populations from which they were drawn, using one of the procedures described in the last chapter. In our study of cholesterol levels, rather than using matched pairs of vegetarians and non-vegetarians, we could use a random sample of both, which might be easier. The paired design could be time-consuming to set up, given that you have to find out details about each individual and then choose the members of each pair carefully.

The advantage of the matched pairs design derives from the fact that it is more powerful in a statistical sense, that is, it can detect a real difference between the two treatments using a smaller sample size, than would be the case with two independent samples. The reason why these procedures are more powerful is that, as we have seen, they remove some of the background variability. With independent samples this variation between observations or sampling units forms the background against which we are trying to detect a difference between treatments. A measure of the variation in each sample has to be incorporated into the statistical test and it may be so marked as to make it difficult to detect a difference between treatments. With a matched pairs design much of the variation present in independent samples is effectively removed from the experiment and from the subsequent statistical analysis.

Data from matched pairs (i.e. related samples) are analysed using different tests from the ones used with independent samples. These related sample tests work on the same general principles and may involve familiar test statistics, so there is not a great deal new to absorb. The only difference is that they focus on the differences between the members of the pairs, rather than on the individual observations. As you might expect from previous chapters the test depends on the type of variable under investigation.

## The sign test

This is the simplest paired sample test available and as we shall see, it uses an idea that we have already encountered and thus does not involve anything new at all. Its simplicity derives from the fact that it does not require us to put any numerical values to the variable – all we need to be able to do is determine the direction of the difference in each pair. This makes it particularly useful for variables such, as the severity of a skin infection, which are difficult to quantify, that is, to measure or count. All we would need to decide was whether the symptoms were more or less severe with drug A than with drug B. I have chosen to illustrate how the test works by using an example on the efficiency of two different procedures for staining bacteria.

The Gram stain is used to distinguish between Gram-negative and Gram-positive bacteria. This involves staining cells with a solution of crystal violet which enters both types of bacteria. They are then immersed in alcohol which has no effect on the stain in Gram-positive bacteria but removes the stain from Gram-negative bacteria. However, the standard concentration of crystal violet can produce such intense staining in some Gram-negative bacteria that they do not de-stain properly. One suggested modification is to use a weaker solution of crystal violet, and Box 10.1 gives some data which compare the modified and unmodified method.

Each student prepared two smears of the same Gram-negative organism taken from the same cultures. One smear was stained with crystal violet at the standard concentration and one with a reduced concentration. The results were assessed by simply noting whether the intensity of the blue stain in the organisms was higher (indicated by a (+)) or lower (–) in the standard treatment when compared with the reduced treatment. Out of a class of 34 students, 26 recorded (+) signs and 8 recorded (–) signs. Is there any evidence that the modification is an improvement?

The most important sources of variability in an experiment such as this are likely to be differences in the strain of organisms used, in their culture conditions, in the make-up of different batches of stain solution, in the technique used by each student and in the "measurement" of the result. The material for the two treatments came from the same culture, was stained (with stain solutions from the same bottles) by the same student, who also assessed the results for each treatment using the same microscope. Each sampling unit in

---

**BOX 10.1 THE SIGN TEST FOR PAIRED SAMPLES: THE EFFECT OF TWO DIFFERENT CONCENTRATIONS OF CRYSTAL VIOLET ON THE INTENSITY OF THE GRAM STAIN**

Null hypothesis:   No difference in stain intensity in smears stained using two different concentrations. Median difference between members of a pair is zero.

Alternative hypothesis:   Staining lighter in smears stained with less concentrated stain. Median difference between members of a pair is positive.

Significance level:   5%, one tailed.

Intensity of staining in standard treatment compared to intensity in modified treatment.

|  | Higher (+) | Lower (−) |
|---|---|---|
| Frequency | 26 | 8 |

Test statistic:   The number of (+) or (−) signs which is predicted to be the larger

Calculated value of $S$ = 26

Number of pairs ($n$) = 34

Critical value of $s$ = 23

Decision rule:   Reject null hypothesis if calculated value which is predicted to be larger is equal to, or greater than critical value.

Decision:  Highly significant evidence for the alternative hypothesis ($P < 0.01$). Modified treatment destains more effectively.

---

one treatment was therefore matched with one in the other treatment with respect to these sources of variation. The experimental design is paired.

## The test statistic S

We proceed, as in any statistical test, from the null hypothesis that the treatment has no effect on the variable, which in this case is intensity of staining. If this were the case, we would not necessarily expect to see equal intensity of staining in any particular pair, because there will still be variation due to experimental error; sometimes the standard method would be recorded as giving more intense staining (+) and sometimes as giving less intense staining (−).

Under the null hypothesis, these two outcomes, that is, the (+) and (−) results, should be equally probable and this is equivalent to saying that the median difference in staining intensity between members of a pair is zero. If this is so we would expect the numbers of the two signs to be roughly equal (roughly, of course, because of the effects of chance). An imbal-

257

ance in the numbers of (+) and (–) signs, that is very few of one relative to the other, would be an unlikely occurrence under the null hypothesis. This suggests that the number of one of the signs might make an appropriate test statistic and this is in fact what we use.

The test statistic $S$ is obtained by counting up the total number of (+) signs and the total number of (–) signs and, for a two-tailed test, taking the larger of the two totals. In our case a one-tailed alternative hypothesis is appropriate for two reasons. First, we are only interested in whether the modification is an improvement. Second, if the weaker solution does anything, it must surely give a weaker stain; it is difficult to see how it could give a more intense stain. With a one-tailed test, we have to use the value of $S$ which relates to the sign which should be the most common if the alternative hypothesis is correct. In this case it would be the number of (+) signs, so $S = 26$.

## The sampling distribution of $S$

The sampling distribution of $S$ is known; in fact, if you think back you will probably realize that we have already dealt with it at some length in Chapter 2. The sampling distribution of $S$ is a binomial distribution. In the sign test, there are only two possible outcomes for each pair, that is, for each difference. It can be either a (+) or a (–) and, under the null hypothesis, these are equally probable. In a well designed experiment, these two outcomes should also be independent, that is, the direction of one difference should not influence the direction of the difference in subsequent pairs. Incidentally, this condition of independence might not hold true with a class of students doing the staining experiment. It would be easy for the results obtained by quicker working students to be communicated to the rest of the class and to influence the decisions made by slower working students. An easy way to ensure that this cannot happen is to make it a blind experiment which in this case would require that the identity of the two slides was not revealed to the experimenters until all the results had been obtained.

If you remember, when there are two independent outcomes, then the distribution of possible results will be a binomial distribution. In the case of the sign test, where the probability of a (+) and a (–) is always equal, the sampling distribution is always a binomial with $p = 0.5$. Critical values of $S$ for sample sizes ($n$) between 1 and 100 are given in Table A14 for different significance levels and for one-tailed and two-tailed tests. The notation is identical to the other tables we have used. Our test is one-tailed, so looking down the column under $\alpha_1 = 5\%$ we find that for $n$ (the number of pairs) = 34 the critical value is 23. We have already seen why large values of calculated values of $S$ are unlikely if the null hypothesis is true and so the decision rule is as usual. We reject the null hypothesis if the calculated value is equal to, or larger than, the critical value. (Note: most other statistics books, tables and programs use the smaller value of $S$ as the test statistic and the decision rule is then to reject the null hypothesis if the calculated value is less than or equal to the critical value. I have decided to use the alternative form because it brings this decision rule into line with almost every other one we shall use. Should you need to convert from one value of $S$ to the other, the relationship is simply $S_{larger} = n - S_{smaller}$.)

The table of critical values of $S$ is related to the probability tables for the binomial for $p = 0.5$ that we have already used. You can check this by looking up the critical values for a sample size of 17, which was the number of trials in the woodlouse experiment. At the 5% significance level, for a two-tailed test, the critical value of $S$ is 13, which is the same value as we obtained from the binomial tables.

There is one complication which you may meet in practice, and which you may already have thought of. What happens if there is no detectable difference between the two members of a pair, in other words, a situation in which we cannot assign either a (+) or a (−)? This actually happened in the example above; four students could not see any difference. I used one of the two conventions for dealing with this, I ignored those pairs where there was no difference. They were simply excluded from the analysis. The other strategy which is recommended, especially if the number of such pairs is large, is to divide the "no difference" pairs up equally between the (+) and (−) categories. This is not as devious it might sound! The justification is that if the variable is continuous, as it is in our example, then if we could measure the outcome with sufficient precision we would record a difference for every pair. A "no difference" score is therefore an artefact caused by our imprecise description of the result. If the null hypothesis is true, then half of these "no difference" scores will actually be (+) scores and half will be (−), but it does not matter which are which.

## Wilcoxon's signed rank test

The sign test uses information on the *direction* of the differences between the members of each pair, which makes it ideal for situations where it is difficult to quantify the difference. However, it makes no use of the *magnitude* of the differences. If these can be quantified, then we ought to use them because the more we know about the differences, the easier it will be to tell whether the treatment has an effect.

Imagine that you have done another staining experiment and that you have recorded nine (+) and three (−) results. If you check against the critical values of $S$ this gives you no evidence against the null hypothesis at the 5% significance level (for a one-tailed test). But suppose that you notice that on the slides where the score is negative the difference in staining intensity is very slight, whereas the difference in intensity is much more marked on those slides with a positive score. This additional information, if it could be incorporated into a statistical test, would clearly go in the balance as evidence for the alternative hypothesis, because small negative differences should not count as much as big positive ones. Wilcoxon's signed rank test is the simplest test for paired samples where both the direction and the magnitude of the differences are known. As its name suggests, it does not use the differences themselves but the ranks of the differences.

Box 10.2 has some paired data from an ecological survey. The question of interest was whether the population density of mayfly larvae (family *Ephemeridae*) in streams in a certain area was different at the edge of streams compared to the middle. We could take two independent samples by randomly locating one set of sampling units in the middle of the streams and another set at the edges, but a major problem would be the variable nature of the different streams. They will almost certainly differ in factors such as water quality, substrate type and water depth, which may also have an influence on the numbers of mayfly larvae. Random samples taken from the middle and the edge of the different streams are likely to be very variable and this variability within the two habitats is likely to obscure differences between them. To overcome this, the sampling sites were matched. Twelve sites were chosen at random (so as to achieve a representative sample of the streams), but at each site animals were collected from both the middle and the edge of the stream. At each site

## BOX 10.2 WILCOXON'S SIGNED RANK TEST FOR PAIRED SAMPLES: NUMBERS OF MAYFLY LARVAE IN THE MIDDLE AND AT THE EDGES OF STREAMS

Null hypothesis:   There is no difference in the numbers of mayfly larvae in the middle and the edges of streams. Median difference between members of a pair is zero.

Alternative hypothesis:   There is a difference in the numbers of mayfly larvae in the middle and the edges of streams. Median difference between members of a pair is not zero.

Significance level:   5%, two tailed.

| Numbers of mayfly larvae | | Sign of difference | Size of difference ($d$) | Rank of difference | Signed rank |
|---|---|---|---|---|---|
| Middle | Edge | | | | |
| 0 | 8 | + | 8 | 5 | +5 |
| 6 | 0 | − | 6 | 4 | −4 |
| 8 | 3 | − | 5 | 3 | −3 |
| 4 | 23 | + | 19 | 9 | +9 |
| 5 | 19 | + | 14 | 8 | +8 |
| 1 | 2 | + | 1 | 1 | +1 |
| 23 | 47 | + | 24 | 10 | +10 |
| 11 | 7 | − | 4 | 2 | −2 |
| 8 | 20 | + | 12 | 7 | +7 |
| 6 | 15 | + | 9 | 6 | +6 |
| 0 | 27 | + | 27 | 12 | +12 |
| 2 | 28 | + | 26 | 11 | +11 |

Sum of positive ranks: $T+ve = 69$
Sum of negative ranks: $T-ve = 9$

Test statistic:   The larger sum of like signed ranks.

Calculated value of $T = 69$

Number of pairs ($n$) $= 12$

Critical value of $T = 65$

Decision rule:   Reject the null hypothesis if calculated value is equal to, or greater than, critical value.

Decision:   Significant evidence for the alternative hypothesis ($P < 0.05$). Mayfly larvae are more numerous at the edges of streams.

the stream has two edges which might be different so, just to be on the safe side, the edge to be sampled was chosen at random at each site.

## The test statistic *T*

To carry out the test, we first note for each pair the direction of the difference and then calculate the size of the difference (*d*). These values are given in the third and fourth columns. The next stage is to ignore the signs in column three and to focus exclusively on the absolute values of *d*. We rank these from smallest to largest, giving the smallest rank (1) to the smallest difference (irrespective of its sign). So, the difference of (+)1 is taken as the smallest value and given the rank of 1, while the difference of (–)4 is taken as the next largest value and is given the rank of 2. These ranks are given in the fifth column. The next step is to reincorporate the information on the direction of each difference by giving each rank the same sign as the difference to which it refers. These are the signed ranks and are given in the sixth column.

If there is no difference in the numbers in the two habitats, we would not expect the two members of a pair to be identical. By chance some pairs would have larger numbers in the edge, others would have larger numbers in the middle, but overall the directions and the sizes of the differences should be more or less equal. Pairs giving rise to large positive differences (and ranks) should be balanced by other pairs giving large negative differences (and ranks) and similarly for pairs with small differences. One way to summarize all this information would be to calculate separately the sum of the positive ranks and the sum of the negative ranks, that is the sums of the like-signed ranks.

Under the null hypothesis, the sums of the like-signed ranks should be more or less equal. If the sum of one of the like-signed ranks is either very small or very large, then this would be evidence for the alternative hypothesis. This suggests that one of these sums could be used as a test statistic. As with the sign test, most books use the smaller of these two sums as the test statistic, but we shall use the larger, which in this case is the sum of the ranks with positive signs. The decision rule will then be in the standard form; if a calculated value of *T* is equal to, or greater than, the critical value, then we reject the null hypothesis. Now we are back to the familiar problem that we need to know the sampling distribution of *T* under the null hypothesis, so that we can assess the probability associated with our calculated value.

## The sampling distribution of *T*

We could produce a sampling distribution of *T* for our example of 12 pairs by the same sort of procedure that we used in the Mann–Whitney *U* test, but it is very laborious so we won't. We shall, however, look briefly at the underlying theory.

With 12 pairs there are 12 differences (we are ignoring, for the moment, the possibility of the two members of a pair having the same number of larvae). If the null hypothesis is true, then for any pair, a difference in one direction will be just as likely as a difference in the opposite direction (as was the case in the sign test). At one extreme the differences could all be positive, at the other all negative and in between there could be any combination of the two signs. Similarly, any of the ranks could belong to any of the pairs if the null hypothesis is true and this means that any of the 12 ranks can have either of the two signs. Another way of stating the null hypothesis is that the median difference between members of pairs is zero.

We could list all the possible allocations of signs to ranks and for each one calculate the value of $T$. If we have 12 pairs, then there will be $2^{12} = 4096$ different allocations (hence the decision not to look at this in detail!). If we work with $T$ as the sum of the positive ranks, then, when all the differences are positive $T = 1+2+3+4+5+6+7+8+9+10+11+12 = 78$. When the 11 largest differences are positive and the smallest difference is negative, then $T = 2+3+4+5+6+7+8+9+10+11+12 = 77$, etc. When they are all negative, $T = 0$. If we worked through all the 4096 allocations, we would find that different values of $T$ occurred with different frequencies. The more extreme ones, e.g. $T = 78$, only occur once, whereas intermediate values are more frequent. We could produce a frequency distribution of these values of $T$ and this of course is the sampling distribution of $T$ under the null hypothesis. Table A15 gives critical values for one-tailed and two-tailed tests for different significance levels; the notation is the same as in the other tables, as is the decision rule.

Our calculated value of 69 is larger than the critical value of 65 and so we can reject the null hypothesis. Note how this test is more powerful than the sign test. Out of the 12 differences, nine are positive, so in the sign test, $S$ would be 9. This is less than the critical value of 10, so that we would have had no evidence against for the alternative hypothesis if we had used the sign test. Wilcoxon's test is more powerful and the reason is that it takes account of both the signs and the sizes of the differences, which, in this case means that the negative differences, because they are small, do not carry as much weight as the positive differences. The test also has the virtue of being distribution free so it does not depend on the population of differences from which the sample is drawn being normally distributed. Thus it can be used in any situation where we are not sure about the underlying distribution of the variable. Later on we shall consider the equivalent test for a variable that is normally distributed.

## One-tailed tests with tied observations

The biological issue dealt with in Box 10.3 is one that we have already met in Chapter 9; the effect of drinking on urea concentration. It has been included because it illustrates how a paired experimental design and statistical analysis can be more powerful than one based on independent samples. It also illustrates two "technical" points about the test, both of which we have met elsewhere.

In the previous experiment on drinking and urea concentration, subjects were allocated to the two treatments at random, so we had two independent samples. In this example, the design is paired. Each subject was both a drinker and an abstainer, with the two treatments being administered sequentially 1 week apart but on the same day of the week and at the same time of day. Half of the subjects were chosen at random to be drinkers in week 1 and abstainers in week 2, while the sequence was reversed for the rest. This was to control for any differences that there might be between the two times. The null hypothesis is that there is no difference in urea concentration in the two treatments, which means that the average difference between the pairs should be zero. As before, the alternative hypothesis is one-tailed.

The test proceeds as usual. We calculate the differences for each pair and rank them without regard to their sign. One difference is zero and this is discarded from the analysis in

---

**BOX 10.3 WILCOXON'S SIGNED RANK TEST FOR PAIRED SAMPLES: UREA CONCENTRATION IN DRINKERS AND NON-DRINKERS**

Null hypothesis:   There is no difference in urea concentration in urine of drinkers and non-drinkers. Median difference between members of a pair is zero.

Alternative hypothesis:   Urea concentration is higher in non-drinkers compared to drinkers. Median difference between members of a pair is greater than zero.

Significance level:   5% , one tailed.

| Urea concentration (mg/ml) Non-drinkers | Drinkers | Sign of difference | Size of difference | Rank of difference | Signed rank |
|---|---|---|---|---|---|
| 4.4 | 5.0 | − | 0.6 | 1 | −1 |
| 18.6 | 17.2 | + | 1.4 | 2 | +2 |
| 9.0 | 9.0 | | 0 | | |
| 20.0 | 24.0 | − | 4.0 | 3.5 | −3.5 |
| 31.5 | 18.5 | + | 13.0 | 8 | +8 |
| 26.0 | 21.5 | + | 14.5 | 9 | +9 |
| 17.0 | 7.6 | + | 9.4 | 5 | +5 |
| 17.2 | 5.8 | + | 11.4 | 6 | +6 |
| 20.2 | 7.6 | + | 12.6 | 7 | +7 |
| 11.5 | 7.5 | + | 4.0 | 3.5 | +3.5 |

Sum of positive ranks: T +ve = 40.5
Sum of negative ranks: T −ve = 4.5

---

Test statistic:   Sum of the like signed ranks which should be the largest under the alternative hypothesis (in this case T +ve).

Calculated value = 40.5

Number of pairs with non-zero difference ($n$) = 9

Critical value of $T$ = 37

Decision rule:   Reject the null hypothesis if calculated value is equal to, or larger than, critical value.

Decision:   Significant evidence for the alternative hypothesis. Urea concentration higher in non-drinkers ($P < 0.05$).

the same way as in the sign test. Two differences have the same absolute value (4.0) and they share ranks 3 and 4. They are both allocated the average of their shared ranks, that is 3.5. The test statistic for a one-tailed test is not necessarily the sum of the like signed ranks which is the largest. It is the sum which is predicted to be the largest, if the alternative hypothesis is correct. Because the differences have been calculated as "non-drinkers minus drinkers", we expect most of them to be positive if the alternative hypothesis is true. Consequently it is the sum of the positive ranks which should be largest in this case, so this is the calculated value we need. The critical value is based on the number of pairs with non-zero differences so $n = 10 - 1 = 9$, giving a critical value of 37. The calculated value of 40.5 is larger than this so we can reject the null hypothesis.

With the two independent samples in Box 9.6, we found no evidence for the alternative hypothesis, even though it was based on a larger number of observations and despite the fact that the sets of data look similar in the two cases. This is due to the greater power of the paired design and the paired test. You can see that the paired test has greater power if you analyse (incorrectly) the data in Box 10.3 using the Mann–Whitney $U$ test. The result comes out as non-significant.

## The *t* test for paired samples

Wilcoxon's test makes no assumptions about the shape of the distribution of the differences between the pairs. When we are sure that the sample of differences is drawn from a normally distributed population, then we can use a modification of the $t$ test. This test uses the actual differences to calculate the mean difference and the standard error of the difference. It then examines whether the mean difference departs significantly from zero.

The land snail (*Cepaea nemoralis*) is polymorphic for shell colour, which means that several different shell colours are found in most (biological) populations. These yellow, pink and brown colour forms are genetically controlled and their frequencies vary from one population to another. One possible explanation for this frequency variation is that natural selection favours different shell colours in different environments and you may be familiar with studies on selective predation in this species. Selective mortality due to climatic factors has also been suggested to account for the observation that dark coloured shells are more common in cooler, less sunny places, while pale shells are at a higher frequency where it is warmer and sunnier. The reasoning is that dark colours absorb more radiant thermal energy than do light colours, so dark shelled snails would be more prone to overheat and die in sunny places, than would pale shelled snails.

To test whether shells of different colours did reach different temperatures in sunlight, pairs of mercury-filled shells of the same size and shape but of contrasting colours were exposed to sunlight at the same time, side by side and in the same orientation. A thermistor inside each shell measured the temperature. The null hypothesis is that, on average, there is no temperature difference between the members of a pair, and a one-tailed alternative hypothesis is probably in order. This is because we could predict from the laws of physics that if there is a temperature difference, then it will be that darker shells are hotter.

A set of results for 10 pairs of brown and yellow shells is given in Box 10.4. The third column gives the temperature difference, with a negative sign indicating that the lighter coloured shell was hotter than the darker coloured shell. Our observations are not now the 20 temperatures but the 10 differences in temperature and all we need to do is to calculate their mean and standard deviation. This can be done using the standard deviation function on your calculator, but a word of warning. You must include the sign of the difference, so negative differences have to be entered as such. This can be done by pressing the $+/-$ button after you have keyed in the number, but before you have entered it.

If the null hypothesis is true then we can imagine a sampling distribution of mean differences between pairs, based on samples of ten pairs of shells. This distribution will have a mean of zero and a spread which, as usual, is related to the sample size and the variability in the population ($\sigma$). The population now is a population of temperature differences be-

---

**BOX 10.4 *t* TEST FOR PAIRED SAMPLES: THE EFFECT OF SHELL COLOUR ON INTERNAL TEMPERATURE IN THE LANDSNAIL (*CEPAEA NEMORALIS*).**

Null hypothesis:   No difference in temperature between brown and yellow shells. Mean difference between members of a pair is zero.

Alternative hypothesis:   Brown shells have higher internal temperatures than yellow shells. Mean difference between members of a pair (brown-yellow) is greater than zero.

Significance level:   5%, one tailed

| Temperature (°C) | | Difference, *d* |
| Brown | Yellow | (°C) |
|---|---|---|
| 25.5 | 25.6 | −0.1 |
| 27.5 | 27.8 | −0.3 |
| 27.3 | 26.3 | +1.0 |
| 27.3 | 25.9 | +1.4 |
| 29.2 | 28.0 | +1.2 |
| 25.3 | 25.4 | −0.1 |
| 26.4 | 24.6 | +1.8 |
| 28.5 | 28.9 | −0.4 |
| 28.1 | 27.2 | +0.9 |
| 26.4 | 26.0 | +0.4 |

Mean difference ($\bar{d}$) = + 0.58
Standard deviation (*s*) = 0.783

$$t = \frac{\bar{d}}{s/\sqrt{n}} = \frac{0.58}{0.2476} = 2.343$$

Degrees of freedom = Numbers of pairs − 1 = 10 − 1 = 9

Critical value of *t* = 1.812

Decision rule:   Reject null hypothesis if calculated value is equal to, or larger, than critical value.

Decision:   Significant evidence for alternative hypothesis ($P < 0.05$). Temperature higher in brown shells.

tween pairs of shells. The standard error (remember this is what describes the spread of a sample statistic) is $\sigma/\sqrt{n}$, where $n$ is the number of pairs. We can estimate this by $s/\sqrt{n}$. Furthermore, if the differences follow a normal distribution so will the means of the differences.

We now can ask the familiar question: "How far away is our mean difference from the mean of the sampling distribution in relation to the size of the standard error?". This is what the formula in Box 10.4 is doing. Since the mean of the sampling distribution is zero we can ignore it and simply calculate how big the mean difference is in relation to the standard error. The answer will be a value of $t$ (because we are taking small samples from a normally distributed population) which can be compared with a critical value in exactly the same way as before. The degrees of freedom are $(n - 1)$, where $n$ is the number of pairs.

With a two-tailed test the sign of the difference is immaterial, but we are doing a one-tailed test so the sign has to be considered. Because our observations are differences calculated as "brown temperatures minus yellow temperatures" the mean difference is going to be positive if the alternative hypothesis is correct. It follows that large positive values of $t$ are going to provide evidence for the alternative hypothesis, so the critical value will lie in the upper, right-hand tail of the sampling distribution. We have significant evidence for the alternative hypothesis.

## Why paired tests are more powerful

We can use this last example to examine why paired tests are more powerful. First, we can re-analyse it using the $t$ test for independent samples. It is actually quite incorrect to do this because the samples are not independent but it is informative. The result (Box 10.5) is non-significant, in the same way that the Mann–Whitney $U$ test would give us a non-significant result on the paired urea data in Box 10.3.

Why does a different test on the same data give such a different conclusion? The answer is clear if we look at the quantities at the final stage in the calculation of $t$ in the two cases. It cannot be due to the quantity in the numerator (the top line) because it is the same in both cases, i.e. 0.58. Indeed it can be shown mathematically that the mean difference between $n$ pairs will always be equal to the difference between the means of two samples of size $n$. What is different is the bottom line, which is the standard error of the difference. With the paired samples it is less than half the size.

The standard error is a measure of the chance variation in a statistic, so this tells us that the extent of chance variation is less with the paired design than with the independent design. At first sight this is rather surprising because of the difference in the value of $n$. In the paired experiment $n = 10$ (the number of pairs), whereas in the independent design the total sample size is 20. Surely the standard errors should be the other way round, that is, larger in the experiment based on the smaller sample size?

The standard error is smaller because of the way we have done the experiment, that is, because of the paired experimental design. The temperature inside a shell may be affected by many factors besides its colour. Factors such as shell shape and size, orientation to the sun and the incident energy from the sun at the time, could all have an effect. The temperature is therefore the result of a number of factors, some with positive effects and some with negative effects. If, for instance, we take a random sample of yellow shells and expose

**BOX 10.5  INCORRECT ANALYSIS OF TEMPERATURE DIFFERENCES BETWEEN PAIRS OF YELLOW AND BROWN SHELLS USING $t$ TEST FOR INDEPENDENT SAMPLES (DATA FROM BOX 10.4)**

|  | Mean temperature (°C) | |
|---|---|---|
|  | Yellow | Brown |
| Mean | 26.57 | 27.15 |
| Standard deviation | 1.35 | 1.26 |
| Variance | 1.825 | 1.596 |

$$t = \frac{27.15 - 26.57}{\sqrt{\dfrac{9(1.825) + 9(1.596)}{18}\left[\dfrac{1}{10} + \dfrac{1}{10}\right]}}$$

$$= \frac{0.58}{0.5849} = 0.9917$$

Degrees of freedom $= (n_1 - 1) + (n_2 - 1) = n_1 + n_2 - 2 = 18$

Critical value of $t = 1.7341$
($\alpha = 5\%$, one tailed)

Decision rule:   Reject the null hypothesis if calculated value is equal to, or greater than, critical value.

Decision:   No evidence that brown shells have higher temperatures.

them to the sun one at a time, there will be a lot of variability in temperature between shells because of these factors. In a randomized design the difference in mean temperature between yellow and brown shells has to be judged against this large background variation in temperature due to differences in shell shape, shell size and incident sunlight. This will lead to a large standard error which is undesirable, so we could describe this variability as being unwanted. It stops us from getting a reliable estimate of the temperature.

The purpose of matching in pairs is to remove or at least to minimize this unwanted "background" variation by matching sampling units. If a yellow and a brown shell are exposed to the sun at the same time, then none of the difference in temperature between them can be due to differences in the amount of incident energy. Similarly, if the two members of the pair are the same size, then none of the temperature difference can be due to a size difference. Since these sources of variation for each pair have been removed all we are left with is any temperature difference due to colour. Because we have removed these sources of unwanted variability in the design of the experiment, the measure of the variability in the statistical test is reduced. The size of the mean temperature difference between pairs can be judged against the variability in the temperature differences, rather than against the variability of the individual temperatures. It is therefore easier to detect differences between treatments. The power of the test, that is, its ability to reject the null hypothesis when it is false, is increased.

Three final points. If we analyse these results using Wilcoxon's signed rank test, the calculated value of $T$ comes out as 44.5, which is just below the critical value of 45 and is therefore not formally significant. The parametric $t$ test is therefore more powerful than the non-parametric signed rank test which is something we saw in the previous chapter. This increased power is due to the fact that measuring gives a more detailed description than ranking and makes it easier to detect differences. It also shows that we can change the conclusion by changing the test. We could therefore put a set of data into a variety of tests and we might be able to find one which gives the answer we want. This would not be fair! It is another reason why you ought to have decided what test you are going to use before you obtain the data. Finally, whatever test we use, the population we have sampled is all the differences which we could have obtained under these conditions and this means that we need to be careful about extrapolating from this study to a more natural situation. It does not follow from this result that brown shelled snails will get hotter – they may have some physiological or behavioural means for reducing their temperature.

# 11

# Tests using three or more samples: do three or more populations differ?

We now move on to the analysis of data from more complex experiments and sampling programmes, that is, situations where we have more than two levels of a treatment or where we have taken samples from more than two habitats, places or times. As result we shall have more than two samples.

One possibility for analysing data of this type may have occurred to you. Can we not simply use the two-sample tests which we already know about and compare every pair of samples? So if we had three samples, we could compare 1 *vs* 2, 1 *vs* 3 and 2 *vs* 3. This type of approach is both time consuming, inefficient and incorrect. For a start, the number of comparisons which are needed goes up rapidly as the number of samples increases. With three samples, there are three comparisons which can be made, but with four samples there are six (1 *vs* 2, 1 *vs* 3, 1 *vs* 4, 2 *vs* 3, 2 *vs* 4 and 3 *vs* 4). In general, if we have $n$ samples then there are $n(n - 1)/2$ comparisons to be made; so if we had grown plants under six different conditions and wanted to test their means, we would be involved in 15 $t$ tests!

Furthermore, we know that if even if the null hypothesis is true, from time to time we will get a result which is so extreme that it leads us to believe that the alternative hypothesis is correct. We saw this with the maize kernels, but it can happen in any test. Indeed, if we are working at the 5% significance level, then one out of every 20 comparisons between two samples will be of this type. We shall make a type 1 error and conclude that the null hypothesis is false when it is true. This means that if we have an experiment with several treatments, so that we have to do a large number of two-sample tests on the data, a few of them will be judged to be significant even when the treatment has no effect. Needless to say, there are ways around this and they form the subject of most of this chapter.

As you can perhaps judge by some of the formulae and calculations in this chapter, the statistical analysis of such data is more involved than the two sample case. The ideas, however, should be familiar because most of the tests described in this chapter are simple extensions of ideas which we have already covered. As for the calculations, you will find that they are not as complex as they seem at first sight. Most can be carried out quite easily on a calculator with a memory and a standard deviation function.

As you might expect from the preceding chapters, the type of test depends on the type and distribution of variable and on the question to be answered. Following our usual pattern of dealing with the simpler situations first, we shall begin with the test for nominal variables. This is very easy because the appropriate test should by now be very familiar. We use contingency tables and $\chi^2$.

# The chi-squared test for differences in proportions in three or more populations

There is, in fact, nothing new to learn here, so we will work through these examples rather quickly. They are mainly intended to illustrate a variety of different applications of the procedure.

## Contingency tables with several columns and two rows

Our first example (Box 11.1) is similar to the one in Box 9.3, where we had two samples of invertebrates with 14 categories in each, giving a contingency table with two columns and 14 rows. Here we have five samples with two categories in each, giving a table with five columns and two rows.

The data come from an observational study into the possible effects of trampling on the abundance of plant species on a footpath. Five lines were set out parallel to the path at varying distances from the centre. One ran down the centre, one was positioned at each edge of the path and the last two were on either side, completely off the path itself. A point frame with a single pin, was positioned at 100 random distances along each line and the number of points touching each species was counted. The sampling unit is a point and it can either touch or not touch a given species, so the variable is a nominal variable which falls into only two categories. The sample is the set of 100 points taken along each line, so there are five samples.

The data for one species can then be put in the form of a contingency table, with the five columns for the five distances. The first set of data in Box 11.1a are the results for bird's-foot trefoil (*Lotus corniculatus*). We can treat this table in exactly the same way as the example in Box 9.3, because a contingency table with five columns and two rows is the same mathematically as one with two columns and five rows. When sample sizes are equal, as they are here, the expected frequencies are the same for all the cells in a row, which means that only one in each row needs to be calculated. However, when sample sizes are equal, we could get away without even doing this, because we do not need the expected frequencies to help with the interpretation of a significant result; the equal sample sizes mean that any patterns in the data are immediately apparent. We could use the short-cut formula given in Box 9.3, labelling the values in the first row as $a_1$, $a_2$, etc., and the values in the second row as $b_1$, $b_2$, etc.

The calculated value far exceeds the critical value at the 5% significance level. Indeed, the probability associated with the calculated value is vanishingly small so we have almost conclusive evidence for the alternative hypothesis that the proportion of pins hitting this species is not the same at the different distances across the path. All this tells us is that there are some differences somewhere in the data, but it does not identify which proportions are different from which.

In Boxes 6.5 and 9.3 we had the same sort of result and we saw that we had to be careful about advancing the interpretation too far and trying to pinpoint where the differences were. In this case our interpretation can safely be pushed further, because of the clear pattern in the data. It would be reasonable to conclude that bird's-foot trefoil is more common on the path and becomes less abundant with increasing distance from the path. Note though, that there is possibly another pattern superimposed on this. Although this species is more common on than off the path, it looks as though it might be less common on one side of the path than on the other. This would need further investigation.

**BOX 11.1 CHI-SQUARED TEST FOR A DIFFERENCE IN PROPORTIONS IN THREE OR MORE SAMPLES WHEN THERE ARE ONLY TWO CATEGORIES: THE EFFECT OF TRAMPLING ON THE ABUNDANCE OF (a) BIRD'S-FOOT TREFOIL (*LOTUS CORNICULATUS*) (b) RESTHARROW (*ONONIS REPENS*)**

Null hypothesis:   Proportion of pin hits the same at different distances across a  footpath.

Alternative hypothesis:   Proportion of pin hits not the same at different distances across a footpath.

Significance level:   5%.

**(a)  Bird's-foot trefoil**

| | \multicolumn{6}{c}{Distance across path (m)} |
| | 0 | 1 | 2 | 3 | 4 | Total |
|---|---|---|---|---|---|---|
| Frequency of pin hits: observed and (expected) | 26(39.2) | 52(39.2) | 75(39.2) | 23(39.2) | 20(39.2) | 196 |
| Frequency of pin misses: observed and (expected) | 74(60.8) | 48(60.8) | 25(60.8) | 77(60.8) | 80(60.8) | 304 |
| Total | 100 | 100 | 100 | 100 | 100 | 500 |

$$X^2 = \sum \frac{O^2}{E} - N = 94.44$$

Degrees of freedom = $(c - 1)(r - 1) = 4$

Critical value of $\chi^2 = 9.488$

Decision rule:   Reject the null hypothesis  if calculated value of  is equal to, or greater than, critical value.

Decision:   Almost conclusive evidence for alternative hypothesis ($P < 0.001$). Bird's-foot trefoil more abundant in centre of path, less abundant at edges.

**(b)  Restharrow**

| | \multicolumn{5}{c}{Distant across path (m)} |
| | 0 | 1 | 2 | 3 | 4 |
|---|---|---|---|---|---|
| Frequency of pin hits observed | 30 | 44 | 25 | 31 | 38 |
| Frequency of pin misses observed | 70 | 56 | 75 | 69 | 62 |

$X^2 = 9.915$

Critical value of $\chi^2 = 9.488$

Decision:   Significant evidence that trampling affects abundance ($P < 0.05$).

The problems of interpretation of significant results is highlighted in the second set of data in Box 11.1 which is from the same study. There is significant evidence that the frequency of hits varies with distance, but the pattern is less clear.

## A more complex example with several rows

This test can be easily extended to cover situations in which there are several columns (i.e. samples) and where each sampling unit can be put into one of several categories. The results can be entered in an $r \times c$ contingency table in the same way as in Box 8.3.

---

**BOX 11.2 CHI-SQUARED TEST FOR DIFFERENCE IN PROPORTIONS IN THREE OR MORE SAMPLES WHEN THERE ARE SEVERAL CATEGORIES: PROPORTION OF CELLS IN DIFFERENT STAGES OF MITOSIS IN ROOT TIPS OF *ALLIUM* IN SAMPLES SCORED BY FOUR STUDENTS**

Null hypothesis:   Proportion of cells at each stage of mitosis the same in the four populations.

Alternative hypothesis:   Proportion of cells at each stage of mitosis not the same in the four populations.

Significance level:   5%.

|  | Student 1 | 2 | 3 | 4 | Row total |
|---|---|---|---|---|---|
| Interphase | 250 | 204 | 189 | 227 | 870 |
| Prophase | 23 | 27 | 18 | 15 | 83 |
| Metaphase | 13 | 11 | 8 | 7 | 39 |
| Anaphase | 5 | 5 | 3 | 5 | 18 |
| Telophase | 8 | 7 | 8 | 7 | 30 |
| Column total | 299 | 254 | 226 | 261 | 1040 |

$$X^2 = \sum \frac{O^2}{E} - N = 6.699$$

Degrees of freedom $= (r - 1)(c - 1) = (5 - 1)(4 - 1) = 12$

Critical value of $\chi^2 = 21.026$

Decision rule:   Reject null hypothesis if calculated value is equal to, or larger than, critical value.

Decision:   No evidence for alternative hypothesis.

---

Box 11.2 presents some data on the frequency of cells at different stages of mitosis, gathered by four students. The sampling units are individual cells and the variable is the stage of mitosis of the cell (actually this is another continuous variable which has been categorized). The purpose of the exercise was to produce a reliable estimate of the proportion of cells at each stage, something which requires as large a sample as possible. An efficient way to do this is to combine the results from several sources (i.e. several microscope slides and students). However, if the samples have been drawn from different populations, then adding the results together could give a misleading overall estimate and obscure possibly interesting features, so before we combine results in this way, we always ought to check whether the samples are homogeneous. This illustrates an alternative use for statistical analysis. In most of the previous tests, we have been hoping to find evidence for differences in the variable, either between experimental treatments or between naturally occurring situations which are contrasted in some way. Here we hope for no evidence of a difference between "treatments".

The principle of calculating the value of $X^2$ is exactly the same as before, with the exception that you do have to calculate the expected frequencies. Remember that for a given cell, the expected frequency is the product of the marginal totals divided by the grand total. As each one of these is obtained it can be used immediately to calculate $O^2/E$. These values can be accumulated in the memory. In this example the calculated value lies far below the critical value so we have no evidence that the four samples are drawn from different populations and we can justifiably pool them and use the row totals to calculate the proportions of cells at each stage. Note, however, that it would be incorrect to say that we have shown that the samples have been taken from the same population, i.e. that we have proved the null hypothesis. As we saw in Chapter 6 the null hypothesis is probably never true and even if it was a non-significant result would not prove it.

# The Kruskal–Wallis test for differences in the medians of three or more populations

Just as there was a two sample test for ordinal variables, the Mann–Whitney $U$ test, so there is a corresponding test for three or more samples. Like the two-sample test, it can be used on data which were gathered as ordinal data, or it can be used on counts and measurements which are subsequently converted to ranks. This test is called the Kruskal–Wallis test and, like the Mann–Whitney $U$ test, it is a non-parametric, distribution-free test, which tests for differences in medians. It does not require estimates of parameters such as the mean and the variable does not have to have a specified (e.g. a normal) distribution. This last property means that the test can be used when we either lack information about the distribution of the variable and feel that a normal distribution may not be appropriate, or when we know that the distribution is not a normal one.

Box 11.3 (p. 278) presents some data which are suitable for this test. Many plant species consist of what are known as ecotypes. Ecotypes are phenotypically distinct forms of a species, which are genetically different and which have arisen through adaptation to differing local environments. In this study, 10 random plants of ribwort plantain (*Plantago lanceolata*) were taken from five sites and the dry weight of their shoots and roots measured. The variable of interest is the ratio of shoot weight to root weight. Is there any evidence that average shoot/root ratios are different at the five sites?

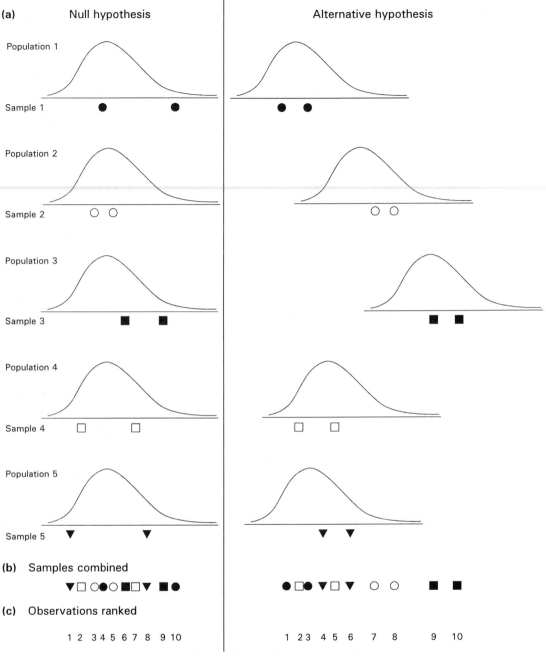

**(a)** Null hypothesis — Alternative hypothesis

Population 1
Sample 1

Population 2
Sample 2

Population 3
Sample 3

Population 4
Sample 4

Population 5
Sample 5

**(b)** Samples combined

**(c)** Observations ranked

1 2  3 4 5 6 7 8  9 10     1 2 3  4 5  6  7  8   9  10

**(d)** Sum of ranks for each sample $R_1$, $R_2$, $R_3$, $R_4$, $R_5$

Rather similiar     Quite different

**(e)** $H = \dfrac{n_1\left(\bar{R}_1 - \bar{R}\right)^2 + n_2\left(\bar{R}_2 - \bar{R}\right)^2 + n_3\left(\bar{R}_3 - \bar{R}\right)^2 + n_4\left(\bar{R}_4 - \bar{R}\right)^2 + n_5\left(\bar{R}_5 - \bar{R}\right)^2}{\left(N^2 - 1\right)/12} \times \dfrac{(N-1)}{N}$

**(f)** Formula for calculation, $H = \dfrac{12}{N(N+1)} \sum \dfrac{R^2}{N} - 3(N+1)$

**Figure 11.1** The Kruskal–Wallis test: how it works.

The test is in two stages. Initially we test the null hypothesis that the five samples have been taken from populations which have the same median shoot/root ratio. (Note: the populations are populations of measurements not populations of plants!) If we find no evidence against this hypothesis, then we proceed no further. If, on the other hand, we reject this hypothesis (i.e. we have evidence that the samples have come from populations with different medians), we move on to the second stage. This involves finding out which of the populations are different.

## Testing whether the samples have been taken from populations with the same median: how does the test work?

We can illustrate this using the diagrams in Figure 11.1. We start as always from the null hypothesis which is that the five populations all have the same location, which is equivalent to them all having the same median. The test is based on the assumption that the distribution of the variable has the same shape in all the populations. Both these points are shown by the curves on left in Figure 11.1a. The symbols under each curve represent the observations in a random sample from each population; for simplicity in this explanation we shall work with sample sizes of two. The test is similar to the Mann–Whitney $U$ test in that the observations are ranked and the sums of the ranks are used to calculate the test statistic.

The observations from all the samples are first of all combined and placed in order with the origin of each observation (i.e. the sample to which it belongs) being noted (Fig. 11.1b). The position of each observation in the order is then described by a numerical rank (Fig. 11.1c). The sum of the ranks ($R$) is then determined separately for each sample ($R_1$, $R_2$, etc.). If the null hypothesis is true, then there should be a great deal of overlap between the samples. As a result, high and low ranks should be scattered fairly evenly across all the samples and no sample should have a preponderance of high (or low) ranks. The sum of the ranks for each sample ($R_1$, $R_2$, etc.) should therefore be rather similar, with the only differences being due to sampling variation.

If the alternative hypothesis is true then there are differences between the medians of some (at least two) of the populations, as shown on the right of Figure 11.1. Not all the populations have to have different medians, so the pattern of differences shown is only one of a number of patterns which there could be. Under these circumstances, the lower ranks will tend to be concentrated in some of the samples, the higher ranks in others. The sums of the ranks for the different samples will now tend to be quite different from one another. From this we can see that the variation in $R$ (the sums of the ranks) should be large if the alternative hypothesis is correct and small if the null hypothesis is true. This suggests that some measure of the variation in $R$ could be used as a test statistic, although it will have to have a known sampling distribution under the null hypothesis.

A suitable test statistic ($H$) is given by the formula in Figure 11.1e and, as you can see, things are starting to get a bit more complicated! I shall try and unpack this for you so you can see how the test works, but if you just want to get on and carry out the test, then go straight to the next section. As is often the case, the most convenient formula for calculating the test statistic is not the easiest one to use if you want to understand what the test is doing, so an alternative form is presented in Figure 11.1f. This second formula is quite easy to handle on an ordinary calculator.

So what is this formula doing and what does it tell us about how the test works? If the null hypothesis is true then as we saw the sums of the ranks of the samples ($R_1$, $R_2$, etc.) should be similar. However, using the sums of the ranks will be misleading if the sample sizes are not the same, because a larger sample will automatically tend to have a larger value of $R$. To overcome this, we work with mean ranks of each sample $\bar{R}_1$, $\bar{R}_2$, etc. Under the null hypothesis these mean ranks should all be close to one another and to the overall mean rank $R$ denoted by $\bar{R}$.

A useful measure of this closeness can be obtained by finding out the deviation of the mean rank of each sample from the overall mean and summing the deviations for each sample. You will perhaps recognize this type of procedure from elsewhere and it is fairly obvious that some of these deviations will be negative and some will be positive. In fact, their sum will be zero. As you might expect, to overcome this we shall need to square the deviations before we add them and this explains the terms in the brackets, e.g. $(\bar{R}_1 - \bar{R})^2$, in the top line of the formula. These deviations are not necessarily all equally useful, because their reliability depends on the sample size and we ought to take proportionally more notice of those based on large samples. We can do this by weighting the squared deviations by the sample sizes, which is why we multiply each deviation by its appropriate $n$, before adding them. Finally, we have to bear in mind that the size of this sum will be affected by the total number of observations ($N$) involved. As $N$ gets larger, so we have to use larger ranks. Mean ranks will increase and so will the sum of the squared deviations. To take this into account, we use the familiar device of standardization which, as elsewhere, involves dividing through by a measure of the variability. In this case we use the variance which can be shown to be $(N^2-1)/12$.

What this all amounts to is that $H$ is a familiar type of test statistic. Like $\chi^2$ it is a sum of squared standardized deviates. In fact, if sample sizes are not too small, the sampling distribution of $H$ is near enough to the distribution of $\chi^2$ and we can use critical values from $\chi^2$ tables – but more of that later.

## The sampling distribution of $H$

The basis for calculating the sampling distributions of $H$ is the same as that used for the sampling distribution of $U$, namely putting the observations into all the possible orders and calculating the values of $H$ that result. We won't attempt to do it because with our five samples there would be over $3 \times 10^6$ possible orders! You only need to know that sampling distributions can be produced.

A sampling distribution tells us all the possible values of $H$ which we could obtain if the null hypothesis were true and their probabilities of occurrence. We would find that extreme large values of $H$ were improbable so large values of $H$ are evidence for the alternative hypothesis. (Large values of $H$ arise under these conditions because the mean ranks of the different samples will be very different from one another. They will therefore be widely spread around the overall mean rank.) We could then use our sampling distribution to set critical values of $H$ which cut off known proportions of the distribution. The critical value of $H$ depends upon three things, the significance level, the sample sizes and the number of samples.

In a test with more than two samples we cannot have a one-tailed test; the alternative hypothesis is simply that the populations differ in median. As we discussed in relation to Figure 11.1, if the null hypothesis is not true, there are many patterns of differences between medians which could be present. At one extreme only two of the populations could have different medians while all the others are the same; at the other extreme they could all be different. Because of this we cannot specify the direction of the difference, as we would need to do for a one-tailed test. The distinction between one-tailed and two-tailed tests disappears so that the significance level can be simply referred to as $\alpha$.

## Calculating the value of $H$

This is fairly straightforward, but it does pay to set the results out clearly to begin with (Box 11.3). First we rank the observations from the smallest to the largest, irrespective of the sample to which they belong. Tied values are treated in exactly the same way as in the Mann–Whitney $U$ test. This is the stage at which things are most likely to go wrong, particularly if there are several ties. You may find it helpful to set out all the observations in order, in a single column. The most common mistake is to get "out of step" with tied values. In this example two observations (0.566) share ranks 10 and 11 so they are both allocated rank 10.5. The next observation (0.584) has rank 12 (not 11, which has already been allocated). The rank of the largest observation should be equal to $N$ (the total number of observations) unless it is a tied value, so this is a useful check, although it is no guarantee that the ranking is correct.

The formula for calculating $H$ is easy to use on a calculator. Clear the memory and sum the ranks to get a value of $R$ for each sample. (If you add these values of $R$, the answer $(\Sigma R)$ should be equal to $N(N+1)/2$, so this is another check on the ranking). As you calculate each value of $R$, square it, divide it by the sample size to get $R^2/n$ and add this into the memory to get $\Sigma(R^2/n)$. Calculate $12/N(N+1)$ and multiply it by what is in the memory, then subtract $3(N+1)$. The answer is $H$.

Strictly, if there are tied values, then we ought to apply a modified formula which corrects the value of $H$ for the presence of ties. Unless there are a lot of ties this correction makes little difference to the calculated value; in our example it makes the value 20.686. The correction always increases the calculated value, which means that the uncorrected value is conservative, so we will not bother with the correction.

## Critical values of $H$

Critical values of $H$ for equal sample sizes are given in Table A16. We have five samples so $k = 5$, and the number of observations in each sample is 10, so $n = 10$. The critical value at the 5% significance level is 9.200. The decision rule has the standard form, because it is large calculated values which are unlikely if the null hypothesis is true. Our value is larger even than the 1% critical value, so we have highly significant evidence for the alternative hypothesis.

The maximum number of samples $(k)$ in Table A16 is 6. If you have more than six samples and the sample sizes are 5 or above, the distribution of $H$ is quite close to a $\chi^2$ distribution and appropriate critical values can be found from tables of critical values of $\chi^2$. The significance level used is as before and the degrees of freedom are one less than the number of samples $(k)$.

We can illustrate this with our example. The degrees of freedom will be $5 - 1 = 4$, giving an approximate critical value of $\chi^2$ of 9.488. (These values are given in the last row of

**BOX 11.3 KRUSKAL–WALLIS TEST FOR DIFFERENCES IN MEDIAN OF THREE OR MORE POPULATIONS: SHOOT/ROOT RATIOS IN RIBWORT PLANTAIN (*PLANTAGO LANCEOLATA*) FROM FIVE SITES**

Null hypothesis:   Median shoot/root ratios the same at all five sites.

Alternative hypothesis:   Median shoot/root ratios not the same at all five sites.

Significance level:   5%.

| Shoot/root ratio (bracketed figures are ranks) | | | | |
|---|---|---|---|---|
| Site 1 | Site 2 | Site 3 | Site 4 | Site 5 |
| 0.768 (23) | 3.377 (50) | 0.750 (20.5) | 1.055 (33) | 0.262 (41) |
| 0.854 (27) | 2.163 (47) | 0.833 (26) | 0.566 (10.5) | 0.149 (1) |
| 1.833 (44) | 2.654 (48) | 1.199 (37) | 0.730 (18) | 0.907 (30) |
| 1.103 (35) | 0.566 (10.5) | 0.518 (7) | 1.521 (41) | 0.716 (16) |
| 0.741 (19) | 1.095 (34) | 0.821 (25) | 0.642 (15) | 0.216 (3) |
| 1.683 (43) | 1.002 (32) | 1.340 (39) | 0.865 (28) | 0.724 (17) |
| 0.920 (31) | 0.896 (29) | 1.543 (42) | 0.565 (9) | 0.614 (14) |
| 1.127 (36) | 1.491 (40) | 1.211 (38) | 0.761 (22) | 0.210 (2) |
| 1.929 (46) | 0.584 (12) | 0.562 (8) | 0.772 (24) | 0.435 (6) |
| 1.837 (45) | 3.168 (49) | 0.589 (13) | 0.750 (20.5) | 0.352 (5) |

$R =$   349   $+$   351.5   $+$   255.5   $+$   221   $+$   98    $\Sigma R = 1275$

$n =$   10   $+$   10   $+$   10   $+$   10   $+$   10    $N = 50$

$\dfrac{R^2}{n} = 12180.10 + 12355.23 + 6528.02 + 4884.10 + 960.40$   $\sum \dfrac{R^2}{N} = 36907.85$

Check:   $\Sigma R = N(N+1)/2$
$$1275 = 50 \times 51/2$$

$$H = \frac{12}{N(N+1)} \sum \frac{R^2}{n} - 3(N+1)$$

$$= \left( \frac{12}{50 \times 51} \times 36907.85 \right) - (3 \times 51)$$

$H = 20.864$  (uncorrected for ties)

  (= 20.686 corrected for ties)

Number of samples being compared $(k) = 5$

Critical value of $H = 9.200$

Decision rule:   Reject null hypothesis if calculated value is equal to, or greater than, critical value.

Decision:   Highly significant evidence ($P < 0.01$) for a difference in medians between sites.

Table A16.) You can see that this is larger than the correct critical value of $H$ ($= 9.200$), which means that this approximation is conservative. Because it errs on the side of caution some differences which would be significant using the exact critical value of $H$ will be declared non-significant using the approximate critical values based on $\chi^2$. As a result, using the approximate critical values, you will fail to find significant differences slightly more often than you should.

The test can be used on quite small sample sizes, for example, three samples of size 2, four samples of sizes 3, 2, 1 and 1 and five samples of sizes 3, 2, 1, 1 and 1 can yield results which are significant at the 5% level. With these small samples, critical values have to be tabulated separately for all combinations of sample sizes and this makes such tables too lengthy for this book. So if you have unequal sample sizes below 5, then you may need to find a set of these tables (e.g. Neave, 1985). The alternative is to use the approximation described above but it is very conservative under these conditions. For example, with samples of sizes 3, 2, 2, 1 and 1, and a significance level ($\alpha$) of 5% the exact critical value of $H$ is 6.800, while the approximate critical value using $\chi^2$ is 9.488.

Returning to our example the highly significant difference that we have found merely tells us that it is very unlikely that the five samples have been taken from populations with the same median shoot/root ratio. The next question is which populations are different from which others. To answer this we need to move to the second stage of the test which involves multiple comparisons.

## Finding which populations are different: multiple comparisons between samples

There are several different procedures available but we shall only deal with one, which has the advantages of being simple, suitable for unequal sample sizes and conservative in its outcome. As we saw earlier, the Kruskal–Wallis test actually uses mean ranks, although they do not appear as such in the formula. This second stage (Box 11.4) uses the difference in mean ranks of a pair of samples as a convenient measure of the difference between two populations. It works like the $z$ test, by examining the difference between each pair of samples in turn and asking a familiar question: "How big is the difference between them in terms of the standard error of the difference?". To put this test into operation we shall need to make a modification to our ideas about significance levels and also meet a new test statistic.

## The test statistic $Q$

Suppose that the median shoot/root ratio is the same in a pair of populations and imagine a sampling distribution of the difference in mean rank between pairs of samples (one from each population). It can be shown that this distribution will have a mean of zero (i.e. on average, there is no difference between mean ranks) and a standard deviation given by the formula in Box 11.4a. This is a sampling distribution of a statistic, so, as usual, we shall call this standard deviation a standard error. The distribution will also be approximately normal in shape.

Now we are back on familiar territory. We can examine an actual difference between a pair of mean ranks in the light of the likely magnitude of sampling variation measured by

---

**BOX 11.4  KRUSKAL–WALLIS TEST: MAKING MULTIPLE COMPARISONS**

(a) For any two samples A and B the distribution of the differences in mean rank $(\bar{R}_A - \bar{R}_B)$:

(1) Has a mean of 0.

(2) Has a standard error of $\sqrt{\dfrac{N(N+1)}{12}\left(\dfrac{1}{n_A} + \dfrac{1}{n_B}\right)}$.

(3) Is approximately a normal distribution.

(b) For each pair of samples calculate:

$$\frac{\text{Difference in mean rank}}{\text{Standard error}} = \frac{\bar{R}_A - \bar{R}_B}{\sqrt{\dfrac{N(N+1)}{12}\left(\dfrac{1}{n_A} + \dfrac{1}{n_B}\right)}} = Q$$

(c) Compare calculated value of $Q$ to critical value in Table A17.

---

the standard error of the difference (Box 11.4b). Dividing the differences in mean ranks by the standard error is exactly the same procedure that we used in the $z$ test. There, you remember, the test statistic was a value of $z$ because the samples were large enough to give us a very reliable estimate of the population standard deviation. Here, the formula for the standard error only involves the sample sizes and as a result we actually know what the population standard deviation is. Calculated values based on this expression will also be values of $z$. A large value will result when the difference in mean ranks (of a pair of samples) is large relative to the standard error and this will be indicative of a real difference between the pair of populations. So these calculated values can act as a test statistic.

However, we cannot refer a calculated value of this test statistic to a table of critical values of $z$ for the following reason. We know that if we set the significance level at 5%, then on average one comparison in 20 will by chance yield a value of the test statistic which would lead us to reject the null hypothesis, even when it is true. We saw this happening with the maize kernels in Box 5.1.

When we are making a single comparison we are prepared to take this risk, because we can argue that our particular comparison is not very likely to be one of those which gives the "wrong" answer. In fact we only have a 5% chance of getting it wrong. When we are contemplating making several comparisons at the 5% level, then we obviously increase the chance that one or more of them will lead to an erroneous conclusion. If we make 20 comparisons, then the odds are very much in favour of one of them being of this type; by chance, two or three of them could lead us to say that there was a difference, even though in fact there was not.

To overcome this problem we have to have a more stringent criterion for rejection, that is, we have to set a smaller significance level for each individual comparison, while retaining the same overall significance level for the set of comparisons. This involves sharing out the overall risk (say, 5%) equally amongst the multiple comparisons. In our case we have 10 comparisons to make so that the appropriate figure for each is $5\%/10 = 0.5\%$. This means that we have to use a larger value of $z$ for our critical value, one which cuts of 0.5% in the two tails of the distribution or 0.25% in each tail.

We could obtain this value directly from Table A4, where you will find it to be 2.81, but it is easier to use a special table of critical values (Table A17). These are values of a new test statistic, $Q$, which have been derived from the tables of $z$ using the logic we have just worked through. The critical value depends on the total number of samples to be compared ($= k$) and the overall significance level ($\alpha$) for the whole set of comparisons. For $k = 5$ and $\alpha = 5\%$, the critical value of $Q$ is 2.807 (our value from Table A4 is slightly inaccurate because of rounding errors).

## Calculating the values of $Q$

Calculating a value of $Q$ is easy, but there are 10 different values to find. The following is the most efficient way to deal with this on a simple calculator.

---

**BOX 11.5  KRUSKAL–WALLIS TEST: MULTIPLE COMPARISONS OF DIFFERENCES IN SHOOT/ROOT RATIOS IN RIBWORT PLANTAIN (*PLANTAGO LANCEOLATA*) FROM FIVE SITES**

Null hypothesis:   A pair of populations (A and B) have same median.

Alternative hypothesis:   A pair of populations (A and B) have different medians.

Significance level:   5%, over all comparisons.

Standard error for comparisons of samples of size $n_A = 10$, $n_B = 10$:

$$\sqrt{\frac{50 \times 51}{12}\left(\frac{1}{10} + \frac{1}{10}\right)} = 6.52$$

| Site | $\bar{R}$ | | | | | |
|------|-----------|---|---|---|---|---|
| | | Values are calculated values of $Q$ | | | | |
| 2 | 35.15 | | | | | |
| 1 | 34.90 | | | | | 0.038 |
| 3 | 25.55 | | | | 1.442 | 1.473 |
| 4 | 22.10 | | | 0.529 | 1.963 | 2.001 |
| 5 | 9.80 | | 1.886 | 2.416 | 3.850 | 3.889 |
| | | 9.80 | 22.10 | 25.55 | 34.90 | 35.15  $\bar{R}$ |
| | | 5 | 4 | 3 | 1 | 2  Site |

Critical value:   = 2.807

Decision rule:   Reject the null hypothesis for a given pair of samples if calculated value is equal to, or larger than, critical value.

Decision:   Median shoot/root ratios higher at sites 1 and 2 than at site 5.

---

Set out the mean ranks in order of increasing size along the two sides of a table as shown in Box 11.5 and include the sample sizes if they are unequal. Look up the critical value of $Q$. With equal sample sizes the standard error will be the same for all comparisons and only needs to be calculated once and stored in the memory. Then calculate the difference between each pair of mean ranks, divide by the standard error and write the value of $Q$ in the table. If you work systematically through the comparisons dealing with the largest differences first you may not need to calculate every value of $Q$. Once the value entered in a column is less than the critical value you know that any smaller difference cannot be significant.

With unequal sample sizes, the standard error will vary from comparison to comparison so you will have to calculate several. I suggest you calculate $N(N+1)/12$ and write the answer down somewhere and then calculate the standard error for a given pair of sample sizes and store it. Then work through all the pairwise comparisons with these size samples, again writing the values of $Q$ in the table. Repeat for the other combinations of sample size.

I have calculated all the values of $Q$ in our example and you can see that only two of them are larger than the critical value. We can conclude that the median shoot/root ratio is higher at sites 1 and 2 than it is at site 5. There are no significant differences between other sites.

## One-way analysis of variance for differences in means of several populations of a normally distributed variable

Analysis of variance often abbreviated to anova, is the parametric analogue of the Kruskal–Wallis test and has the same relationship to it as the $t$ test has to the Mann–Whitney $U$ test. It is a parametric test because it involves the estimation of parameters and one of the assumptions on which is based is that the variable has a normal distribution. Like the Kruskal–Wallis test it is a two-stage procedure; we first test the hypothesis that the populations have the same mean and, if this is rejected, we go on to perform multiple comparisons to see where the differences are.

We shall be using this test on some data on mean enzyme activity at four different pH values, which are given in Box 11.6 (p. 287). It will be a one-way analysis of variance because we are using it to investigate the effect of one factor, i.e. pH (of which there are four different levels). Although our data are based on samples of equal size, a one-way analysis of variance can also be carried out when samples are of unequal size. At first sight a test called "analysis of variance" seems quite inappropriate, because we are interested in differences between means. Surely means are measures of location and variances are measures of spread, so how can we use variances to answer a question about differences in means? To see how, we need to understand the principle of the test.

### How does analysis of variance work?

Because the ideas behind this test are more difficult we shall work through the explanation twice, first using diagrams and then putting the ideas into symbols. Figure 11.2 is similar in layout to Figure 11.1. The four curves on the left represent the distribution of enzyme activity at each of the four different pH values, assuming that the null hypothesis is true. If this is

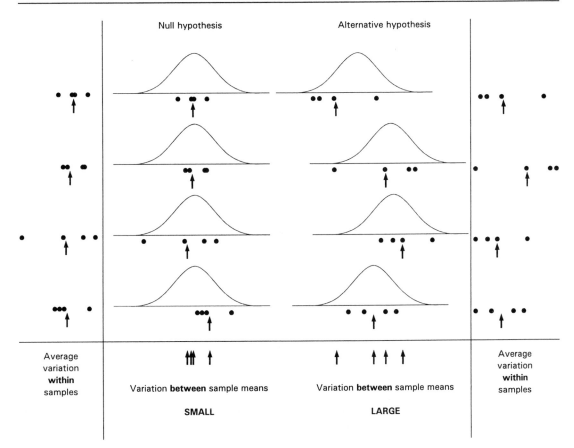

**Figure 11.2**  Analysis of variance: how it works.

the case, then pH has no effect on mean enzyme activity, which is why all four distributions are centred on the same value. Note that they are all drawn as normal distributions because this is one of the assumptions of the analysis of variance. They are also shown as having the same degree of spread, which means that they all have the same variance. This is another assumption of the test.

In practice, we have a random sample from each of these populations and the observations in each sample are represented by the dots. As we would expect, there is chance variation in the make-up of each sample and as a result there are differences between the four samples. These differences are reflected in differences in the four sample means (indicated by the arrows). This variation in the four sample means is shown clearly when we put all four together as has been done at the bottom of the diagram. However, because the four population means are the same we would not expect too much variation in the sample means.

The right-hand set of distributions show how things would be when the null hypothesis is not true. Now the means of the populations in the four treatments are not all the same, although, as before, the distributions are normal and have equal variances. The four sample means will again broadly reflect the means of their populations. Because the population

283

means are different now, there will be greater variation between the sample means than when the null hypothesis is true. The amount of variation between sample means is therefore indicative of the amount of variation between population means. Perhaps we could measure it and use the result to discriminate between the two hypotheses?

However, it is not quite this simple. There will be variation between the means of samples even if the populations are identical, because of the vagaries of chance. The extent of this sampling variation is related to two things, the inherent variation in the material being studied and the size of the samples used. It follows that the amount of variation between sample means is of little use on its own; it has to be seen in the context of the amount of chance variation which could occur anyway in the course of sampling.

To gauge the amount of sampling variation which could occur, we need a measure of the inherent variation of the material. We can get this measure by looking at the variation within each sample, that is, the variation of the individual observations around their own sample mean which is shown at both sides of the diagram. Our best measure of this inherent variation would be to use information from all the samples, that is, the average variation within samples. Now we have a yardstick against which to judge the variation between sample means. The variation between sample means should be small relative to the variation within samples if the null hypothesis is true. If the variation between sample means is large while the variation within samples is small, then this would be evidence for the alternative hypothesis.

We need a way of discriminating between these two scenarios and the logic used is as follows. You will need to keep in mind that the extent of chance variation in sample means is related both to the underlying variability and the sample size. Also, instead of talking loosely about the amounts of variation we will use the proper measure of variability – the variance. We ask the following question: "Is the variance between the means of the four samples too big to be reasonably attributed to chance, given what we know of the underlying variability estimated from the variance within each sample and taking into account the sizes of the samples?".

If the answer to this question is yes, then what we are saying is that the variance in sample means is too big to be due to sampling variation. If this is so, then the only thing that can be causing it to be so large is a difference in population means. Or to put it another way, if the null hypothesis is true then we are not likely to obtain a variance between sample means which is very large in relation to the variance within samples.

Note that this question does not mention differences between pairs of means, but instead uses the variance among a set of means and this is examined in the context of the variance within samples. This is why it is called analysis of variance. The way this analysis is done involves using the data to get two independent estimates of the variance which are then compared. One is based on the variation of the individual observations within each sample around the sample mean and the other is based on the variation between the sample means. To see how this works we shall have to use symbols.

## An explanation using symbols

So that you can keep your bearings, the diagram in Figure 11.2 is reproduced in Figure 11.3. We shall work through it again, starting with the null hypothesis, but now using symbols.

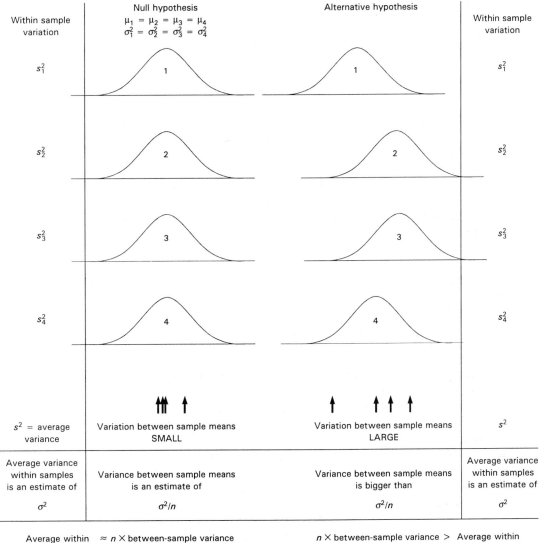

**Figure 11.3** Analysis of variance: how it works using symbols.

The variance of each of the four populations is denoted by $\sigma^2$, with a subscript to identify which population it refers to. For each sample we can calculate a value of the variance, labelled $s_1^2$, $s_2^2$, etc., which quantifies the variation of the observations around the sample mean. These are shown at the extreme left of the diagram. Each of these will be an estimate of $\sigma^2$. Since $\sigma^2$ is assumed to be the same for all populations, we can get a much better estimate of it by combining the values of the variance from each sample and taking the average. This value derived from the variation within samples, is called the **within-sample variance** and is denoted by $s^2$. It is used as one estimate of $\sigma^2$. The other estimate is based on the variance in sample means around their joint mean (which is the same as the grand or

285

overall mean for all the observations). It is called the **between-sample variance**. In practice, this will involve calculating a variance using the four sample means as the observations (rather than the five individual observations) and we shall see how to do this later.

We now focus on the variation in the sample means shown in the second column. We already know from Chapter 7 that if we repeatedly take samples of size $n$ from a normally distributed population, then these sample means will form a normal distribution around the population mean. Here, we are considering taking one sample from each of four populations, but this will be equivalent to taking four samples from one population because we are assuming that the four populations are exactly the same. (Remember this argument is based on the null hypothesis.) As a result our four sample means should belong to a normal distribution of sample means, centred on $\mu$.

We also saw in Chapter 7 how the spread of this distribution of sample means was related to the population standard deviation ($\sigma$) and the sample size ($n$); its standard deviation was given by $\sigma/\sqrt{n}$. For our present purposes, we need to modify this slightly so that we are working with variances ($\sigma^2$) rather than standard deviations ($\sigma$), which we can do by simply squaring both the top and the bottom. The variance of the distribution of sample means is then given by $\sigma^2/n$. So when we calculate the variance of our four sample means, that is, the between-sample variance, it will be an estimate of $\sigma^2/n$. It follows that if we multiply the between-sample variance by $n$ then the answer will be another estimate of $\sigma^2$.

If the null hypothesis is true then our two calculated values, "the average within-sample variance" and "$n \times$ the between-sample variance" are estimates of the same thing ($\sigma^2$) obtained in two different ways. The two values should therefore be more or less the same, give or take some chance variation due to sampling. If the alternative hypothesis is true the calculated value of the within-sample variance will still be an estimate of $\sigma^2$ as shown at the extreme right. However, the sample means will not have been drawn from the same population and the variability between sample means will be increased. Under these conditions the value of "$n \times$ the between-sample variance" will be much bigger than $\sigma^2$. This is the crux of the analysis.

So all we need is a statistical test which will tell us whether the difference between our two estimates of $\sigma^2$ can reasonably be accounted for by sampling variation. We already know how to do this; we can calculate the ratio of the between-sample variance to the within-sample variance and this will be a value of $F$ which we can compare with a critical value.

In case you have lost sight of the purpose of the exercise, I shall sum this all up by repeating the question we are seeking to answer: "Can the amount of variation between mean enzyme activities in samples of this size be reasonably accounted for by chance, given what we know about the inherent variability within samples?". Having dealt with the theory we can now move on to the calculations.

## Preliminary calculations

We shall use the data on enzyme activities at the four pH values. The calculations are rather more complicated than in the Kruskal–Wallis test, but they can all be handled quite efficiently on a calculator with a standard deviation function. However, it will pay you to set the data out in a neat tabular form (Box 11.6a) with plenty of space at the bottom and at the left to write down the various intermediate stages of the calculations. Some of the values

---

**BOX 11.6 ANALYSIS OF VARIANCE FOR DIFFERENCE IN MEANS OF THREE OR MORE
POPULATIONS: ACTIVITY OF ACID PHOSPHATASE AT PH 3, 5, 7 AND 9.
CALCULATING THE SUMS OF SQUARES**

Null hypothesis:   Mean enzyme activity the same at all four pH values.

Alternative hypothesis:   Mean enzyme activity not the same at all four pH values.

Significance level:   5%.

**(a)**

| | Enzyme activity (μM/min) at pH | | | |
|---|---|---|---|---|
| | 3 | 5 | 7 | 9 |
| | 11.1 | 12.0 | 11.2 | 5.6 |
| | 10.0 | 15.3 | 9.1 | 7.2 |
| | 13.3 | 15.1 | 9.6 | 6.4 |
| | 10.5 | 15.0 | 10.0 | 5.9 |
| | 11.3 | 13.2 | 9.8 | 6.3 |

**(b)**

| | | | | | | | | | Totals |
|---|---|---|---|---|---|---|---|---|---|
| (1) Sample size | $n$ | 5 | + | 5 | + | 5 | + | 5 | = 20 |
| (2) Sample mean | $\bar{Y}$ | 11.24 | | 14.12 | | 9.94 | | 6.28 | |
| (3) Sample standard deviation | $s$ | 1.26 | | 1.45 | | 0.78 | | 0.61 | |
| (4) Sample variance | $s^2$ | 1.588 | | 2.117 | | 0.608 | | 0.367 | |
| (5) Within sample sum of squares | $s^2(n-1)$ | 6.35 | + | 8.47 | + | 2.43 | + | 1.47 | = 18.72 |
| (6) | $\Sigma Y$ | 56.2 | + | 70.6 | + | 49.7 | + | 31.4 | = 207.9 |
| (7) | $(\Sigma Y)^2$ | 3158.44 | | 4984.36 | | 2470.09 | | 985.96 | |
| (8) | $(\Sigma Y)^2/n$ | 631.69 | + | 996.87 | + | 494.02 | + | 197.19 | = 2319.77 |
| (9) | $\Sigma Y^2$ | 638.04 | + | 1005.34 | + | 496.45 | + | 198.66 | = 2338.49 |

**(c)**

| | Totals | Correction term (CT) | | Sums of squares |
|---|---|---|---|---|
| $N =$ | 20 | | | |
| | 18.72 | | = 18.72 | Within samples |
| Grand total (GT) = | 207.9 | = $GT^2/N$ = $(207.9)^2/20$ | | |
| | | = 2161.025 | | |
| | 2319.77 | − 2161.025 | = 158.6495 | Between samples |
| | 2338.49 | − 2161.025 | = 177.3695 | Total |

calculated have to be used again so do not be tempted to round them off at this stage in your own calculations, because it is important to retain sufficient accuracy. (They have been rounded here to aid the presentation). If you have access to a computer program, then all these calculations will be done for you but your understanding of how the test works will be enhanced if you follow through the next section.

If you think back to the calculation of a sample variance (Ch. 4) you will remember that there were two stages. First, we obtained the sum of the squared deviations of each observation from the sample mean. This was called the sum of squares, or SS for short, and it measures the total of the squared deviations of all the items in the sample from the sample mean. Then, to obtain the variance we divided by $n - 1$, the degrees of freedom. We follow the same general procedure here.

Our first task is to calculate a number of different quantities based on each sample (Box 11.6b), so, to begin with, each column of figures is dealt with separately. With your calculator in standard deviation mode enter the five values from sample 1. Although you know $n$ already, it is a good idea to check that you have entered the correct number of items into the calculator by finding $n$ from the calculator if you can. Then find the mean $\bar{Y}$ and the sample standard deviation, $s_1$. Remember to press the correct button here – we want the sample standard deviation which is the result of dividing the sum of squares by $(n - 1)$. It is the larger of the two values which you can calculate. Then square $s_1$ to get $s_1^2$ the sample variance. We will also need the sum of squares for each sample and this is something which we can easily get this at this stage. The variance is the sum of squares divided by the degrees of freedom $(n - 1)$ so it follows that if we have the variance and we multiply it by $(n - 1)$ we end up with the sum of squares. This has been done in line 5. This sum of squares is a measure of the variation in the sample and is called a **within-sample sum of squares**.

Many calculators enable you to get the next four values very easily. $\Sigma Y$ is the sum of the values of $Y$, which you can get with the $\Sigma x$ button. You need to note this value. You then need to square this and divide it by the sample size $n$ to get line 8, which again we will need later. The last line, $\Sigma Y^2$, is what you get if you square each value of $Y$ and add the answers together and can be obtained with the $\Sigma x^2$ button. If your calculator does not have these two buttons then you can get by without. If you multiply the mean (line 2) by the sample size you get $\Sigma Y$, which you can then use to get line 8. Strictly, the value in the last row, $\Sigma Y^2$, is not needed, although later it does serve as a check on the calculations. If you want to use it for this purpose, the only way you could obtain it is to enter each value of $Y$, square it and add the results together.

Once you have repeated these steps for each sample, the worst of the calculation is over. The final stage of these preliminary calculations is to add up the values in lines 1, 5, 6, 8 and 9 to get the figures in the column on the right headed totals. If you were actually doing an analysis of variance then you would pause at this stage to check something. Remember from Figure 11.2 that one assumption of the procedure is that the population variances are the same and this is something we can check using our four sample variances. If they are not the same, then we might decide that there was no point in continuing with the analysis. We shall see how to do this later, but for the moment we shall carry on with the calculations.

## Calculating the sums of squares

There are three of these. The within-sample sum of squares is based on all the samples and is given by the total of line 5. The next step is to calculate the **between-sample sum of squares**. These two sums of squares are needed to calculate the two variances, the within-sample variance and the between-sample variance. We shall also calculate a third sum of squares the **total sum of squares**, which describes the total variability; the reasons for doing this will become apparent later.

In this book I have tried, wherever possible, to explain where formulae come from and what they are doing, so that you can get a understanding of how they relate to the test. At this point I shall have to abandon this principle because the formulae which underlie these calculations are complex and in any case they have been substantially rearranged to make the calculations easy. If you have an algebraic turn of mind, you will find accounts of all of this in Sokal & Rohlf (1981), but for our purposes all you need to bear in mind is the outline of the procedure. If you have set out the intermediate answers as suggested, then this stage is very easy.

The steps are shown in Box 11.6c. They involve the totals from section (b), which are shown again on the left. We already have the within-sample sum of squares; it is the total in line 5. The sums of squares in line 5 measure the variation of the items around the sample mean for each sample separately. We want to use the information from all the samples so we simply add the four sums of squares together to get 18.72

Obtaining the between-samples sum of squares involves two steps. First we calculate something called the **correction term (CT)** which involves the total from line 6. This is actually the grand total (GT) of the 20 original variates. The correction term is given by $(GT)^2/N = 2161.025$. In the second stage we take the total in line 8 and subtract from this the value of the correction term. The answer, 1586.6495, is the required between-sample sum of squares. To calculate the total sum of squares, we simply take the total in line 9 and subtract the correction term to get 177.3695.

As we saw earlier, we are primarily interested in the within-sample and the between-sample sums of squares, as preliminaries to calculating within-sample and between-sample variances, so why do we bother with this last sum of squares? The answer is that these three sums of squares have an interesting property which is clear from our example. Numerically the total sum of squares is equal to the sum of the within-sample and the between-sample sums of squares. The value of this is that if we calculate any two of the sums of squares using the long methods just described, we can always easily find the value of the third one. When these calculations had to be done without a calculator, it was easiest to work out the total sum of squares and the between-sample sum of squares and then to obtain the within-sample sum of squares by subtracting one from the other. You could do the same by omitting line 5 of the calculations, but I suggest that you retain this step. It is very quick and it provides a check on your subsequent calculations.

Having found the sums of squares we are ready to move on to the final stage which is to convert them to variances and calculate the value of $F$. After we have done this we shall see how to check one of the assumptions involved in the analysis of variance.

## From sums of squares to variances and the value of $F$

Remember that our analysis is an analysis of variance, so what we need to do is convert the sums of squares into variances. The procedure is the same as the one we used in Chapter 4, we simply divide the sums of squares by the degrees of freedom. These are related to the number of "observations" contributing to the sum of squares and the number of pieces of information which have been used as estimates of parameters. Conventionally, the results of the analysis are set out in a table as illustrated in Box 11.7.

The sums of squares are simply copied from Box 11.6c and the degrees of freedom are calculated as follows. For the between-sample sum of squares the degrees of freedom are

BOX 11.7 ANALYSIS OF VARIANCE FOR DIFFERENCE IN MEAN OF THREE OR MORE SAMPLES. ACTIVITY OF ACID PHOSPHATASE AT PH 3, 5, 7 AND 9. FROM SUMS OF SQUARES TO VARIANCES AND THE $F$ TEST

| Source of variation | Sum of squares (SS) | Degrees of freedom (df) | Mean square (MS) | $F$ |
|---|---|---|---|---|
| Between samples | 158.6495 | $k - 1 = 4 - 1 = 3$ | 52.883 | 45.199 |
| Within samples | 18.72 | $N - k = 20 - 4 = 16$ | 1.17 | |
| Total | 177.3695 | $N - 1 = 19$ | | |

given by the number of samples ($k$) minus one, so df $= (4-1) = 3$. The reason is that this sum of squares measures the variation of the four sample means around the grand mean. Each mean is used as an observation, so we only have four observations. To calculate a sum of squares we need first to calculate the mean of the observations which is used as an estimate of the population mean. So we need one piece of information from the data so we lose one degree of freedom.

The within-sample sum of squares is based on the variation of the items in each sample around their own mean and it has 16 degrees of freedom. The easy way to remember this is that the within-sample degrees of freedom are given by the total sample size minus the number of treatments $20-4 = 16$. Put simply, the reason is that this sum of squares is based on 20 observations, but required the use of four pieces of information calculated from the data. These are the four sample means which enter into the calculation of the sums of squares for each sample and which are estimates of the four population means. The overall sum of squares measures the variation of all 20 items around the grand mean and has 19 degrees of freedom because we have to use the grand mean to calculate it.

Dividing the between-sample and the within-sample sums of squares by their appropriate degrees of freedom gives us our two estimates of the variance; in the terminology of analysis of variance, they are called mean squares (MS) rather than variances. The final column gives us the calculated $F$ value, which is the ratio of the two variances. It is obtained by simply dividing the between-sample sum of squares by the within-sample sum of squares. This calculated value can then be compared with the appropriate critical value in the table of $F$ values.

## Critical values of $F$

We saw in Chapter 9 that the critical value of $F$ depended on two sets of degrees of freedom, well the same thing applies here. We find the appropriate subtable of $F$ values by choosing the one with the correct degrees of freedom at the top. These are the degrees of freedom associated with the top line (the between-sample mean square) in the analysis of variance table. This mean square is also the one that is on the top, i.e. in the numerator, in the calculation of $F$. The degrees of freedom for the within-sample mean square are to be found down the side of the table of critical values.

In the context of analysis of variance the $F$ test is always a one-tailed test and the reason for this was touched on earlier. If the null hypothesis is true, then the sample means are likely to be close to the grand mean, the two estimates of the variance are likely to be similar and $F$ will be around 1. If the null hypothesis is not true and the population means are different then the sample means are likely to be more widely spread around the grand mean. If the sample means are more widely spread, then the between-sample variance will be larger and so will the value of $F$. So it is large values of $F$ which suggest that the population means are different. As a result, we need the critical values which cut off the appropriate sized fraction of the area of the sampling distribution in the right-hand tail, that is $\alpha_1^R$.

Our calculated value of 45.199 is far larger than the 5% critical value and even exceeds the critical value of 9.01 at $\alpha = 0.1\%$. We can conclude that we have virtually conclusive evidence for the alternative hypothesis, which is that the means of the four treatments are not the same. We could put this another way which highlights why it is an analysis of variance. If the treatments have no effect (so that the four population means are the same) then, given the variability observed within samples, we would be very unlikely to draw four samples (of this size) which had as much variation between their means as that which we observed.

However, it does not tell us which mean or means are different from which others; to answer this requires a multiple comparison test. Before we do this we shall see how to check one of the assumptions of the analysis of variance. We should really have checked this before proceeding with the analysis but I have left it until this stage so as not to interrupt the flow of the main argument.

## Checking that the population variances are not different; the $F_{max}$ test

As we saw in Figure 11.2, the analysis of variance is based on the assumption that the population variances are equal, i.e. $\sigma_1^2 = \sigma_2^2 = \sigma_3^2 = \sigma_4^2$. We can easily check this using the four sample variances in line 4 of Box 11.6, which are estimates of the four population variances.

We face a familiar problem. Even if the four population variances are the same, the four sample variances are likely to be different because of chance, although we would not expect them to be too different. We need a statistical test which will tell us whether the differences between the sample variances are too big to be reasonably attributable to sampling variation. We shall use a quick and easy test – the $F_{max}$ test (Box 11.8).

All we need to do is divide the largest sample variance by the smallest sample variance to get a variance ratio, $F_{max}$. It is called $F_{max}$ because it is obviously the largest of several $F$ values which we could calculate from pairwise combinations of our sample variances. The null hypothesis is that the population variances are all the same and the sampling distribution of $F_{max}$ based on samples of equal size under these conditions is known. It should be obvious that large values (implying big differences in variances) are less likely than small values. Accordingly, the decision rule will be to reject the null hypothesis if the calculated value is equal to or greater than the critical value.

The critical value depends on three things: the significance level, the number of samples involved ($k$), and the degrees of freedom associated with the two variances being compared. In our case the 5% significance level will be appropriate and we have four samples so $k = 4$. Each of the two variances is based on four degrees of freedom so the critical value from Table A18 is 20.6. Our calculated value is much smaller than this so we have no evidence that our samples have been drawn from populations with different variances. We were quite safe in proceeding with the analysis of variance.

---

**BOX 11.8  THE $F_{max}$ TEST FOR EQUALITY OF VARIANCES IN THE FOUR POPULATIONS (DATA FROM BOX 11.6)**

Null hypothesis:   The two samples with the largest and smallest variance have been taken from populations with the same variance.

Alternative hypothesis:   The two samples with the largest and smallest variance have not been taken from populations with the same variance.

Significance level:   5%.

Largest variance = 2.117          Smallest variance = 0.367

Calculated value of $F_{max}$  =  2.117/0.367  =  5.768

Total number of samples being compared in the analysis of variance ($k$)  =  4

Degrees of freedom  = $n - 1$ = 4

Critical value of $F_{max}$  = 20.6

Decision rule:   Reject null hypothesis if calculated value is equal to, or more extreme than, critical value.

Decision:   No evidence that samples have been drawn from populations with different variance.

---

If we had found evidence that the variances were different, we would have to reconsider what to do and this is something which is covered in the final section. If you are actually embarking on an analysis of variance, it obviously makes sense to pause and make this check straight after calculating line 4 and before starting on the sums of squares. Should you have samples of unequal size then you cannot find exact critical values from the table. What should you do in this the case? The answer is that you can get an approximate, conservative test by using the critical value appropriate to the smaller of the two samples (i.e. by using the smaller of the two degrees of freedom).

## Multiple comparisons between means

When the $F$ test in the main analysis of variance is significant, it implies that there are differences between some of the population means. Under these circumstances we can proceed to a multiple comparison test.

For simplicity, we shall only deal with one of these tests which can be used with either equal or unequal sample sizes. It is called the Tukey–Kramer test and produces a test statistic $q$. When sample sizes are unequal there are several other tests available, but they all work on the same general principle which should be familiar. They involve calculating a test statistic by dividing the difference between each pair of means by the standard error, just as we did in the non-parametric multiple comparison test. The only difference between the various tests is that they have different ways of estimating the standard error and produce different test statistics which have different tables of critical values. The calculated value of the test statistic is then compared with a critical value which takes into account the number of comparisons being made so that the overall significance level (for all the comparisons) is at the specified level.

The details are set out in Boxes 11.9 and 11.10; the logic and the procedure are the same as for the non-parametric multiple comparisons. You will need to calculate the standard error of the difference using the formula. Like any other standard error, this one is the square root of a variance (here called a mean square) divided by the sample size. Note that the equation uses the within-sample mean square which is based on all of the samples, not just the two particular samples being compared. The test will work with unequal and equal sample sizes, but, in the latter case, you can use a simplified version of the equation. Obviously, if all the samples are the same size, the standard error will be the same for all comparisons and can usefully be calculated once and stored in the memory. If sample sizes are different then you will have to calculate several standard errors, one for each pair of sample sizes.

The appropriate critical value of $q$ (Table A19) depends on the total number of samples ($k$) in the analysis of variance, the significance level and the degrees of freedom. The degrees of freedom are those relating to the within-sample mean square. In this example, $k = 4$, $\alpha = 5\%$ and df = 16, so the critical value is 4.046. Set out the sample means in order of increasing size, along the two sides of a table, as illustrated. Then work systematically through the pairs, column by column, calculating the difference and dividing by the appropriate standard error to get the calculated values of $q$ which can be entered in the table.

---

**BOX 11.9 ANALYSIS OF VARIANCE: MULTIPLE COMPARISONS USING THE TUKEY–KRAMER TEST**

(1) For any two sample means calculate the standard error of the difference between the means ($\bar{Y}_A - \bar{Y}_B$), using the formula below.

(a) General formula:

$$\text{Standard error (s.e.)} = \sqrt{\frac{\text{mean square within samples}}{2}\left(\frac{1}{n_A} + \frac{1}{n_B}\right)}$$

(b) Formula for equal sample size:

$$\text{Standard error (s.e.)} = \sqrt{\frac{\text{mean square within samples}}{n}}$$

(2) Calculated value of $q = (\bar{Y}_A - \bar{Y}_B)/\text{s.e.}$

(3) Compare calculated value of $q$ to critical value.

---

---

**BOX 11.10  ANALYSIS OF VARIANCE: MULTIPLE COMPARISONS OF DIFFERENCES IN MEAN ACTIVITY OF ACID PHOSPHATASE AT PH 3, 5, 7 AND 9 USING THE TUKEY–KRAMER TEST**

Null hypothesis:   A pair of populations (A and B) have the same means.

Alternative hypothesis:   A pair of populations (A and B) have different means.

Significance level:   5%, over all comparisons.

$$\text{Standard error (s.e.)} = \sqrt{\frac{\text{mean square within samples}}{n}}$$

$$= \sqrt{\frac{1.17}{5}} = 0.484$$

Table gives calculated values of $q$ for all pairwise combinations of sample means.

| pH | Sample mean µM/min | | | |
|----|------|------|------|------|
| 5 | 14.12 | | | |
| 3 | 11.24 | | | 5.954 |
| 7 | 9.94 | | 2.708 | 8.641 |
| 9 | 6.28 | 7.566 | 10.254 | 16.207 |
| | | 6.28 | 9.94 | 11.24 | 14.12 | Sample mean µM/min |
| | | 9 | 7 | 3 | 5 | pH |

Critical  value:   4.076

Decision rule:   Reject null hypothesis if calculated value is equal to or greater than critical value.

Decision:   Mean activity at pH 5 higher than at pH 3, 7 and 9.
Mean activity at pH 3 higher than at pH 9
Mean activity at pH 7 higher than at pH 9

As you might expect from the formula for calculating $q$, it is large values which are indicative of differences in population means, so the decision rule is to reject the null hypothesis if the calculated value of $q$ is equal to, or larger than, the critical value. If you deal with the largest differences first then after you have encountered one non-significant result in a column, you may not need to calculate the remaining values of $q$. (This will not be true if the smaller differences in the column happen to be related to larger samples, so that the standard error is smaller.)

## What to do if your data do not meet the assumptions

Some of the assumptions which underlie the analysis of variance were mentioned when we examined the rationale of the test but there are others. In full they are that samples should be random, independent and from normally distributed populations with equal variance. We have already seen how to test for equality of variances using the $F_{max}$ test, but we left unanswered the question about what to do if the data failed this test. Neither have we discussed the other assumptions.

As I have emphasized before, the issue of random, independent samples is related to experimental design. If our data have not been obtained in the correct way then the statistical analysis tells us nothing about the question of interest. We are interested in the effects of pH on enzyme activity, so we must ensure that all the other factors which could affect the variable are randomized. Before reading the next paragraph, you might like to produce a list of other factors which could affect the enzyme activity and which would, therefore, need to be taken into account in the experimental design.

One possibility is that the enzyme gradually loses its activity over time when in a solution. Another is that the temperature in the laboratory will change over time and so may aspects of your technique and the properties of the spectrophotometer. Finally, the use of different batches of enzyme or other reagents for the measurements at different pH values could affect the result. Were any of these complications not taken care of by the experimental design there is no way that we can use the statistical test to disentangle their possible effects. The quality of the answer you get out will only be as good as the quality of the data that you have put in. Obviously the bigger and more complex the experiment, the more scope there is for unwanted effects like this to creep in and of course the greater the loss of time and effort if they do! So the message is that you need to think very carefully about the design of your experiment or sampling programme before starting work.

The analysis of variance, like the $t$ test, appears not to be greatly affected by inequalities in variance or non-normality in the distribution of the variable if the sample sizes are large and more or less equal. With small samples we need to be cautious. As we saw earlier, it is difficult to tell "by eye" whether a small sample follows a normal distribution and we have not considered any tests for small samples to check whether this is so. Our best bet is to go back to the biology and to think about whether a normal distribution is reasonable. We shall probably be fairly safe with biochemical data of the type dealt with here and with biochemical, physiological and morphological variables measured on individual organisms which are of the same sex and age and from the same environment. We should be more wary if our experimental subjects are heterogeneous with respect to these factors.

It would be unwise to use analysis of variance on variables such as count data from Poisson or contagious distributions or on binomial proportions, all of which tend to show marked departures from normality. With variables such as these, one solution would be to find a suitable transformation of the type described in Chapter 7. There they were used to make the distribution closer to a normal distribution, but they often have the effect of simultaneously equalising the variances. The analysis of variance can then be conducted on the transformed values. However, transformation makes the calculations more long-winded and a simpler option may be to analyse the data using the non-parametric Kruskal–Wallis test which is the distribution-free alternative.

The disadvantage of using the Kruskal–Wallis test is that it is less powerful than the parametric analysis of variance, in the same way that the Mann–Whitney $U$ test is less powerful than the $t$ test. For example, the data on shoot/root ratios are not really suitable for an analysis of variance because they fail the $F_{\max}$ test, but if you go ahead and do the analysis anyway you will find that it detects a greater number of significant differences between sites than does the Kruskal–Wallis test.

## The value of analysis of variance

As a result of reading the last section you might feel that analysis of variance is so hedged about with difficulties and involves so much calculation that you would be better off avoiding it and concentrating on simple experiments involving two treatments, or analysing more complex experiments using distribution-free tests. That would be a mistake!

Examining several treatments simultaneously is very efficient in terms of time and effort. If you wanted to carry out the acid phosphatase experiment, two treatments at a time, you would have to conduct six separate experiments to cover every pairwise combination of pH values. With the same number of replicates per treatment, you would have to do a total of 60 determinations as opposed to 20. You would then have to calculate six separate $t$ tests with the increased risk of making a type 1 error.

Finally, analysis of variance can be used for the analysis of much more complicated experiment. In general distribution-free tests cannot be used for these more complex cases. For example, it is possible to design experiments in which the effects of two or more different factors are investigated and each factor can be present at several different levels. We could, for instance, investigate the joint effects of two factors such as pH and temperature on acid phosphatase activity in one experiment, using four pHs and four temperatures. The results could be analysed by a two-way analysis of variance which would enable us to see what the effect of pH was and what the effect of temperature was. We could also see whether the effect of temperature was different at different pH values, that is whether there was an interaction between the two factors. The field experiment in Plate 5 which involved two levels of three factors (temperature, rainfall and nutrients) was analysed with a three-way analysis of variance. You will find accounts of these tests in more advanced textbooks (e.g. Sokal & Rohlf 1981).

# Linear regression: describing the relationship between an independent and a dependent variable

In this final section we look at a rather different procedure for analysing samples from several populations. The other tests described in this chapter have been used to detect differences between populations. Linear regression is designed to detect and describe a linear relationship between the values of $Y$ in a series of populations

## When is regression used?

Figure 11.4 has some data on the organic content of the soil at different distances across the width of a coastal shingle ridge. The data have a familiar appearance, in that they resemble

| Distance from back of ridge (m) | 0 | 10 | 20 | 30 | 40 | 50 | 60 | 70 | 80 | 90 |
|---|---|---|---|---|---|---|---|---|---|---|
| Organic content of soil (%) | 4.98 | 3.56 | 2.53 | 2.76 | 1.57 | 1.15 | 1.20 | 0.82 | 0.08 | 0.12 |

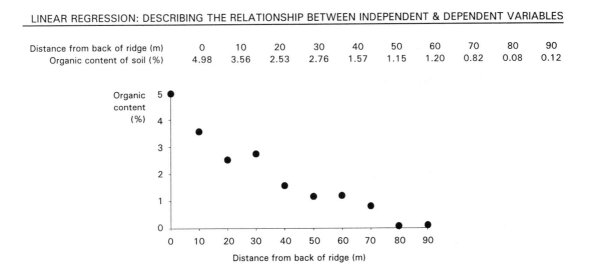

**Figure 11.4**   Relationship between soil organic content (%) and distance across a shingle ridge.

the bivariate data we met when discussing correlation in Chapter 7. There we took sampling units at random and described them with respect to two variables. Here the sampling unit is a quantity of soil and we have two pieces of information about it; the organic content and the place where it was collected from, measured as the distance (from the back of the shingle ridge). This similarity is deceptive because there is a fundamental difference.

In the case of correlation the sampling units were chosen at random with respect to both variables. For example, in Box 8.3, before we chose the subjects we knew nothing about either the amount of physical exercise they took or their index of physical efficiency so they were sampled at random with respect to both of them. This was not the case for the data in Figure 11.4. The values of one of the variables, distance, were deliberately chosen by the investigator. You can see this because the distances are equally spaced and each value occurs once. In practice, this was achieved by choosing stations every 10 m across the ridge and at each station placing a tape measure running parallel to the ridge. A random position was chosen along each tape measure. Values of one of the variables (distance) are said to be fixed (by the investigator), while the values of the other variable (organic content) are chosen at random.

The difference between the two sets of data has another facet. We saw in the case of correlation that all we are looking for is a tendency for the two variables to vary together. It was not an attempt to investigate cause and effect, which is why we did not label the variables as dependent and independent. This is not the case for the data in Figure 11.4. Now there is an independent variable (distance) which, it is suggested, may be causing change in a dependent variable (organic content). This distinction between the two cases is highlighted by the graphical presentation, which again takes the form of a scatter diagram (Fig. 11.4). In the examples in Chapter 8, where there was no distinction between dependent and independent variables, either variable could go along either axis. In this example distance is clearly the independent variable and so conventionally it goes along the horizontal ($x$) axis. To convey this important distinction we use $X$ as the symbol for values of the independent variable and $Y$ for values of the dependent variable.

Collecting data in this way with values of one of the variables fixed by the investigator is very common. For example, we might measure the size of fungal colonies grown at a number of different temperatures. Temperature is the independent variable and the different temperatures are fixed by us. When we have data of this type they are not analysed using correlation. Instead we use a technique called regression.

The idea of regression is to describe the nature of the relationship between the independent and dependent variables by means of an equation. The independent variable is one which is thought to be implicated in causing changes in the dependent variable. The equation allows values of the dependent variable to be predicted from a knowledge of the independent variable.

## How does regression work?

There are several different sorts of regression, but all we shall deal with is simple rectilinear regression. It is simple because there is only one independent variable and it is rectilinear because the equation produced describes a straight line.

The theory of regression, which underlies the calculations used to obtain the equation, is based on a number of assumptions. One assumption is that the values of one of the variables, the independent variable $(X)$, are fixed or deliberately chosen by the investigation so that they are not subject to random variation. Values of the other variable $(Y)$ are chosen at random. In this sense the conditions under which regression is used are similar to those for analysis of variance. There we had an independent variable (pH) which had a number of levels (3, 5, 7 and 9) which were fixed by the experimenter. Values of $Y$ (enzyme activity) were sampled at random from a population of values. In our current example, each population is a population of soil organic contents at a fixed distance across the ridge and we have a single sampling unit from each. However, in terms of the calculations involved, regression is much closer to correlation and as we shall see we do not need to learn any new terms.

We can see how it works in Figure 11.5a. Regression analysis produces the equation of a straight line which runs through the data points on the scatter diagram. This line has a special property, which arises from the way that its equation is calculated and which is summarized by its name. It is the **line of best fit**. The vertical distance (marked $d$) between any point and the line is called the deviation (from the regression line). The position of the regression line is such that the sum of the squared deviations for all the points is as small as it can be. What this means is that overall the line is as near to all the points as it can be, hence the name, line of best fit.

The position of this line can be described with respect to two features. First, the direction and steepness of its slope and second its height, that is its position with respect to the y axis. These can be described by an equation of the form given in Figure 11.5a, which enables us to calculate a value of $Y$ corresponding to a given value of $X$. The equation for a particular set of data has two constants $a$ and $b$. Constant $a$ is the intercept, which is the value of $Y$ when $X = 0$. It corresponds to the point at which the line crosses the y axis and is a measure of the overall position of the line. Constant $b$ is the slope or gradient of the line, which tells us how much $Y$ changes by for a specified increase in $X$.

We shall see how to find the equation for a given set of data in the next section, but we should look first at the other assumptions which underlie the theory. Two of these are shown in Figure 11.5b. We assume that for each value of $X$ the values of $Y$ have a normal distribution and that for different values of $X$ the variances of these distributions are the same. In addition we assume that the sampling is at random.

**(a)  Straight line of best fit for data in Figure 11.4**

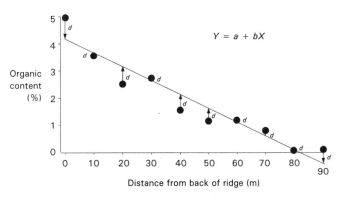

**(b)  Assumptions of linear regression. Values of _Y_ at each value of _X_ are normally distributed with equal variances**

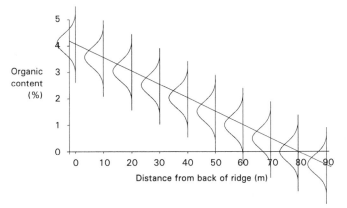

**Figure 11.5**  Linear regression. The line of best fit and the assumptions involved.

## Calculating the line of best fit

To do this all we need to do is to calculate the intercept and the slope. We have now reached the stage where it becomes impossible to see how the formulae used in the calculations relate to the biology. Because of this it is not possible to explain why the formula results in a line of best fit.

Many calculators have a linear regression function, but even if yours does not it is not difficult to do the calculations (Box 11.11). Calculating the slope (_b_) is the worst part but the steps are similar to those involved in correlation, so they do not involve anything new. If you look towards the bottom of Box 11.11, you will see that the slope is given by dividing the sum of products by the sum of squares of _X_, both of which we saw how to calculate in correlation. (There is only a slight difference in notation; now one of the variables is called _X_, the other _Y_, to distinguish the dependent and independent variables.) The five rows of figures at the bottom of the column of _X_ values are easily obtained with your calculator in

299

**BOX 11.11  LINEAR REGRESSION FOR CALCULATION OF EQUATION FOR LINE OF BEST FIT TO A SERIES OF DATA POINTS: ORGANIC CONTENT OF SOIL AS A FUNCTION OF DISTANCE ACROSS A COASTAL SHINGLE RIDGE.**

| $X$ Distance from back of ridge (m) | $Y$ Organic content (%) |
|:---:|:---:|
| 0 | 4.98 |
| 10 | 3.56 |
| 20 | 2.53 |
| 30 | 2.76 |
| 40 | 1.57 |
| 50 | 1.15 |
| 60 | 1.20 |
| 70 | 0.82 |
| 80 | 0.08 |
| 90 | 0.12 |

Mean:      $\bar{X} = 45$      $\bar{Y} = 1.88$

Sum of values:      $\Sigma X = 450$      $\Sigma Y = 18.77$      $\Sigma XY = 435.9$

Standard deviation ($s$):      30.27      1.57

Variance ($s^2$):      916.67      2.46

Sum of squares [$s^2(n-1)$]      $SS_X = 8250$      $SS_Y = 22.18$

Calculation of slope:

$$\text{Slope } b = \frac{\text{sum of products}}{\text{sum of squares of } X} = \frac{\Sigma XY - (\Sigma X \Sigma Y / n)}{SS_X}$$

$$= \frac{435.9 - (450 \times 18.77 / 10)}{8250} = -0.0495$$

Calculation of intercept:

$$\text{Intercept } a = \bar{Y} - b\bar{X} = 1.88 - (-0.0495 \times 45) = 4.11$$

Equation for line of best fit:

$$Y = 4.11 - 0.0495X$$

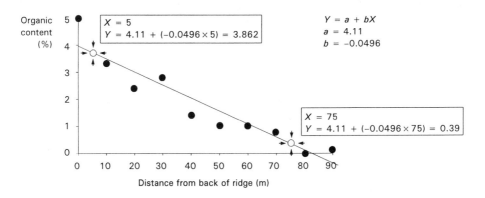

**Figure 11.6**  Drawing the line of best fit.

standard deviation mode and the same goes for those at the bottom of the $Y$ column. To obtain $\Sigma XY$ you multiply each value of $X$ by the corresponding value of $Y$ and add the values for all the pairs of $X$ and $Y$ together in the memory. You can then easily find the slope as shown.

The value of the intercept a is much simpler to obtain (Box 11.11). We rearrange the equation and substitute the values of $\bar{X}$, $\bar{Y}$ and $b$ into it and solve for $a$. Having calculated the intercept and the slope it is easy to draw the straight line (Fig. 11.6). We simply take a value of $X$ which lies near to the $y$ axis and put this into the equation (along with $a$ and $b$) to find the corresponding value of $Y$. We mark this on the graph and then repeat the procedure with a value of $X$ which lies a long way from the $y$ axis. Joining up the two points gives us the regression line.

## Testing whether the slope is significantly different from zero

We saw in Chapter 8 how, by chance, a sample can show a correlation even when there is not one in the population and the same thing can happen here. Just because we can describe a relationship in a sample it does not follow that the relationship is present in the population. If there is no relationship in the populations, so that organic matter content does not change with distance, then the slope of the line (in the populations) will be zero. This population slope is of course a parameter, so we call it $\beta$ to distinguish it from $b$ which is a sample statistic. Even if the null hypothesis is true so that $\beta = 0$, the value of $b$ is unlikely to be zero because $b$ is subject to sampling variation. We therefore need to test whether $b$ is significantly different from zero.

The basis of the test is simple. The sampling distribution of $b$ under the null hypothesis is known. It has a mean of zero because $\beta = 0$ and we can estimate the standard error from the data. We ask how far away our value of $b$ is from the mean of the sampling distribution in terms of the standard error. The answer is a value of $t$ which can be compared with a critical value. The calculations again look rather complex (Box 11.12), but they only involve manipulating the quantities we have already calculated.

The calculated value of $t$ has $(n-2) = 8$ degrees of freedom because to calculate it we

301

**BOX 11.12  LINEAR REGRESSION FOR FITTING A STRAIGHT LINE THROUGH A SERIES OF POINTS: TEST OF SIGNIFICANCE (DATA FROM BOX 11.11).**

Null hypothesis:   There is no relationship between distance and % organic matter; the slope in the populations ($\beta$) is zero.

Alternative hypothesis:   There is a relationship between distance and % organic matter; the slope in the populations ($\beta$) is not zero.

Significance level:   5%, two tailed

How far is calculated value of $b$ away from zero in relation to standard error?

(1)  Calculate residual variance:

$$\text{Residual variance} = \frac{1}{n-2} \times \left[ SS_Y - \frac{SP^2}{SS_X} \right]$$

$$= \frac{1}{7} \times \left[ 22.18 - \frac{(-408.75)^2}{8250} \right] = 0.276$$

(2)  Calculate standard error of the slope $b$:

$$\text{Standard error of } b = \sqrt{\frac{\text{Residual variance}}{\text{Sum of squares of } X}}$$

$$= \sqrt{\frac{0.276}{8250}} = 0.0058$$

(3)  Calculate  $t$:

$$t = \frac{b}{\text{standard error}}$$

$$= \frac{-0.0496}{0.0058} = -8.54$$

$$\text{df} = n - 2 = 8$$

Critical value:   2.3060

Decision rule:   Reject the null hypothesis if calculated value (ignoring sign) is equal to, or larger than, critical value.

Decision:   Very highly significant evidence for the alternative hypothesis ($P < 0.001$).

used two estimates of population parameters; $\bar{X}$ was used as an estimate of $\mu_X$ and $\bar{Y}$ was used as an estimate of $\mu_Y$. Our alternative hypothesis is two-tailed and so both extreme positive and negative values of $t$ are evidence for the alternative hypothesis. We can ignore the sign of our calculated value and compare it to the critical value in the $\alpha_2$ column. It far exceeds this value so we have very highly significant evidence for the alternative hypothesis. Note, however, that the demonstration of a significant relationship does not tell us that change in distance directly causes change in soil organic content. The reason is that this was an observational study not a manipulative experiment. The actual chain of causal connection is more complex. Distance directly affects micro-climate and the physical properties of the substrate which in turn affect plant biomass. It is changes in the latter which cause the change in soil organic matter.

## Further uses of regression

Regression not only tells us whether there is a relationship, it also describes it. This enables us to predict what the value of $Y$ will be for a given value of $X$. A classic example of this was a study of the relation between distance (from an industrial area) and relative survival of the dark form of the peppered moth (*Biston betularia*). This was measured at seven different distances and the regression line calculated. From this it was possible to predict how well dark moths should do at other distances and, therefore, how common they should be.

The simple ideas of regression can be extended. We can calculate regression lines for two or more sets of data, for example to describe the relationship between drug dose and blood pressure for two different drugs or for two different types of patient. We can then statistically compare the two regression lines to see if there is any evidence that the relationship is different in the two groups. We can also look for a relationship between a dependent variable and two or more independent variables, using multiple regression.

# 12

# Checklists for the design of sampling programmes and experiments and a key to statistical tests

This chapter will draw together and summarize all the ideas on the design of sampling programmes and experiments covered earlier in the book. It is impossible to give detailed instructions on how to conduct a particular experiment or sampling programme because of the wide variety of possible biological investigations. As a result, this summary is of necessity general.

There are many elements in common to a well-designed sampling programme and a well-designed experiment, however it is more convenient to separate the two types of practical work at this stage. Since these issues of design should be dealt with *before* you begin the practical work, they are set out in the form of a checklist. Each checklist is presented as a linear sequence but remember that in practice this is not how the design process works; you may have to backtrack or jump ahead, you may even go round in circles for a time before you settle on a good design.

There is also a key to statistical tests which you can use to identify an appropriate test for the data you are planning to obtain. Again you should ensure that there is an appropriate statistical test available *before* you start the practical work. You should also check the details of the test in the appropriate chapter so that you know that your data will fulfil all the necessary conditions.

## Checklist for the design of a sampling programme for an observational study

Observational studies are designed to find out whether differences in the variable of interest are associated with naturally occurring variation in the factor(s) of interest.

1. *Have you made a clear statement of the objectives of your study?*
   The answers you get will only be as good as the questions you have asked. Break the problem down into a number of questions and make each question precise. Rephrase each question in the form of a statement (i.e. a research hypothesis) and then make each of these statements more precise by defining the method to be used. Define what the sampling unit is and ensure that you know what sort(s) of variable(s) will be involved.

2. *Does your study involve "treatments", e.g. different times, genotypes, species, habitats or locations?*

If so, how many "treatments" will you have and, given the constraints of time and money, will you be able to take sufficient replicates from each population (see (4)).

3. *Have you defined the population(s)?*

You will be making inferences about populations from your samples, so make sure that the populations that you will be sampling are the ones which you wish to make inferences about. Statistics does not enable you to extrapolate from one population to another; all you can rely on then is your knowledge of the subject. Remember that the population sampled is determined partly by the method used, so you need to be aware of any sources of bias in the method.

4. *Have you considered the number of sampling units (replicates) to be taken from each population?*

Differences between populations can only be detected if you know something about the variability within populations, so you must take more than one sampling unit from each population. The number required will depend on the size of the difference between the populations and the variability within populations. Larger samples are better than small ones and equal sized samples are better than unequal. A preliminary study will give you an idea of the variability within populations and will also allow you to assess the feasibility of what you intend to do. It will also give you an idea of how your variable is distributed.

5. *Do you have a clear procedure for choosing sample units at random from the population(s)?*

Remember that all statistical tests assume that the sampling units have been taken at random. You cannot correct for non-random sampling after the data have been collected. You should not try to achieve random sampling by any means other than by random numbers. Ensure that your sampling units are independent (but see (6)).

6. *Is it possible to match the sampling units into pairs (or into blocks)?*

This will reduce the variability between members of each pair with respect to other factors, which might affect the variable of interest and allow the use of more powerful statistical tests. Pairs will need to be representative of the population.

7. *Are there any other possible sources of unwanted variation such as changes in sampling technique or sampling efficiency and changes due to deterioration of sampled material prior to analysis?*

If the answer is yes, then either fully randomize the order in which sample units are taken, processed and the results recorded or use randomized blocks.

8. *Could there be a subjective element in your assessment of the results?*

If the answer is yes, you should consider incorporating a blind element into the design.

9. *Is there an appropriate statistical test available for the analysis of your results?*
You need to know what type of variable you will have and what sort of distribution it will follow. If you think that you have an appropriate test, check that your variable and your sampling design meet any assumptions of the test. If the data do not meet some of the assumptions (e.g. a normal distribution) which underlie a statistical test (e.g. the *t* test), you may be able to correct this (e.g. by transformation). Check for restrictions on the minimum sample sizes. By this stage you should be able to state the null hypothesis and decide whether a one-tailed or a two-tailed test is appropriate. Set the significance level. If any of these points are problematical, then sort out where the difficulty lies before you carry out the practical work. If necessary, seek advice. Remember that statistics cannot recover useful information from a badly designed sampling programme.

10. *Have you drawn up data recording sheets?*
Doing this can clarify the objectives and the details of the investigation. It will also help you to record your data accurately and unambiguously.

11. *Remember that an observational study can never prove a causal connection between the factor of interest and the variable!*
You may be able to use other evidence to support your argument, but, in order to prove a causal connection, you must do a manipulative experiment.

## Checklist for the design of a manipulative experiment

What defines a manipulative experiment is the fact that you deliberately alter one or more factors while holding all the others constant. Any change in the variable of interest should then be directly attributable to the factor manipulated, provided that the experiment is well designed. Cause and effect can then be linked unambiguously. Manipulative experiments can be done in the field as well as in the laboratory.

1. *Have you made a clear statement of the objectives of your study?*
The answers you get will only be as good as the questions you have asked. Break the problem down into a number of questions and make each question precise. Rephrase each question in the form of a statement (i.e. a research hypothesis) and then make each of these statements more precise by defining the method to be used. Define what the sampling unit is and ensure that you know what sort(s) of variable(s) will be involved. Decide how many treatments you want. Efficient experiments can be designed to investigate the effects of several levels of two or more factors at the same time, but remember that each treatment has to be replicated and so a complex experiment can rapidly become unwieldy.

2. *Have you defined the population(s)?*
You will be making inferences about populations from your samples, so make sure that the population(s) that you will be sampling are the one(s) which you wish to make inferences about. Statistics does not enable you to extrapolate from one popula-

tion to another; all you can rely on then is your knowledge of the subject. Remember that the population sampled is determined by the experimental material and by the method used, so you need to be aware of any sources of bias in the method.

3. *Have you considered the number of replicates in each treatment?*
Differences between populations can only be detected if you know something about the sampling variation within populations so you must have more than one replicate in each treatment. The number required will depend on the size of the difference between the treatments and the variability within treatments. Larger samples are better than small ones and equal sized samples are better than unequal. A preliminary experiment will give you an idea of the variability within populations and will also allow you to assess the feasibility of what you intend to do. It will also give you an idea of how your variable is distributed.

4. *Have you considered reducing the variability within treatments by using uniform material?*
The ability to detect an effect of the treatment depends on how different the treatments are and the variability within each treatment, both of which can be manipulated. However, it may not be biologically realistic to make the treatments very different. Reducing the variability within treatments by using uniform experimental material (e.g. a single strain or vegetatively propagated plants) means that possible inferences are restricted.

5. *Do you have an clear procedure for assigning experimental material (subjects) at random to the different treatments?*
Remember that all statistical tests assume that the sampling units have been taken at random. You cannot correct for non-random allocation after the data have been collected. You should not try to achieve random allocation by any means other than by random numbers. Ensure that your replicates will be independent (but see (6)).

6. *Is it possible to match the experimental units into pairs (or into blocks)?*
This will reduce the variability between members of each pair with respect to other factors, which might affect the variable of interest and allow the use of more powerful statistical tests. Pairs will need to be representative of the population.

7. *Have you drawn up a clearly defined set of instructions for the experiment, detailing how the treatments are to be applied?*
Following these instructions will reduce variation between replicates and make it easier to detect differences. Make sure that experimental and control material receives identical treatment apart from that related to the factor of interest.

8. *Are there other possible factors which might affect the outcome of the experiment, such as variation in physical conditions across the bench, in the incubator or waterbath? Is there a possibility that after the experiment the material from different treatments may be inadvertently subjected to different conditions?*
If the answer is yes, either fully randomize the layout of the experiment and the order in which experimental units are examined or use randomized blocks. If there is a pos-

sibility of interference between replicates then take this into account in the layout of the experiment.

9. *Is there any possibility that the experimenter (or the subject), knowing the purpose of the experiment, could affect the result?*
This is particularly likely where there is a subjective element in the way in which the result is recorded. If this is likely, consider a blind experimental design.

10. *Is there an appropriate statistical test available for the analysis of your results?*
You need to know what type of variable you will have and what sort of distribution it will follow. If you think that you have an appropriate test, check that your variable and your experimental design meet any assumptions of the test. If data do not meet some of the assumptions (e.g. a normal distribution) which underlie a statistical test (e.g. the *t* test ) you may be able to correct this (e.g. by transformation). Check for restrictions on the minimum sample sizes. By this stage you should be able to state the null hypothesis and decide whether a one-tailed or a two-tailed test is appropriate. Set the significance level. If you any of these points are problematical, then sort out where the difficulty lies before you carry out the practical work. If necessary, seek advice. Remember that statistics cannot recover useful information from a badly designed experiment.

11. *Have you drawn up data recording sheets?*
Doing this can clarify the objectives and the details of the investigation. It will also help you to record your data accurately and unambiguously.

12. *Remember that a manipulative experiment is always somewhat unnatural and involves a degree of simplification!*
Although it is the only way to demonstrate unambiguously whether the factor of interest causes change in the variable of interest, you may need to be cautious about extrapolating from the experimental results to a natural system.

# Key to statistical procedures and tests

To use this key you will need to know:
(a) The number of samples you have.
(b) The type of variable involved.

**How many samples will you have?**
(a) One ..................................................................... *go to* **A**
(b) Two ..................................................................... *go to* **B**
(c) More than two ......................................................... *go to* **C**

## A. Statistical procedures for single samples

1. How many variables will be used to describe each sampling unit (p. 46)?
(a) One – univariate data ..................................................... *go to* **2**
(b) Two – bivariate data ...................................................... *go to* **12**

2. What will you want to use the sample for?
(a) To estimate population parameters from sample statistics .............. *go to* **3**
(b) To test whether it has been taken from some specified population .... *go to* **5**
(c) To quantify the reliability of the estimates .............................. *go to* **10**

3. What sorts of variables have you used to describe the sampling units?
(a) Nominal – find mode (p. 74) or use **sample proportion** ($\hat{p}$) as estimate of **population proportion** ($p$) (p. 86) and **sample standard deviation** ($s$) as estimate of **population standard deviation** ($\sigma$) (p. 86).
(b) Ordinal – find **median** score (p. 74).
(c) Discontinuous, continuous derived ...................................... *go to* **4**

4.
(a) Distribution of variable asymmetrical – find **median** (p. 74).
(b) Distribution of variable more or less symmetrical – Calculate **sample mean** ($\bar{Y}$) (p. 76) and sample standard deviation ($s$) (p. 79) as estimates of **population mean** ($\mu$) and **population standard deviation** ($\sigma$).

5. What sort of variable have you used to describe the sampling unit?
(a) Nominal .................................................................. *go to* **6**
(b) Discontinuous, continuous or derived ................................... *go to* **7**

6. How many categories will be involved?
(a) Two, use
**Binomial test** (p. 123).
Null hypothesis: sample taken from population with specified proportions in the two categories.
or
**Chi-squared goodness-of-fit test** (p. 135).
Null hypothesis: sample has been taken from population with specified proportions in

the two categories.
(b) More than two, use **chi-squared goodness-of-fit test** (p. 141).
Null hypothesis: sample taken from population with specified proportions in each category.

7. What sort of distribution do you want to test for?
(a) Poisson distribution as model for random dispersion counts of objects/events in space or time ................................................................. *go to* **8**
(b) Normal distribution (for other discontinuous variables, continuous and derived variables) ................................................................. *go to* **9**

8. How big is your sample?
(a) A single observation:
Use **tables** of **Poisson distribution** or calculate **probabilities** (p. 94).
Null hypothesis: sample taken from a randomly distributed population with specified mean ($\mu$).
(b) Small (say, less than 30):
Use **chi-squared test of dispersion** (p. 150).
Null hypothesis: sample taken from population with mean = variance
(c) Large (say, 30 or more):
Use **chi-squared goodness-of-fit test** to Poisson distribution (p. 144).
Null hypothesis: sample taken from a population which follows a Poisson distribution.

9. How big will your sample be?
(a) A single observation:
Calculate $z$ or $t$ **value** (p. 117).
Null hypothesis: sample taken from a population which follows a normal distribution with specified mean ($\mu$) and standard deviation ($\sigma$)
(b) Large:
Use **chi-squared goodness-of-fit test** for a normal distribution (p. 148).
Null hypothesis: sample taken from a population which follows a normal distribution.

10. What sort of variable have you used to describe the sampling units?
(a) Nominal: Calculate **standard error** (p. 158) or find **confidence limits** (p. 176) for proportion.
(b) Other types of variable ................................................................. *go to* **11**

11. What sort of distribution will the variable have?
(a) More or less normal:
Calculate **standard error** (p. 163) or find **confidence limits** (p. 168, 174).
(b) Poisson:
Calculate **standard error** (p. 163) or find **confidence limits** (p. 180).
(c) Some other shape, e.g. from contagious dispersion:
**Transform** data (p. 183) then calculate **standard error** or **confidence limits**.

12. Statistical tests for a single sample of bivariate data. (Note: the tests here are appro-

priate if sampling units were chosen at random with respect to both variables. If you have deliberately chosen or fixed the values of one of the variables (the independent variable) then go to Section C, 4(c).)

What sort of variables have you used to describe the sampling unit?

(a) Nominal:

Use **chi-squared test of association** (p. 187).

Null hypothesis: sample taken from population in which there is no association between the two variables.

(b) Ordinal:

Calculate **Spearman's rank correlation coefficient** (p. 196).

Null hypothesis: sample taken from population in which there is no correlation between the two variables.

(c) Discontinuous, continuous or derived .................................... *go to* **13**

13. What sort of distribution will the variables have?

(a) Not normal ................................................................. *go to* **14**

(b) More or less normal ...................................................... *go to* **15**

14. Are suitable transformations available to make either or both variables more or less normally distributed (p. 183)?

(a) Yes:

Transform data ............................................................ *go to* **15**

(b) No:

**Convert data to ranks** calculate **Spearman's rank correlation coefficient** (p. 196).

Null hypothesis: sample taken from population in which there is no correlation between the two variables.

15. Estimate the direction and strength of correlation between the two variables:

Calculate **Pearson's product–moment correlation coefficient** (p. 205).

Null hypothesis: sample taken from population in which there is no correlation between the two variables.

## B. Statistical test for two samples

1. How were the sampling units allocated to treatments or chosen?

(a) At random, i.e. samples independent ..................................... *go to* **2**

(b) As matched pairs, i.e. samples related ................................. *go to* **6**

2. What sort of variable have you used to describe each sampling unit?

(a) Nominal:

Use **chi-squared test for difference in proportions** (p. 216).

Null hypothesis: Samples taken from populations with the same proportions.

(b) Ordinal:

Use **Mann–Whitney *U* test** (p. 222).

Null hypothesis: Samples take from populations with the same median.

(c) Discontinuous/continuous/derived ........................................ *go to* **3**

3. What sort of distribution will the variable have?
(a) Not normal (e.g. binomial, Poisson or from contagious dispersion) .. *go to* **4**
(b) Normal ................................................................... *go to* **5**

4. Is a suitable transformation available to make the distribution more or less normal (p. 183)?
(a) Yes – **transform data** ...................................................... *go to* **5**
(b) No:
    **Convert data to ranks** and use **Mann–Whitney *U* test** (p. 222).
    Null hypothesis: Samples taken from populations with same median.

5. What sort of difference between the populations will you want to detect?
(a) Differences in dispersion:
    Use *F* **test** (p. 239).
    Null hypothesis: samples take from populations with same variance.
(b) Differences in location (small samples):
    Use *t* **test for two independent samples** (p. 242).
    Null hypothesis: samples taken from populations with the same mean.
(c) Differences in location (large samples):
    Use *z* **test for two independent samples** (p. 232).
    Null hypothesis: samples taken from populations with the same mean.

6. What information will you have on the differences between members of each pair?
(a) Only the direction of the difference:
    Use **sign test** (p. 256).
    Null hypothesis: samples taken from a population in which median difference between members of pairs is zero.
(b) The direction and the magnitude of the difference ...................... *go to* **7**

7. What sort of distribution will the differences have?
(a) Not normal or not known:
    Use **Wilcoxon's signed rank test** (p. 259).
    Null hypothesis: Samples taken from a population in which median difference between members of pairs is zero.
(b) Normal:
    Use *t* **test for paired samples** (p. 264).
    Null hypothesis: samples taken from a population in which mean difference between members of pairs is zero.

## C. Statistical tests for three or more samples

1. What sort of variable have you used to describe each sampling unit?
(a) Nominal:
    Use **chi-squared test for difference in proportions** (p. 270).
    Null hypothesis: samples taken from populations with the same proportions.
(b) Ordinal:
    Use **Kruskal–Wallis test** (p. 273).

Null hypothesis: samples taken from populations with the same median. If null hypothesis rejected use **multiple comparisons test** (p. 279).

(c) Discontinuous, continuous, derived ..................................... *go to* **2**

2. What sort of distribution will the variable have?

(a) Not normal (e.g. binomial, Poisson or from contagious dispersion) .. *go to* **3**

(b) More or less normal ........................................................ *go to* **4**

3. Is a suitable transformation available to make distributions more or less normal (p. 183)?

(a) Yes – **transform data** ...................................................... *go to* **4**

(b) No:

Convert data to ranks and use **Kruskal–Wallis test** (p. 273).

Null hypothesis: samples taken from populations with the same median.

If null hypothesis rejected use **multiple comparisons test** (p. 279).

4. What will you want to use the data for?

(a) To detect differences in dispersion:

Use $F_{max}$ **test** (p. 291).

Null hypothesis: samples with largest and smallest variance taken from populations with same variance.

(b) To detect differences in mean:

Use **one-way analysis of variance** (p. 282).

Null hypothesis: samples taken from populations with the same mean.

If null hypothesis rejected use **multiple comparisons test** (p. 292).

(c) To produce an equation for predicting values of dependent variable from values of independent variable:

Calculate **linear regression equation** (p. 296). Test **slope** for **significance** (p. 302).

Null hypothesis: samples taken from populations in which slope of relationship is zero.

# Appendix

315

## Table A1 Random numbers.

A random sequence of digits can be obtained by reading along rows or columns in either direction. For convenience digits are grouped into blocks and rows are numbered in the shaded margin.

| | | | | | | | | | | |
|---|---|---|---|---|---|---|---|---|---|---|
| 1 | 75523 | 70787 | 04873 | 98093 | 57054 | 26622 | 88599 | 26925 | 19605 | 11720 |
| 2 | 23399 | 42116 | 44736 | 21504 | 92732 | 96640 | 56980 | 59465 | 54342 | 02903 |
| 3 | 34142 | 65196 | 53174 | 75802 | 01310 | 56580 | 16431 | 79507 | 44423 | 05190 |
| 4 | 42872 | 62003 | 72242 | 30428 | 57935 | 06491 | 96314 | 29103 | 59548 | 86793 |
| 5 | 44614 | 58545 | 26468 | 33215 | 25384 | 16767 | 18290 | 07341 | 11075 | 52477 |
| 6 | 39706 | 70442 | 98647 | 23491 | 63110 | 49549 | 23325 | 15959 | 64607 | 74123 |
| 7 | 70578 | 39869 | 84197 | 64011 | 54126 | 84183 | 37242 | 86338 | 59538 | 22130 |
| 8 | 78274 | 87404 | 37371 | 59291 | 96595 | 28047 | 23352 | 52081 | 69298 | 44783 |
| 9 | 13416 | 02651 | 94042 | 78811 | 01319 | 85712 | 08773 | 45852 | 23535 | 45052 |
| 10 | 99984 | 89216 | 03611 | 23532 | 83348 | 47649 | 65257 | 49268 | 23029 | 75625 |
| 11 | 78054 | 04644 | 99556 | 01201 | 45570 | 40365 | 53975 | 47102 | 79367 | 07426 |
| 12 | 99606 | 90902 | 58005 | 34574 | 30457 | 56847 | 76717 | 02982 | 12922 | 76773 |
| 13 | 87823 | 53256 | 10686 | 41497 | 34560 | 68314 | 73474 | 75120 | 92638 | 37130 |
| 14 | 44503 | 13769 | 85619 | 36577 | 67255 | 52439 | 55403 | 16113 | 06744 | 37208 |
| 15 | 04778 | 24884 | 43285 | 56072 | 02936 | 33322 | 46265 | 60012 | 35288 | 53021 |
| 16 | 69328 | 15666 | 80382 | 63848 | 27829 | 98564 | 16683 | 05034 | 29066 | 50739 |
| 17 | 23332 | 57255 | 99287 | 46441 | 91561 | 67740 | 45041 | 13962 | 44044 | 92153 |
| 18 | 53206 | 35845 | 30111 | 72106 | 85895 | 06326 | 43281 | 46571 | 30792 | 29143 |
| 19 | 39551 | 80772 | 94732 | 65467 | 01105 | 92384 | 28019 | 88926 | 69408 | 12635 |
| 20 | 95974 | 75400 | 34361 | 41920 | 87278 | 92499 | 58548 | 78176 | 98774 | 75134 |
| 21 | 88395 | 94730 | 85887 | 90822 | 09925 | 32216 | 58026 | 42839 | 37863 | 56508 |
| 22 | 01344 | 80820 | 70162 | 82256 | 61933 | 08449 | 25084 | 53172 | 23471 | 48572 |
| 23 | 06563 | 61938 | 14078 | 24506 | 90709 | 96611 | 96811 | 20392 | 78088 | 21606 |
| 24 | 49126 | 23009 | 74448 | 47332 | 20783 | 49648 | 31957 | 63933 | 27819 | 68735 |
| 25 | 81745 | 00890 | 69134 | 91862 | 09460 | 09284 | 12782 | 38845 | 19203 | 63677 |
| 26 | 10219 | 16922 | 77315 | 12864 | 52029 | 08072 | 90548 | 48805 | 57491 | 66749 |
| 27 | 81628 | 73061 | 63065 | 58921 | 78387 | 35354 | 57328 | 21121 | 59326 | 26309 |
| 28 | 26933 | 47287 | 41129 | 03723 | 13490 | 70667 | 47193 | 02618 | 14949 | 47721 |
| 29 | 69753 | 27266 | 19860 | 40674 | 05683 | 26356 | 61697 | 91372 | 99891 | 91779 |
| 30 | 49143 | 50370 | 20827 | 97462 | 03243 | 64844 | 66905 | 06144 | 95728 | 83852 |
| 31 | 84909 | 82246 | 29160 | 80441 | 03099 | 50673 | 51625 | 85936 | 17788 | 62450 |
| 32 | 19285 | 18230 | 43639 | 70335 | 29429 | 79438 | 00723 | 72324 | 96843 | 85350 |
| 33 | 84511 | 60998 | 89383 | 48864 | 16103 | 53652 | 24439 | 33922 | 26594 | 53108 |
| 34 | 44309 | 52438 | 78214 | 11486 | 70259 | 71669 | 55348 | 96918 | 93946 | 53476 |
| 35 | 50683 | 49541 | 52570 | 54526 | 36832 | 18352 | 38320 | 39091 | 26165 | 21611 |
| 36 | 01246 | 74379 | 41711 | 62841 | 78316 | 68828 | 81002 | 20542 | 60405 | 59429 |
| 37 | 34947 | 12802 | 02375 | 87679 | 23599 | 73150 | 23982 | 03796 | 42277 | 17124 |
| 38 | 38668 | 33888 | 87536 | 12703 | 07474 | 47753 | 05436 | 65603 | 67860 | 69965 |
| 39 | 72645 | 33248 | 30667 | 40475 | 58583 | 11712 | 04368 | 59954 | 66283 | 50184 |
| 40 | 74020 | 23580 | 66247 | 88423 | 78143 | 16075 | 89691 | 28970 | 32007 | 69196 |
| 41 | 72650 | 47588 | 65016 | 52769 | 30875 | 22068 | 07947 | 61186 | 57433 | 37764 |
| 42 | 03264 | 26947 | 91421 | 79146 | 58867 | 96821 | 70811 | 32069 | 96633 | 68969 |
| 43 | 72074 | 11279 | 22044 | 30258 | 10497 | 88538 | 62936 | 86243 | 78959 | 47987 |
| 44 | 41396 | 40803 | 21590 | 04685 | 13495 | 58720 | 40147 | 50456 | 17661 | 29748 |
| 45 | 38926 | 40901 | 76756 | 38647 | 14322 | 18947 | 07705 | 04661 | 04786 | 32288 |
| 46 | 44657 | 23453 | 42103 | 86768 | 66249 | 61807 | 31779 | 35423 | 72061 | 25074 |
| 47 | 23000 | 90903 | 24619 | 41082 | 55179 | 00873 | 61930 | 98669 | 20981 | 76609 |
| 48 | 92484 | 92813 | 46998 | 18603 | 56722 | 43663 | 43795 | 12863 | 33962 | 78059 |
| 49 | 18293 | 35756 | 57446 | 81024 | 17037 | 13674 | 31384 | 09472 | 26977 | 56564 |
| 50 | 08743 | 81986 | 54402 | 37969 | 06725 | 34996 | 51391 | 53458 | 28038 | 81996 |
| 51 | 10735 | 30630 | 74188 | 84526 | 55422 | 58354 | 50319 | 89606 | 39278 | 30645 |
| 52 | 19399 | 80249 | 76343 | 46840 | 23394 | 81483 | 09294 | 74737 | 02239 | 68217 |
| 53 | 42283 | 45339 | 68297 | 01759 | 03573 | 39797 | 60397 | 11789 | 06238 | 87907 |
| 54 | 03923 | 95308 | 33235 | 03379 | 39322 | 88940 | 34421 | 83322 | 99020 | 14373 |
| 55 | 04299 | 54205 | 89675 | 44396 | 76200 | 81035 | 52847 | 50327 | 48322 | 38070 |

(continued on next page)

**Table A1 (continued)  Random numbers.**

| | | | | | | | | | |
|------|-------|-------|-------|-------|-------|-------|-------|-------|-------|
| 56 | 12555 | 10295 | 85500 | 39224 | 52929 | 24010 | 92161 | 56319 | 40392 | 87495 |
| 57 | 76893 | 98786 | 06910 | 47516 | 10720 | 64668 | 10068 | 66210 | 88167 | 11183 |
| 58 | 61081 | 65364 | 79415 | 22573 | 08651 | 21443 | 85776 | 66719 | 63900 | 47366 |
| 59 | 57768 | 47314 | 24213 | 58850 | 22921 | 31585 | 79953 | 88441 | 52513 | 78665 |
| 60 | 91383 | 32299 | 48159 | 00191 | 14326 | 85209 | 91958 | 64519 | 60875 | 32590 |
| 61 | 54353 | 65123 | 65597 | 70398 | 87265 | 25583 | 29217 | 25854 | 99144 | 15348 |
| 62 | 44089 | 98099 | 77052 | 99780 | 74972 | 75047 | 39538 | 69710 | 28796 | 39669 |
| 63 | 14367 | 46479 | 66320 | 31434 | 08218 | 30790 | 58354 | 56905 | 80256 | 74763 |
| 64 | 46624 | 88143 | 49273 | 42610 | 91975 | 87055 | 58176 | 15965 | 97013 | 39584 |
| 65 | 64117 | 34209 | 96508 | 42226 | 20849 | 36920 | 46593 | 63931 | 03215 | 36293 |
| 66 | 28572 | 97373 | 02161 | 40750 | 20992 | 05065 | 12053 | 92779 | 27231 | 90594 |
| 67 | 27363 | 54342 | 99274 | 86005 | 30818 | 86140 | 76448 | 34952 | 14806 | 50831 |
| 68 | 76911 | 45777 | 88068 | 48798 | 26574 | 37554 | 37569 | 38856 | 53766 | 16853 |
| 69 | 91154 | 68393 | 57596 | 28163 | 45334 | 20441 | 17050 | 19135 | 69778 | 02705 |
| 70 | 19482 | 43484 | 45635 | 83278 | 95397 | 51438 | 82061 | 83089 | 92223 | 79890 |
| 71 | 97882 | 22033 | 19699 | 57786 | 36880 | 53462 | 34321 | 00594 | 30869 | 56598 |
| 72 | 89321 | 25870 | 99368 | 61586 | 36175 | 51758 | 75632 | 56473 | 80364 | 54199 |
| 73 | 33265 | 85064 | 43002 | 43395 | 53792 | 13291 | 78643 | 64587 | 72027 | 58604 |
| 74 | 66814 | 30286 | 28564 | 61582 | 61000 | 36566 | 95465 | 93607 | 30356 | 99834 |
| 75 | 50187 | 68880 | 89259 | 65145 | 83710 | 50268 | 62533 | 23737 | 15493 | 19679 |

## Table A2 Expected frequencies for binomial distributions.

This table is used to find the expected relative frequency of samples (of size n) with a specified number of occurences (Y) of an event that has a binomial distribution and a probability (p) of occurring.

1. Find subtable for required sample size.
2. For p between 0.01 and 0.50 find the value of p at the top of the table, and the value of Y at the left of the subtable.
3. For p between 0.50 and 0.99 find the value of p at the bottom of the table and the value of Y at the right of the subtable.
4. Values in the body of the table are expected relative frequencies.

e.g. $n = 6$, $p = 0.35$; expected relative frequency of samples with $Y = 4$ is 0.0951.

| p = | .01 | .02 | .03 | .04 | .05 | .10 | .15 | .20 | .25 | .30 | .333 | .35 | .40 | .45 | .50 | |
|---|---|---|---|---|---|---|---|---|---|---|---|---|---|---|---|---|
| **n = 1** | | | | | | | | | | | | | | | | |
| Y = 0 | .9900 | .9800 | .9700 | .9600 | .9500 | .9000 | .8500 | .8000 | .7500 | .7000 | .6667 | .6500 | .6000 | .5500 | .5000 | 1 |
| 1 | .0100 | .0200 | .0300 | .0400 | .0500 | .1000 | .1500 | .2000 | .2500 | .3000 | .3333 | .3500 | .4000 | .4500 | .5000 | 0 |
| **n = 2** | | | | | | | | | | | | | | | | |
| 0 | .9801 | .9604 | .9409 | .9216 | .9025 | .8100 | .7225 | .6400 | .5625 | .4900 | .4444 | .4225 | .3600 | .3025 | .2500 | 2 |
| 1 | .0198 | .0392 | .0582 | .0768 | .0950 | .1800 | .2550 | .3200 | .3750 | .4200 | .4444 | .4550 | .4800 | .4950 | .5000 | 1 |
| 2 | .0001 | .0004 | .0009 | .0016 | .0025 | .0100 | .0225 | .0400 | .0625 | .0900 | .1111 | .1225 | .1600 | .2025 | .2500 | 0 |
| **n = 3** | | | | | | | | | | | | | | | | |
| 0 | .9703 | .9412 | .9127 | .8847 | .8574 | .7290 | .6141 | .5120 | .4219 | .3430 | .2963 | .2746 | .2160 | .1664 | .1250 | 3 |
| 1 | .0294 | .0576 | .0847 | .1106 | .1354 | .2430 | .3251 | .3840 | .4219 | .4410 | .4444 | .4436 | .4320 | .4084 | .3750 | 2 |
| 2 | .0003 | .0012 | .0026 | .0046 | .0071 | .0270 | .0574 | .0960 | .1406 | .1890 | .2222 | .2389 | .2880 | .3341 | .3750 | 1 |
| 3 | .0000 | .0000 | .0000 | .0001 | .0001 | .0010 | .0034 | .0080 | .0156 | .0270 | .0370 | .0429 | .0640 | .0911 | .1250 | 0 |
| **n = 4** | | | | | | | | | | | | | | | | |
| 0 | .9606 | .9224 | .8853 | .8493 | .8145 | .6561 | .5220 | .4096 | .3164 | .2401 | .1975 | .1785 | .1296 | .0915 | .0625 | 4 |
| 1 | .0388 | .0753 | .1095 | .1416 | .1715 | .2916 | .3685 | .4096 | .4219 | .4116 | .3951 | .3845 | .3456 | .2995 | .2500 | 3 |
| 2 | .0006 | .0023 | .0051 | .0088 | .0135 | .0486 | .0975 | .1536 | .2109 | .2646 | .2963 | .3105 | .3456 | .3675 | .3750 | 2 |
| 3 | .0000 | .0000 | .0001 | .0002 | .0005 | .0036 | .0115 | .0256 | .0469 | .0756 | .0988 | .1115 | .1536 | .2005 | .2500 | 1 |
| 4 | .0000 | .0000 | .0000 | .0000 | .0000 | .0001 | .0005 | .0016 | .0039 | .0081 | .0123 | .0150 | .0256 | .0410 | .0625 | 0 |
| **n = 5** | | | | | | | | | | | | | | | | |
| 0 | .9510 | .9039 | .8587 | .8154 | .7738 | .5905 | .4437 | .3277 | .2373 | .1681 | .1317 | .1160 | .0778 | .0503 | .0313 | 5 |
| 1 | .0480 | .0922 | .1328 | .1699 | .2036 | .3281 | .3915 | .4096 | .3955 | .3602 | .3292 | .3124 | .2592 | .2059 | .1563 | 4 |
| 2 | .0010 | .0038 | .0082 | .0142 | .0214 | .0729 | .1382 | .2048 | .2637 | .3087 | .3292 | .3364 | .3456 | .3369 | .3125 | 3 |
| 3 | .0000 | .0001 | .0003 | .0006 | .0011 | .0081 | .0244 | .0512 | .0879 | .1323 | .1646 | .1811 | .2304 | .2757 | .3125 | 2 |
| 4 | .0000 | .0000 | .0000 | .0000 | .0000 | .0005 | .0022 | .0064 | .0146 | .0284 | .0412 | .0488 | .0768 | .1128 | .1563 | 1 |
| 5 | .0000 | .0000 | .0000 | .0000 | .0000 | .0000 | .0001 | .0003 | .0010 | .0024 | .0041 | .0053 | .0102 | .0185 | .0313 | 0 |
| **n = 6** | | | | | | | | | | | | | | | | |
| 0 | .9415 | .8858 | .8330 | .7828 | .7351 | .5314 | .3771 | .2621 | .1780 | .1176 | .0878 | .0754 | .0467 | .0277 | .0156 | 6 |
| 1 | .0571 | .1085 | .1546 | .1957 | .2321 | .3543 | .3993 | .3932 | .3560 | .3025 | .2634 | .2437 | .1866 | .1359 | .0938 | 5 |
| 2 | .0014 | .0055 | .0120 | .0204 | .0305 | .0984 | .1762 | .2458 | .2966 | .3241 | .3292 | .3280 | .3110 | .2780 | .2344 | 4 |
| 3 | .0000 | .0002 | .0005 | .0011 | .0021 | .0146 | .0415 | .0819 | .1318 | .1852 | .2195 | .2355 | .2765 | .3032 | .3125 | 3 |
| 4 | .0000 | .0000 | .0000 | .0000 | .0001 | .0012 | .0055 | .0154 | .0330 | .0595 | .0823 | .0951 | .1382 | .1861 | .2344 | 2 |
| 5 | .0000 | .0000 | .0000 | .0000 | .0000 | .0001 | .0004 | .0015 | .0044 | .0102 | .0165 | .0205 | .0369 | .0609 | .0938 | 1 |
| 6 | .0000 | .0000 | .0000 | .0000 | .0000 | .0000 | .0000 | .0001 | .0002 | .0007 | .0014 | .0018 | .0041 | .0083 | .0156 | 0 |
| **n = 7** | | | | | | | | | | | | | | | | |
| 0 | .9321 | .8681 | .8080 | .7514 | .6983 | .4783 | .3206 | .2097 | .1335 | .0824 | .0585 | .0490 | .0280 | .0152 | .0078 | 7 |
| 1 | .0659 | .1240 | .1749 | .2192 | .2573 | .3720 | .3960 | .3670 | .3115 | .2471 | .2048 | .1848 | .1306 | .0872 | .0547 | 6 |
| 2 | .0020 | .0076 | .0162 | .0274 | .0406 | .1240 | .2097 | .2753 | .3115 | .3177 | .3073 | .2985 | .2613 | .2140 | .1641 | 5 |
| 3 | .0000 | .0003 | .0008 | .0019 | .0036 | .0230 | .0617 | .1147 | .1730 | .2269 | .2561 | .2679 | .2903 | .2918 | .2734 | 4 |
| 4 | .0000 | .0000 | .0000 | .0001 | .0002 | .0026 | .0109 | .0287 | .0577 | .0972 | .1280 | .1442 | .1935 | .2388 | .2734 | 3 |
| 5 | .0000 | .0000 | .0000 | .0000 | .0000 | .0002 | .0012 | .0043 | .0115 | .0250 | .0384 | .0466 | .0774 | .1172 | .1641 | 2 |
| 6 | .0000 | .0000 | .0000 | .0000 | .0000 | .0000 | .0001 | .0004 | .0013 | .0036 | .0064 | .0084 | .0172 | .0320 | .0547 | 1 |
| 7 | .0000 | .0000 | .0000 | .0000 | .0000 | .0000 | .0000 | .0000 | .0001 | .0002 | .0005 | .0006 | .0016 | .0037 | .0078 | 0 = Y |
| | .99 | .98 | .97 | .96 | .95 | .90 | .85 | .80 | .75 | .70 | .667 | .65 | .60 | .55 | .50 | = p |

(continued on next page)

# Table A2 (continued) Expected frequencies for binomial distributions.

| p = | .01 | .02 | .03 | .04 | .05 | .10 | .15 | .20 | .25 | .30 | .333 | .35 | .40 | .45 | .50 | |
|---|---|---|---|---|---|---|---|---|---|---|---|---|---|---|---|---|
| | | | | | | | **n = 8** | | | | | | | | | |
| Y = 0 | .9227 | .8508 | .7837 | .7214 | .6634 | .4305 | .2725 | .1678 | .1001 | .0576 | .0390 | .0319 | .0168 | .0084 | .0039 | 8 |
| 1 | .0746 | .1389 | .1939 | .2405 | .2793 | .3826 | .3847 | .3355 | .2670 | .1977 | .1561 | .1373 | .0896 | .0548 | .0313 | 7 |
| 2 | .0026 | .0099 | .0210 | .0351 | .0515 | .1488 | .2376 | .2936 | .3115 | .2965 | .2731 | .2587 | .2090 | .1569 | .1094 | 6 |
| 3 | .0001 | .0004 | .0013 | .0029 | .0054 | .0331 | .0839 | .1468 | .2076 | .2541 | .2731 | .2786 | .2787 | .2568 | .2188 | 5 |
| 4 | .0000 | .0000 | .0001 | .0002 | .0004 | .0046 | .0185 | .0459 | .0865 | .1361 | .1707 | .1875 | .2322 | .2627 | .2734 | 4 |
| 5 | .0000 | .0000 | .0000 | .0000 | .0000 | .0004 | .0026 | .0092 | .0231 | .0467 | .0683 | .0808 | .1239 | .1719 | .2188 | 3 |
| 6 | .0000 | .0000 | .0000 | .0000 | .0000 | .0002 | .0002 | .0011 | .0038 | .0100 | .0171 | .0217 | .0413 | .0703 | .1094 | 2 |
| 7 | .0000 | .0000 | .0000 | .0000 | .0000 | .0000 | .0000 | .0001 | .0004 | .0012 | .0024 | .0033 | .0079 | .0164 | .0313 | 1 |
| 8 | .0000 | .0000 | .0000 | .0000 | .0000 | .0000 | .0000 | .0000 | .0000 | .0001 | .0002 | .0002 | .0007 | .0017 | .0039 | 0 |
| | | | | | | | **n = 9** | | | | | | | | | |
| 0 | .9135 | .8337 | .7602 | .6925 | .6302 | .3874 | .2316 | .1342 | .0751 | .0404 | .0260 | .0207 | .0101 | .0046 | .0020 | 9 |
| 1 | .0830 | .1531 | .2116 | .2597 | .2985 | .3874 | .3679 | .3020 | .2253 | .1556 | .1171 | .1004 | .0605 | .0339 | .0176 | 8 |
| 2 | .0034 | .0125 | .0262 | .0433 | .0629 | .1722 | .2597 | .3020 | .3003 | .2668 | .2341 | .2162 | .1612 | .1110 | .0703 | 7 |
| 3 | .0001 | .0006 | .0019 | .0042 | .0077 | .0446 | .1069 | .1762 | .2336 | .2668 | .2731 | .2716 | .2508 | .2119 | .1641 | 6 |
| 4 | .0000 | .0000 | .0001 | .0003 | .0006 | .0074 | .0283 | .0661 | .1168 | .1715 | .2048 | .2194 | .2508 | .2600 | .2461 | 5 |
| 5 | .0000 | .0000 | .0000 | .0000 | .0000 | .0008 | .0050 | .0165 | .0389 | .0735 | .1024 | .1181 | .1672 | .2128 | .2461 | 4 |
| 6 | .0000 | .0000 | .0000 | .0000 | .0000 | .0001 | .0006 | .0028 | .0087 | .0210 | .0341 | .0424 | .0743 | .1160 | .1641 | 3 |
| 7 | .0000 | .0000 | .0000 | .0000 | .0000 | .0000 | .0000 | .0003 | .0012 | .0039 | .0073 | .0098 | .0212 | .0407 | .0703 | 2 |
| 8 | .0000 | .0000 | .0000 | .0000 | .0000 | .0000 | .0000 | .0000 | .0001 | .0004 | .0009 | .0013 | .0035 | .0083 | .0176 | 1 |
| 9 | .0000 | .0000 | .0000 | .0000 | .0000 | .0000 | .0000 | .0000 | .0000 | .0000 | .0001 | .0001 | .0003 | .0008 | .0020 | 0 |
| | | | | | | | **n = 10** | | | | | | | | | |
| 0 | .9044 | .8171 | .7374 | .6648 | .5987 | .3487 | .1969 | .1074 | .0563 | .0282 | .0173 | .0135 | .0060 | .0025 | .0010 | 10 |
| 1 | .0914 | .1667 | .2281 | .2770 | .3151 | .3874 | .3474 | .2684 | .1877 | .1211 | .0867 | .0725 | .0403 | .0207 | .0098 | 9 |
| 2 | .0042 | .0153 | .0317 | .0519 | .0746 | .1937 | .2759 | .3020 | .2816 | .2335 | .1951 | .1757 | .1209 | .0763 | .0439 | 8 |
| 3 | .0001 | .0008 | .0026 | .0058 | .0105 | .0574 | .1298 | .2013 | .2503 | .2668 | .2601 | .2522 | .2150 | .1665 | .1172 | 7 |
| 4 | .0000 | .0000 | .0001 | .0004 | .0010 | .0112 | .0401 | .0881 | .1460 | .2001 | .2276 | .2377 | .2508 | .2384 | .2051 | 6 |
| 5 | .0000 | .0000 | .0000 | .0000 | .0001 | .0015 | .0085 | .0264 | .0584 | .1029 | .1366 | .1536 | .2007 | .2340 | .2461 | 5 |
| 6 | .0000 | .0000 | .0000 | .0000 | .0000 | .0001 | .0012 | .0055 | .0162 | .0368 | .0569 | .0689 | .1115 | .1596 | .2051 | 4 |
| 7 | .0000 | .0000 | .0000 | .0000 | .0000 | .0000 | .0001 | .0008 | .0031 | .0090 | .0163 | .0212 | .0425 | .0746 | .1172 | 3 |
| 8 | .0000 | .0000 | .0000 | .0000 | .0000 | .0000 | .0000 | .0001 | .0004 | .0014 | .0030 | .0043 | .0106 | .0229 | .0439 | 2 |
| 9 | .0000 | .0000 | .0000 | .0000 | .0000 | .0000 | .0000 | .0000 | .0000 | .0001 | .0003 | .0005 | .0016 | .0042 | .0098 | 1 |
| 10 | .0000 | .0000 | .0000 | .0000 | .0000 | .0000 | .0000 | .0000 | .0000 | .0000 | .0000 | .0000 | .0001 | .0003 | .0010 | 0 |
| | | | | | | | **n = 11** | | | | | | | | | |
| 0 | .8953 | .8007 | .7153 | .6382 | .5688 | .3138 | .1673 | .0859 | .0422 | .0198 | .0116 | .0088 | .0036 | .0014 | .0005 | 11 |
| 1 | .0995 | .1798 | .2433 | .2925 | .3293 | .3835 | .3248 | .2362 | .1549 | .0932 | .0636 | .0518 | .0266 | .0125 | .0054 | 10 |
| 2 | .0050 | .0183 | .0376 | .0609 | .0867 | .2131 | .2866 | .2953 | .2581 | .1998 | .1590 | .1395 | .0887 | .0513 | .0269 | 9 |
| 3 | .0002 | .0011 | .0035 | .0076 | .0137 | .0710 | .1517 | .2215 | .2581 | .2568 | .2384 | .2254 | .1774 | .1259 | .0806 | 8 |
| 4 | .0000 | .0000 | .0002 | .0006 | .0014 | .0158 | .0536 | .1107 | .1721 | .2201 | .2384 | .2428 | .2365 | .2060 | .1611 | 7 |
| 5 | .0000 | .0000 | .0000 | .0000 | .0001 | .0025 | .0132 | .0388 | .0803 | .1321 | .1669 | .1830 | .2207 | .2360 | .2256 | 6 |
| 6 | .0000 | .0000 | .0000 | .0000 | .0000 | .0003 | .0023 | .0097 | .0268 | .0566 | .0835 | .0985 | .1471 | .1931 | .2256 | 5 |
| 7 | .0000 | .0000 | .0000 | .0000 | .0000 | .0000 | .0003 | .0017 | .0064 | .0173 | .0298 | .0379 | .0701 | .1128 | .1611 | 4 |
| 8 | .0000 | .0000 | .0000 | .0000 | .0000 | .0000 | .0000 | .0002 | .0011 | .0037 | .0075 | .0102 | .0234 | .0462 | .0806 | 3 |
| 9 | .0000 | .0000 | .0000 | .0000 | .0000 | .0000 | .0000 | .0001 | .0005 | .0012 | .0018 | .0052 | .0126 | .0269 | 2 | |
| 10 | .0000 | .0000 | .0000 | .0000 | .0000 | .0000 | .0000 | .0000 | .0000 | .0001 | .0002 | .0007 | .0021 | .0054 | 1 | |
| 11 | .0000 | .0000 | .0000 | .0000 | .0000 | .0000 | .0000 | .0000 | .0000 | .0000 | .0000 | .0000 | .0002 | .0005 | 0 | |
| | | | | | | | **n = 12** | | | | | | | | | |
| 0 | .8864 | .7847 | .6938 | .6127 | .5404 | .2824 | .1422 | .0687 | .0317 | .0138 | .0077 | .0057 | .0022 | .0008 | .0002 | 12 |
| 1 | .1074 | .1922 | .2575 | .3064 | .3413 | .3766 | .3012 | .2062 | .1267 | .0712 | .0462 | .0368 | .0174 | .0075 | .0029 | 11 |
| 2 | .0060 | .0216 | .0438 | .0702 | .0988 | .2301 | .2924 | .2835 | .2323 | .1678 | .1272 | .1088 | .0639 | .0339 | .0161 | 10 |
| 3 | .0002 | .0015 | .0045 | .0098 | .0173 | .0852 | .1720 | .2362 | .2581 | .2397 | .2120 | .1954 | .1419 | .0923 | .0537 | 9 |
| 4 | .0000 | .0001 | .0003 | .0009 | .0021 | .0213 | .0683 | .1329 | .1936 | .2311 | .2384 | .2367 | .2128 | .1700 | .1208 | 8 |
| 5 | .0000 | .0000 | .0000 | .0001 | .0002 | .0038 | .0193 | .0532 | .1032 | .1585 | .1908 | .2039 | .2270 | .2225 | .1934 | 7 |
| 6 | .0000 | .0000 | .0000 | .0000 | .0000 | .0005 | .0040 | .0155 | .0401 | .0792 | .1113 | .1281 | .1766 | .2124 | .2256 | 6 |
| 7 | .0000 | .0000 | .0000 | .0000 | .0000 | .0000 | .0006 | .0033 | .0115 | .0291 | .0477 | .0591 | .1009 | .1489 | .1934 | 5 |
| 8 | .0000 | .0000 | .0000 | .0000 | .0000 | .0000 | .0001 | .0005 | .0024 | .0078 | .0149 | .0199 | .0420 | .0762 | .1208 | 4 |
| 9 | .0000 | .0000 | .0000 | .0000 | .0000 | .0000 | .0000 | .0001 | .0004 | .0015 | .0033 | .0048 | .0125 | .0277 | .0537 | 3 |
| 10 | .0000 | .0000 | .0000 | .0000 | .0000 | .0000 | .0000 | .0000 | .0000 | .0002 | .0005 | .0008 | .0025 | .0068 | .0161 | 2 |
| 11 | .0000 | .0000 | .0000 | .0000 | .0000 | .0000 | .0000 | .0000 | .0000 | .0000 | .0000 | .0001 | .0003 | .0010 | .0029 | 1 |
| 12 | .0000 | .0000 | .0000 | .0000 | .0000 | .0000 | .0000 | .0000 | .0000 | .0000 | .0000 | .0000 | .0000 | .0001 | .0002 | 0 = Y |
| | .99 | .98 | .97 | .96 | .95 | .90 | .85 | .80 | .75 | .70 | .667 | .65 | .60 | .55 | .50 | = p |

(continued on next page)

# Table A2 (continued) Expected frequencies for binomial distributions.

## n = 13

| p = | .01 | .02 | .03 | .04 | .05 | .10 | .15 | .20 | .25 | .30 | .333 | .35 | .40 | .45 | .50 | |
|---|---|---|---|---|---|---|---|---|---|---|---|---|---|---|---|---|
| Y = 0 | .8775 | .7690 | .6730 | .5882 | .5133 | .2542 | .1209 | .0550 | .0238 | .0097 | .0051 | .0037 | .0013 | .0004 | .0001 | 13 |
| 1 | .1152 | .2040 | .2706 | .3186 | .3512 | .3672 | .2774 | .1787 | .1029 | .0540 | .0334 | .0259 | .0113 | .0045 | .0016 | 12 |
| 2 | .0070 | .0250 | .0502 | .0797 | .1109 | .2448 | .2937 | .2680 | .2059 | .1388 | .1002 | .0836 | .0453 | .0220 | .0095 | 11 |
| 3 | .0003 | .0019 | .0057 | .0122 | .0214 | .0997 | .1900 | .2457 | .2517 | .2181 | .1837 | .1651 | .1107 | .0660 | .0349 | 10 |
| 4 | .0000 | .0001 | .0004 | .0013 | .0028 | .0277 | .0838 | .1535 | .2097 | .2337 | .2296 | .2222 | .1845 | .1350 | .0873 | 9 |
| 5 | .0000 | .0000 | .0000 | .0001 | .0003 | .0055 | .0266 | .0691 | .1258 | .1803 | .2067 | .2154 | .2214 | .1989 | .1571 | 8 |
| 6 | .0000 | .0000 | .0000 | .0000 | .0000 | .0008 | .0063 | .0230 | .0559 | .1030 | .1378 | .1546 | .1968 | .2169 | .2095 | 7 |
| 7 | .0000 | .0000 | .0000 | .0000 | .0000 | .0001 | .0011 | .0058 | .0186 | .0442 | .0689 | .0833 | .1312 | .1775 | .2095 | 6 |
| 8 | .0000 | .0000 | .0000 | .0000 | .0000 | .0000 | .0001 | .0011 | .0047 | .0142 | .0258 | .0336 | .0656 | .1089 | .1571 | 5 |
| 9 | .0000 | .0000 | .0000 | .0000 | .0000 | .0000 | .0000 | .0001 | .0009 | .0034 | .0072 | .0101 | .0243 | .0495 | .0873 | 4 |
| 10 | .0000 | .0000 | .0000 | .0000 | .0000 | .0000 | .0000 | .0000 | .0001 | .0006 | .0014 | .0022 | .0065 | .0162 | .0349 | 3 |
| 11 | .0000 | .0000 | .0000 | .0000 | .0000 | .0000 | .0000 | .0000 | .0000 | .0001 | .0002 | .0003 | .0012 | .0036 | .0095 | 2 |
| 12 | .0000 | .0000 | .0000 | .0000 | .0000 | .0000 | .0000 | .0000 | .0000 | .0000 | .0000 | .0000 | .0001 | .0005 | .0016 | 1 |
| 13 | .0000 | .0000 | .0000 | .0000 | .0000 | .0000 | .0000 | .0000 | .0000 | .0000 | .0000 | .0000 | .0000 | .0000 | .0001 | 0 |

## n = 14

| Y | .01 | .02 | .03 | .04 | .05 | .10 | .15 | .20 | .25 | .30 | .333 | .35 | .40 | .45 | .50 | |
|---|---|---|---|---|---|---|---|---|---|---|---|---|---|---|---|---|
| 0 | .8687 | .7536 | .6528 | .5647 | .4877 | .2288 | .1028 | .0440 | .0178 | .0068 | .0034 | .0024 | .0008 | .0002 | .0001 | 14 |
| 1 | .1229 | .2153 | .2827 | .3294 | .3593 | .3559 | .2539 | .1539 | .0832 | .0407 | .0240 | .0181 | .0073 | .0027 | .0009 | 13 |
| 2 | .0081 | .0286 | .0568 | .0892 | .1229 | .2570 | .2912 | .2501 | .1802 | .1134 | .0779 | .0634 | .0317 | .0141 | .0056 | 12 |
| 3 | .0003 | .0023 | .0070 | .0149 | .0259 | .1142 | .2056 | .2501 | .2402 | .1943 | .1559 | .1366 | .0845 | .0462 | .0222 | 11 |
| 4 | .0000 | .0001 | .0006 | .0017 | .0037 | .0349 | .0998 | .1720 | .2202 | .2290 | .2143 | .2022 | .1549 | .1040 | .0611 | 10 |
| 5 | .0000 | .0000 | .0000 | .0001 | .0004 | .0078 | .0352 | .0860 | .1468 | .1963 | .2143 | .2178 | .2066 | .1701 | .1222 | 9 |
| 6 | .0000 | .0000 | .0000 | .0000 | .0000 | .0013 | .0093 | .0322 | .0734 | .1262 | .1607 | .1759 | .2066 | .2088 | .1833 | 8 |
| 7 | .0000 | .0000 | .0000 | .0000 | .0000 | .0002 | .0019 | .0092 | .0280 | .0618 | .0918 | .1082 | .1574 | .1952 | .2095 | 7 |
| 8 | .0000 | .0000 | .0000 | .0000 | .0000 | .0000 | .0003 | .0020 | .0082 | .0232 | .0402 | .0510 | .0918 | .1398 | .1833 | 6 |
| 9 | .0000 | .0000 | .0000 | .0000 | .0000 | .0000 | .0000 | .0003 | .0018 | .0066 | .0134 | .0183 | .0408 | .0762 | .1222 | 5 |
| 10 | .0000 | .0000 | .0000 | .0000 | .0000 | .0000 | .0000 | .0000 | .0003 | .0014 | .0033 | .0049 | .0136 | .0312 | .0611 | 4 |
| 11 | .0000 | .0000 | .0000 | .0000 | .0000 | .0000 | .0000 | .0000 | .0000 | .0002 | .0006 | .0010 | .0033 | .0093 | .0222 | 3 |
| 12 | .0000 | .0000 | .0000 | .0000 | .0000 | .0000 | .0000 | .0000 | .0000 | .0000 | .0001 | .0001 | .0005 | .0019 | .0056 | 2 |
| 13 | .0000 | .0000 | .0000 | .0000 | .0000 | .0000 | .0000 | .0000 | .0000 | .0000 | .0000 | .0000 | .0001 | .0002 | .0009 | 1 |
| 14 | .0000 | .0000 | .0000 | .0000 | .0000 | .0000 | .0000 | .0000 | .0000 | .0000 | .0000 | .0000 | .0000 | .0000 | .0001 | 0 |

## n = 15

| Y | .01 | .02 | .03 | .04 | .05 | .10 | .15 | .20 | .25 | .30 | .333 | .35 | .40 | .45 | .50 | |
|---|---|---|---|---|---|---|---|---|---|---|---|---|---|---|---|---|
| 0 | .8601 | .7386 | .6333 | .5421 | .4633 | .2059 | .0874 | .0352 | .0134 | .0047 | .0023 | .0016 | .0005 | .0001 | .0000 | 15 |
| 1 | .1303 | .2261 | .2938 | .3388 | .3658 | .3432 | .2312 | .1319 | .0668 | .0305 | .0171 | .0126 | .0047 | .0016 | .0005 | 14 |
| 2 | .0092 | .0323 | .0636 | .0988 | .1348 | .2669 | .2856 | .2309 | .1559 | .0916 | .0599 | .0476 | .0219 | .0090 | .0032 | 13 |
| 3 | .0004 | .0029 | .0085 | .0178 | .0307 | .1285 | .2184 | .2501 | .2252 | .1700 | .1299 | .1110 | .0634 | .0318 | .0139 | 12 |
| 4 | .0000 | .0002 | .0008 | .0022 | .0049 | .0428 | .1156 | .1876 | .2252 | .2186 | .1948 | .1792 | .1268 | .0780 | .0417 | 11 |
| 5 | .0000 | .0000 | .0001 | .0002 | .0006 | .0105 | .0449 | .1032 | .1651 | .2061 | .2143 | .2123 | .1859 | .1404 | .0916 | 10 |
| 6 | .0000 | .0000 | .0000 | .0000 | .0000 | .0019 | .0132 | .0430 | .0917 | .1472 | .1786 | .1906 | .2066 | .1914 | .1527 | 9 |
| 7 | .0000 | .0000 | .0000 | .0000 | .0000 | .0003 | .0030 | .0138 | .0393 | .0811 | .1148 | .1319 | .1771 | .2013 | .1964 | 8 |
| 8 | .0000 | .0000 | .0000 | .0000 | .0000 | .0000 | .0005 | .0035 | .0131 | .0348 | .0574 | .0710 | .1181 | .1647 | .1964 | 7 |
| 9 | .0000 | .0000 | .0000 | .0000 | .0000 | .0000 | .0001 | .0007 | .0034 | .0116 | .0223 | .0298 | .0612 | .1048 | .1527 | 6 |
| 10 | .0000 | .0000 | .0000 | .0000 | .0000 | .0000 | .0000 | .0001 | .0007 | .0030 | .0067 | .0096 | .0245 | .0515 | .0916 | 5 |
| 11 | .0000 | .0000 | .0000 | .0000 | .0000 | .0000 | .0000 | .0000 | .0001 | .0006 | .0015 | .0024 | .0074 | .0191 | .0417 | 4 |
| 12 | .0000 | .0000 | .0000 | .0000 | .0000 | .0000 | .0000 | .0000 | .0000 | .0001 | .0003 | .0004 | .0016 | .0052 | .0139 | 3 |
| 13 | .0000 | .0000 | .0000 | .0000 | .0000 | .0000 | .0000 | .0000 | .0000 | .0000 | .0000 | .0001 | .0003 | .0010 | .0032 | 2 |
| 14 | .0000 | .0000 | .0000 | .0000 | .0000 | .0000 | .0000 | .0000 | .0000 | .0000 | .0000 | .0000 | .0000 | .0001 | .0005 | 1 |
| 15 | .0000 | .0000 | .0000 | .0000 | .0000 | .0000 | .0000 | .0000 | .0000 | .0000 | .0000 | .0000 | .0000 | .0000 | .0000 | 0 |

## n = 16

| Y | .01 | .02 | .03 | .04 | .05 | .10 | .15 | .20 | .25 | .30 | .333 | .35 | .40 | .45 | .50 | |
|---|---|---|---|---|---|---|---|---|---|---|---|---|---|---|---|---|
| 0 | .8515 | .7238 | .6143 | .5204 | .4401 | .1853 | .0743 | .0281 | .0100 | .0033 | .0015 | .0010 | .0003 | .0001 | .0000 | 16 |
| 1 | .1376 | .2363 | .3040 | .3469 | .3706 | .3294 | .2097 | .1126 | .0535 | .0228 | .0122 | .0087 | .0030 | .0009 | .0002 | 15 |
| 2 | .0104 | .0362 | .0705 | .1084 | .1463 | .2745 | .2775 | .2111 | .1336 | .0732 | .0457 | .0353 | .0150 | .0056 | .0018 | 14 |
| 3 | .0005 | .0034 | .0102 | .0211 | .0359 | .1423 | .2285 | .2463 | .2079 | .1465 | .1066 | .0888 | .0468 | .0215 | .0085 | 13 |
| 4 | .0000 | .0002 | .0010 | .0029 | .0061 | .0514 | .1311 | .2001 | .2252 | .2040 | .1732 | .1553 | .1014 | .0572 | .0278 | 12 |
| 5 | .0000 | .0000 | .0001 | .0003 | .0008 | .0137 | .0555 | .1201 | .1802 | .2099 | .2078 | .2008 | .1623 | .1123 | .0667 | 11 |
| 6 | .0000 | .0000 | .0000 | .0000 | .0001 | .0028 | .0180 | .0550 | .1101 | .1649 | .1905 | .1982 | .1983 | .1684 | .1222 | 10 |
| 7 | .0000 | .0000 | .0000 | .0000 | .0000 | .0004 | .0045 | .0197 | .0524 | .1010 | .1361 | .1524 | .1889 | .1969 | .1746 | 9 |
| 8 | .0000 | .0000 | .0000 | .0000 | .0000 | .0001 | .0009 | .0055 | .0197 | .0487 | .0765 | .0923 | .1417 | .1812 | .1964 | 8 |
| 9 | .0000 | .0000 | .0000 | .0000 | .0000 | .0000 | .0001 | .0012 | .0058 | .0185 | .0340 | .0442 | .0840 | .1318 | .1746 | 7 |
| 10 | .0000 | .0000 | .0000 | .0000 | .0000 | .0000 | .0000 | .0002 | .0014 | .0056 | .0119 | .0167 | .0392 | .0755 | .1222 | 6 |
| 11 | .0000 | .0000 | .0000 | .0000 | .0000 | .0000 | .0000 | .0000 | .0002 | .0013 | .0032 | .0049 | .0142 | .0337 | .0667 | 5 |
| 12 | .0000 | .0000 | .0000 | .0000 | .0000 | .0000 | .0000 | .0000 | .0000 | .0002 | .0007 | .0011 | .0040 | .0115 | .0278 | 4 |
| 13 | .0000 | .0000 | .0000 | .0000 | .0000 | .0000 | .0000 | .0000 | .0000 | .0000 | .0001 | .0002 | .0008 | .0029 | .0085 | 3 |
| 14 | .0000 | .0000 | .0000 | .0000 | .0000 | .0000 | .0000 | .0000 | .0000 | .0000 | .0000 | .0000 | .0001 | .0005 | .0018 | 2 |
| 15 | .0000 | .0000 | .0000 | .0000 | .0000 | .0000 | .0000 | .0000 | .0000 | .0000 | .0000 | .0000 | .0000 | .0001 | .0002 | 1 |
| 16 | .0000 | .0000 | .0000 | .0000 | .0000 | .0000 | .0000 | .0000 | .0000 | .0000 | .0000 | .0000 | .0000 | .0000 | .0000 | 0 = Y |

| | .99 | .98 | .97 | .96 | .95 | .90 | .85 | .80 | .75 | .70 | .667 | .65 | .60 | .55 | .50 | = p |

# Table A2 (continued) Expected frequencies for binomial distributions.

| p = | .01 | .02 | .03 | .04 | .05 | .10 | .15 | .20 | .25 | .30 | .333 | .35 | .40 | .45 | .50 | |
|---|---|---|---|---|---|---|---|---|---|---|---|---|---|---|---|---|
| | | | | | | | | n = 17 | | | | | | | | |
| Y = 0 | .8429 | .7093 | .5958 | .4996 | .4181 | .1668 | .0631 | .0225 | .0075 | .0023 | .0010 | .0007 | .0002 | .0000 | .0000 | 17 |
| 1 | .1447 | .2461 | .3133 | .3539 | .3741 | .3150 | .1893 | .0957 | .0426 | .0169 | .0086 | .0060 | .0019 | .0005 | .0001 | 16 |
| 2 | .0117 | .0402 | .0775 | .1180 | .1575 | .2800 | .2673 | .1914 | .1136 | .0581 | .0345 | .0260 | .0102 | .0035 | .0010 | 15 |
| 3 | .0006 | .0041 | .0120 | .0246 | .0415 | .1556 | .2359 | .2393 | .1893 | .1245 | .0863 | .0701 | .0341 | .0144 | .0052 | 14 |
| 4 | .0000 | .0003 | .0013 | .0036 | .0076 | .0605 | .1457 | .2093 | .2209 | .1868 | .1510 | .1320 | .0796 | .0411 | .0182 | 13 |
| 5 | .0000 | .0000 | .0001 | .0004 | .0010 | .0175 | .0668 | .1361 | .1914 | .2081 | .1963 | .1849 | .1379 | .0875 | .0472 | 12 |
| 6 | .0000 | .0000 | .0000 | .0000 | .0001 | .0039 | .0236 | .0680 | .1276 | .1784 | .1963 | .1991 | .1839 | .1432 | .0944 | 11 |
| 7 | .0000 | .0000 | .0000 | .0000 | .0000 | .0007 | .0065 | .0267 | .0668 | .1201 | .1542 | .1685 | .1927 | .1841 | .1484 | 10 |
| 8 | .0000 | .0000 | .0000 | .0000 | .0000 | .0001 | .0014 | .0084 | .0279 | .0644 | .0964 | .1134 | .1606 | .1883 | .1855 | 9 |
| 9 | .0000 | .0000 | .0000 | .0000 | .0000 | .0000 | .0003 | .0021 | .0093 | .0276 | .0482 | .0611 | .1070 | .1540 | .1855 | 8 |
| 10 | .0000 | .0000 | .0000 | .0000 | .0000 | .0000 | .0000 | .0004 | .0025 | .0095 | .0193 | .0263 | .0571 | .1008 | .1484 | 7 |
| 11 | .0000 | .0000 | .0000 | .0000 | .0000 | .0000 | .0000 | .0001 | .0005 | .0026 | .0061 | .0090 | .0242 | .0525 | .0944 | 6 |
| 12 | .0000 | .0000 | .0000 | .0000 | .0000 | .0000 | .0000 | .0000 | .0001 | .0006 | .0015 | .0024 | .0081 | .0215 | .0472 | 5 |
| 13 | .0000 | .0000 | .0000 | .0000 | .0000 | .0000 | .0000 | .0000 | .0000 | .0001 | .0003 | .0005 | .0021 | .0068 | .0182 | 4 |
| 14 | .0000 | .0000 | .0000 | .0000 | .0000 | .0000 | .0000 | .0000 | .0000 | .0000 | .0000 | .0001 | .0004 | .0016 | .0052 | 3 |
| 15 | .0000 | .0000 | .0000 | .0000 | .0000 | .0000 | .0000 | .0000 | .0000 | .0000 | .0000 | .0000 | .0001 | .0003 | .0010 | 2 |
| 16 | .0000 | .0000 | .0000 | .0000 | .0000 | .0000 | .0000 | .0000 | .0000 | .0000 | .0000 | .0000 | .0000 | .0000 | .0001 | 1 |
| 17 | .0000 | .0000 | .0000 | .0000 | .0000 | .0000 | .0000 | .0000 | .0000 | .0000 | .0000 | .0000 | .0000 | .0000 | .0000 | 0 |
| | | | | | | | | n = 18 | | | | | | | | |
| 0 | .8345 | .6951 | .5780 | .4796 | .3972 | .1501 | .0536 | .0180 | .0056 | .0016 | .0007 | .0004 | .0001 | .0000 | .0000 | 18 |
| 1 | .1517 | .2554 | .3217 | .3597 | .3763 | .3002 | .1704 | .0811 | .0338 | .0126 | .0061 | .0042 | .0012 | .0003 | .0001 | 17 |
| 2 | .0130 | .0443 | .0846 | .1274 | .1683 | .2835 | .2556 | .1723 | .0958 | .0458 | .0259 | .0190 | .0069 | .0022 | .0006 | 16 |
| 3 | .0007 | .0048 | .0140 | .0283 | .0473 | .1680 | .2406 | .2297 | .1704 | .1046 | .0690 | .0547 | .0246 | .0095 | .0031 | 15 |
| 4 | .0000 | .0004 | .0016 | .0044 | .0093 | .0700 | .1592 | .2153 | .2130 | .1681 | .1294 | .1104 | .0614 | .0291 | .0117 | 14 |
| 5 | .0000 | .0000 | .0001 | .0005 | .0014 | .0218 | .0787 | .1507 | .1988 | .2017 | .1812 | .1664 | .1146 | .0666 | .0327 | 13 |
| 6 | .0000 | .0000 | .0000 | .0000 | .0002 | .0052 | .0301 | .0816 | .1436 | .1873 | .1963 | .1941 | .1655 | .1181 | .0708 | 12 |
| 7 | .0000 | .0000 | .0000 | .0000 | .0000 | .0010 | .0091 | .0350 | .0820 | .1376 | .1682 | .1792 | .1892 | .1657 | .1214 | 11 |
| 8 | .0000 | .0000 | .0000 | .0000 | .0000 | .0002 | .0022 | .0120 | .0376 | .0811 | .1157 | .1327 | .1734 | .1864 | .1669 | 10 |
| 9 | .0000 | .0000 | .0000 | .0000 | .0000 | .0000 | .0004 | .0033 | .0139 | .0386 | .0643 | .0794 | .1284 | .1694 | .1855 | 9 |
| 10 | .0000 | .0000 | .0000 | .0000 | .0000 | .0000 | .0001 | .0008 | .0042 | .0149 | .0289 | .0385 | .0771 | .1248 | .1669 | 8 |
| 11 | .0000 | .0000 | .0000 | .0000 | .0000 | .0000 | .0000 | .0001 | .0010 | .0046 | .0105 | .0151 | .0374 | .0742 | .1214 | 7 |
| 12 | .0000 | .0000 | .0000 | .0000 | .0000 | .0000 | .0000 | .0000 | .0002 | .0012 | .0031 | .0047 | .0145 | .0354 | .0708 | 6 |
| 13 | .0000 | .0000 | .0000 | .0000 | .0000 | .0000 | .0000 | .0000 | .0000 | .0002 | .0007 | .0012 | .0045 | .0134 | .0327 | 5 |
| 14 | .0000 | .0000 | .0000 | .0000 | .0000 | .0000 | .0000 | .0000 | .0000 | .0000 | .0001 | .0002 | .0011 | .0039 | .0117 | 4 |
| 15 | .0000 | .0000 | .0000 | .0000 | .0000 | .0000 | .0000 | .0000 | .0000 | .0000 | .0000 | .0000 | .0002 | .0009 | .0031 | 3 |
| 16 | .0000 | .0000 | .0000 | .0000 | .0000 | .0000 | .0000 | .0000 | .0000 | .0000 | .0000 | .0000 | .0000 | .0001 | .0006 | 2 |
| 17 | .0000 | .0000 | .0000 | .0000 | .0000 | .0000 | .0000 | .0000 | .0000 | .0000 | .0000 | .0000 | .0000 | .0000 | .0001 | 1 |
| 18 | .0000 | .0000 | .0000 | .0000 | .0000 | .0000 | .0000 | .0000 | .0000 | .0000 | .0000 | .0000 | .0000 | .0000 | .0000 | 0 |
| | | | | | | | | n = 19 | | | | | | | | |
| 0 | .8262 | .6812 | .5606 | .4604 | .3774 | .1351 | .0456 | .0144 | .0042 | .0011 | .0005 | .0003 | .0001 | .0000 | .0000 | 19 |
| 1 | .1586 | .2642 | .3294 | .3645 | .3774 | .2852 | .1529 | .0685 | .0268 | .0093 | .0043 | .0029 | .0008 | .0002 | .0000 | 18 |
| 2 | .0144 | .0485 | .0917 | .1367 | .1787 | .2852 | .2428 | .1540 | .0803 | .0358 | .0193 | .0138 | .0046 | .0013 | .0003 | 17 |
| 3 | .0008 | .0056 | .0161 | .0323 | .0533 | .1796 | .2428 | .2182 | .1517 | .0869 | .0546 | .0422 | .0175 | .0062 | .0018 | 16 |
| 4 | .0000 | .0005 | .0020 | .0054 | .0112 | .0798 | .1714 | .2182 | .2023 | .1491 | .1093 | .0909 | .0467 | .0203 | .0074 | 15 |
| 5 | .0000 | .0000 | .0002 | .0007 | .0018 | .0266 | .0907 | .1636 | .2023 | .1916 | .1639 | .1468 | .0933 | .0497 | .0222 | 14 |
| 6 | .0000 | .0000 | .0000 | .0001 | .0002 | .0069 | .0374 | .0955 | .1574 | .1916 | .1912 | .1844 | .1451 | .0949 | .0518 | 13 |
| 7 | .0000 | .0000 | .0000 | .0000 | .0000 | .0014 | .0122 | .0443 | .0974 | .1525 | .1776 | .1844 | .1797 | .1443 | .0961 | 12 |
| 8 | .0000 | .0000 | .0000 | .0000 | .0000 | .0002 | .0032 | .0166 | .0487 | .0981 | .1332 | .1489 | .1797 | .1771 | .1442 | 11 |
| 9 | .0000 | .0000 | .0000 | .0000 | .0000 | .0000 | .0007 | .0051 | .0198 | .0514 | .0814 | .0980 | .1464 | .1771 | .1762 | 10 |
| 10 | .0000 | .0000 | .0000 | .0000 | .0000 | .0000 | .0001 | .0013 | .0066 | .0220 | .0407 | .0528 | .0976 | .1449 | .1762 | 9 |
| 11 | .0000 | .0000 | .0000 | .0000 | .0000 | .0000 | .0000 | .0003 | .0018 | .0077 | .0166 | .0233 | .0532 | .0970 | .1442 | 8 |
| 12 | .0000 | .0000 | .0000 | .0000 | .0000 | .0000 | .0000 | .0000 | .0004 | .0022 | .0055 | .0083 | .0237 | .0529 | .0961 | 7 |
| 13 | .0000 | .0000 | .0000 | .0000 | .0000 | .0000 | .0000 | .0000 | .0001 | .0005 | .0015 | .0024 | .0085 | .0233 | .0518 | 6 |
| 14 | .0000 | .0000 | .0000 | .0000 | .0000 | .0000 | .0000 | .0000 | .0000 | .0001 | .0003 | .0006 | .0024 | .0082 | .0222 | 5 |
| 15 | .0000 | .0000 | .0000 | .0000 | .0000 | .0000 | .0000 | .0000 | .0000 | .0000 | .0001 | .0001 | .0005 | .0022 | .0074 | 4 |
| 16 | .0000 | .0000 | .0000 | .0000 | .0000 | .0000 | .0000 | .0000 | .0000 | .0000 | .0000 | .0000 | .0001 | .0005 | .0018 | 3 |
| 17 | .0000 | .0000 | .0000 | .0000 | .0000 | .0000 | .0000 | .0000 | .0000 | .0000 | .0000 | .0000 | .0000 | .0001 | .0003 | 2 |
| 18 | .0000 | .0000 | .0000 | .0000 | .0000 | .0000 | .0000 | .0000 | .0000 | .0000 | .0000 | .0000 | .0000 | .0000 | .0000 | 1 |
| 19 | .0000 | .0000 | .0000 | .0000 | .0000 | .0000 | .0000 | .0000 | .0000 | .0000 | .0000 | .0000 | .0000 | .0000 | .0000 | 0 = Y |
| | .99 | .98 | .97 | .96 | .95 | .90 | .85 | .80 | .75 | .70 | .667 | .65 | .60 | .55 | .50 | = p |

(continued on next page)

**Table A2 (continued) Expected frequencies for binomial distributions.**

| p = | .01 | .02 | .03 | .04 | .05 | .10 | .15 | .20 | .25 | .30 | .333 | .35 | .40 | .45 | .50 | |
|---|---|---|---|---|---|---|---|---|---|---|---|---|---|---|---|---|
| | | | | | | | | n = 20 | | | | | | | | |
| 0 | .8179 | .6676 | .5438 | .4420 | .3585 | .1216 | .0388 | .0115 | .0032 | .0008 | .0003 | .0002 | .0000 | .0000 | .0000 | 20 |
| 1 | .1652 | .2725 | .3364 | .3683 | .3774 | .2702 | .1368 | .0576 | .0211 | .0068 | .0030 | .0020 | .0005 | .0001 | .0000 | 19 |
| 2 | .0159 | .0528 | .0988 | .1458 | .1887 | .2852 | .2293 | .1369 | .0669 | .0278 | .0143 | .0100 | .0031 | .0008 | .0002 | 18 |
| 3 | .0010 | .0065 | .0183 | .0364 | .0596 | .1901 | .2428 | .2054 | .1339 | .0716 | .0429 | .0323 | .0123 | .0040 | .0011 | 17 |
| 4 | .0000 | .0006 | .0024 | .0065 | .0133 | .0898 | .1821 | .2182 | .1897 | .1304 | .0911 | .0738 | .0350 | .0139 | .0046 | 16 |
| 5 | .0000 | .0000 | .0002 | .0009 | .0022 | .0319 | .1028 | .1746 | .2023 | .1789 | .1457 | .1272 | .0746 | .0365 | .0148 | 15 |
| 6 | .0000 | .0000 | .0000 | .0001 | .0003 | .0089 | .0454 | .1091 | .1686 | .1916 | .1821 | .1712 | .1244 | .0746 | .0370 | 14 |
| 7 | .0000 | .0000 | .0000 | .0000 | .0000 | .0020 | .0160 | .0545 | .1124 | .1643 | .1821 | .1844 | .1659 | .1221 | .0739 | 13 |
| 8 | .0000 | .0000 | .0000 | .0000 | .0000 | .0004 | .0046 | .0222 | .0609 | .1144 | .1480 | .1614 | .1797 | .1623 | .1201 | 12 |
| 9 | .0000 | .0000 | .0000 | .0000 | .0000 | .0001 | .0011 | .0074 | .0271 | .0654 | .0987 | .1158 | .1597 | .1771 | .1602 | 11 |
| 10 | .0000 | .0000 | .0000 | .0000 | .0000 | .0000 | .0002 | .0020 | .0099 | .0308 | .0543 | .0686 | .1171 | .1593 | .1762 | 10 |
| 11 | .0000 | .0000 | .0000 | .0000 | .0000 | .0000 | .0000 | .0005 | .0030 | .0120 | .0247 | .0336 | .0710 | .1185 | .1602 | 9 |
| 12 | .0000 | .0000 | .0000 | .0000 | .0000 | .0000 | .0000 | .0001 | .0008 | .0039 | .0092 | .0136 | .0355 | .0727 | .1201 | 8 |
| 13 | .0000 | .0000 | .0000 | .0000 | .0000 | .0000 | .0000 | .0000 | .0002 | .0010 | .0028 | .0045 | .0146 | .0366 | .0739 | 7 |
| 14 | .0000 | .0000 | .0000 | .0000 | .0000 | .0000 | .0000 | .0000 | .0000 | .0002 | .0007 | .0012 | .0049 | .0150 | .0370 | 6 |
| 15 | .0000 | .0000 | .0000 | .0000 | .0000 | .0000 | .0000 | .0000 | .0000 | .0000 | .0001 | .0003 | .0013 | .0049 | .0148 | 5 |
| 16 | .0000 | .0000 | .0000 | .0000 | .0000 | .0000 | .0000 | .0000 | .0000 | .0000 | .0000 | .0000 | .0003 | .0013 | .0046 | 4 |
| 17 | .0000 | .0000 | .0000 | .0000 | .0000 | .0000 | .0000 | .0000 | .0000 | .0000 | .0000 | .0000 | .0000 | .0002 | .0011 | 3 |
| 18 | .0000 | .0000 | .0000 | .0000 | .0000 | .0000 | .0000 | .0000 | .0000 | .0000 | .0000 | .0000 | .0000 | .0000 | .0002 | 2 |
| 19 | .0000 | .0000 | .0000 | .0000 | .0000 | .0000 | .0000 | .0000 | .0000 | .0000 | .0000 | .0000 | .0000 | .0000 | .0000 | 1 |
| 20 | .0000 | .0000 | .0000 | .0000 | .0000 | .0000 | .0000 | .0000 | .0000 | .0000 | .0000 | .0000 | .0000 | .0000 | .0000 | 0 = Y |
| | .99 | .98 | .97 | .96 | .95 | .90 | .85 | .80 | .75. | 70 | .667 | .65 | .60 | .55 | .50 | = p |

## Table A3 Expected relative frequencies for Poisson distributions.

This table is used to find the expected relative frequency of samples with a specified number of occurences ($Y$) of an event which has a Poisson distribution with mean $\mu$.

1. Find required value of $\mu$ at top of table.
2. Find required value of $Y$ at left or right of table.
3. Values in body of table are expected relative frequencies e.g. if $\mu = 1.5$, the expected relative frequency of samples with $Y = 1$ is 0.3347.

| $\mu =$ | .01 | .02 | .03 | .04 | .05 | .06 | .07 | .08 | .09 | .1 | .12 | .14 | .16 | .18 | .2 | $= \mu$ |
|---|---|---|---|---|---|---|---|---|---|---|---|---|---|---|---|---|
| $Y = 0$ | .9900 | .9802 | .9704 | .9608 | .9512 | .9418 | .9324 | .9231 | .9139 | .9048 | .8869 | .8694 | .8521 | .8353 | .8187 | 0 = Y |
| 1 | .0099 | .0196 | .0291 | .0384 | .0476 | .0565 | .0653 | .0738 | .0823 | .0905 | .1064 | .1217 | .1363 | .1503 | .1637 | 1 |
| 2 | .0000 | .0002 | .0004 | .0008 | .0012 | .0017 | .0023 | .0030 | .0037 | .0045 | .0064 | .0085 | .0109 | .0135 | .0164 | 2 |
| 3 | .0000 | .0000 | .0000 | .0000 | .0000 | .0000 | .0001 | .0001 | .0001 | .0002 | .0003 | .0004 | .0006 | .0008 | .0011 | 3 |
| 4 | .0000 | .0000 | .0000 | .0000 | .0000 | .0000 | .0000 | .0000 | .0000 | .0000 | .0000 | .0000 | .0000 | .0000 | .0001 | 4 |

| $\mu =$ | 0.25 | 0.30 | 0.35 | 0.40 | 0.45 | 0.50 | 0.55 | 0.60 | 0.65 | 0.70 | 0.75 | 0.80 | 0.85 | 0.90 | 0.95 | $= \mu$ |
|---|---|---|---|---|---|---|---|---|---|---|---|---|---|---|---|---|
| $Y = 0$ | .7788 | .7408 | .7047 | .6703 | .6376 | .6065 | .5769 | .5488 | .5220 | .4966 | .4724 | .4493 | .4274 | .4066 | .3867 | 0 = Y |
| 1 | .1947 | .2222 | .2466 | .2681 | .2869 | .3033 | .3173 | .3293 | .3393 | .3476 | .3543 | .3595 | .3633 | .3659 | .3674 | 1 |
| 2 | .0243 | .0333 | .0432 | .0536 | .0646 | .0758 | .0873 | .0988 | .1103 | .1217 | .1329 | .1438 | .1544 | .1647 | .1745 | 2 |
| 3 | .0020 | .0033 | .0050 | .0072 | .0097 | .0126 | .0160 | .0198 | .0239 | .0284 | .0332 | .0383 | .0437 | .0494 | .0553 | 3 |
| 4 | .0001 | .0003 | .0004 | .0007 | .0011 | .0016 | .0022 | .0030 | .0039 | .0050 | .0062 | .0077 | .0093 | .0111 | .0131 | 4 |
| 5 | .0000 | .0000 | .0000 | .0001 | .0001 | .0002 | .0002 | .0004 | .0005 | .0007 | .0009 | .0012 | .0016 | .0020 | .0025 | 5 |
| 6 | .0000 | .0000 | .0000 | .0000 | .0000 | .0000 | .0000 | .0000 | .0001 | .0001 | .0001 | .0002 | .0002 | .0003 | .0004 | 6 |
| 7 | .0000 | .0000 | .0000 | .0000 | .0000 | .0000 | .0000 | .0000 | .0000 | .0000 | .0000 | .0000 | .0000 | .0000 | .0001 | 7 |

| $\mu =$ | 1.0 | 1.1 | 1.2 | 1.3 | 1.4 | 1.5 | 1.6 | 1.7 | 1.8 | 1.9 | 2.0 | 2.1 | 2.2 | 2.3 | 2.4 | $= \mu$ |
|---|---|---|---|---|---|---|---|---|---|---|---|---|---|---|---|---|
| $Y = 0$ | .3679 | .3329 | .3012 | .2725 | .2466 | .2231 | .2019 | .1827 | .1653 | .1496 | .1353 | .1225 | .1108 | .1003 | .0907 | 0 = Y |
| 1 | .3679 | .3662 | .3614 | .3543 | .3452 | .3347 | .3230 | .3106 | .2975 | .2842 | .2707 | .2572 | .2438 | .2306 | .2177 | 1 |
| 2 | .1839 | .2014 | .2169 | .2303 | .2417 | .2510 | .2584 | .2640 | .2678 | .2700 | .2707 | .2700 | .2681 | .2652 | .2613 | 2 |
| 3 | .0613 | .0738 | .0867 | .0998 | .1128 | .1255 | .1378 | .1496 | .1607 | .1710 | .1804 | .1890 | .1966 | .2033 | .2090 | 3 |
| 4 | .0153 | .0203 | .0260 | .0324 | .0395 | .0471 | .0551 | .0636 | .0723 | .0812 | .0902 | .0992 | .1082 | .1169 | .1254 | 4 |
| 5 | .0031 | .0045 | .0062 | .0084 | .0111 | .0141 | .0176 | .0216 | .0260 | .0309 | .0361 | .0417 | .0476 | .0538 | .0602 | 5 |
| 6 | .0005 | .0008 | .0012 | .0018 | .0026 | .0035 | .0047 | .0061 | .0078 | .0098 | .0120 | .0146 | .0174 | .0206 | .0241 | 6 |
| 7 | .0001 | .0001 | .0002 | .0003 | .0005 | .0008 | .0011 | .0015 | .0020 | .0027 | .0034 | .0044 | .0055 | .0068 | .0083 | 7 |
| 8 | .0000 | .0000 | .0000 | .0001 | .0001 | .0001 | .0002 | .0003 | .0005 | .0006 | .0009 | .0011 | .0015 | .0019 | .0025 | 8 |
| 9 | .0000 | .0000 | .0000 | .0000 | .0000 | .0000 | .0000 | .0001 | .0001 | .0001 | .0002 | .0003 | .0004 | .0005 | .0007 | 9 |
| 10 | .0000 | .0000 | .0000 | .0000 | .0000 | .0000 | .0000 | .0000 | .0000 | .0000 | .0000 | .0001 | .0001 | .0001 | .0002 | 10 |

| $\mu =$ | 2.5 | 2.6 | 2.7 | 2.8 | 2.9 | 3.0 | 3.1 | 3.2 | 3.3 | 3.4 | 3.5 | 3.6 | 3.7 | 3.8 | 3.9 | $= \mu$ |
|---|---|---|---|---|---|---|---|---|---|---|---|---|---|---|---|---|
| $Y = 0$ | .0821 | .0743 | .0672 | .0608 | .0550 | .0498 | .0450 | .0408 | .0369 | .0334 | .0302 | .0273 | .0247 | .0224 | .0202 | 0 = Y |
| 1 | .2052 | .1931 | .1815 | .1703 | .1596 | .1494 | .1397 | .1304 | .1217 | .1135 | .1057 | .0984 | .0915 | .0850 | .0789 | 1 |
| 2 | .2565 | .2510 | .2450 | .2384 | .2314 | .2240 | .2165 | .2087 | .2008 | .1929 | .1850 | .1771 | .1692 | .1615 | .1539 | 2 |
| 3 | .2138 | .2176 | .2205 | .2225 | .2237 | .2240 | .2237 | .2226 | .2209 | .2186 | .2158 | .2125 | .2087 | .2046 | .2001 | 3 |
| 4 | .1336 | .1414 | .1488 | .1557 | .1622 | .1680 | .1733 | .1781 | .1823 | .1858 | .1888 | .1912 | .1931 | .1944 | .1951 | 4 |
| 5 | .0668 | .0735 | .0804 | .0872 | .0940 | .1008 | .1075 | .1140 | .1203 | .1264 | .1322 | .1377 | .1429 | .1477 | .1522 | 5 |
| 6 | .0278 | .0319 | .0362 | .0407 | .0455 | .0504 | .0555 | .0608 | .0662 | .0716 | .0771 | .0826 | .0881 | .0936 | .0989 | 6 |
| 7 | .0099 | .0118 | .0139 | .0163 | .0188 | .0216 | .0246 | .0278 | .0312 | .0348 | .0385 | .0425 | .0466 | .0508 | .0551 | 7 |
| 8 | .0031 | .0038 | .0047 | .0057 | .0068 | .0081 | .0095 | .0111 | .0129 | .0148 | .0169 | .0191 | .0215 | .0241 | .0269 | 8 |
| 9 | .0009 | .0011 | .0014 | .0018 | .0022 | .0027 | .0033 | .0040 | .0047 | .0056 | .0066 | .0076 | .0089 | .0102 | .0116 | 9 |
| 10 | .0002 | .0003 | .0004 | .0005 | .0006 | .0008 | .0010 | .0013 | .0016 | .0019 | .0023 | .0028 | .0033 | .0039 | .0045 | 10 |
| 11 | .0000 | .0001 | .0001 | .0001 | .0002 | .0002 | .0003 | .0004 | .0005 | .0006 | .0007 | .0009 | .0011 | .0013 | .0016 | 11 |
| 12 | .0000 | .0000 | .0000 | .0000 | .0000 | .0001 | .0001 | .0001 | .0001 | .0002 | .0002 | .0003 | .0003 | .0004 | .0005 | 12 |
| 13 | .0000 | .0000 | .0000 | .0000 | .0000 | .0000 | .0000 | .0000 | .0000 | .0000 | .0001 | .0001 | .0001 | .0001 | .0002 | 13 |

(continued on next page)

## Table A3 (continued)  Expected relative frequencies for Poisson distributions.

| μ = | 4.0 | 4.1 | 4.2 | 4.3 | 4.4 | 4.5 | 4.6 | 4.7 | 4.8 | 4.9 | 5.0 | 5.1 | 5.2 | 5.3 | 5.4 | = μ |
|---|---|---|---|---|---|---|---|---|---|---|---|---|---|---|---|---|
| Y = 0 | .0183 | .0166 | .0150 | .0136 | .0123 | .0111 | .0101 | .0091 | .0082 | .0074 | .0067 | .0061 | .0055 | .0050 | .0045 | 0 = Y |
| 1 | .0733 | .0679 | .0630 | .0583 | .0540 | .0500 | .0462 | .0427 | .0395 | .0365 | .0337 | .0311 | .0287 | .0265 | .0244 | 1 |
| 2 | .1465 | .1393 | .1323 | .1254 | .1188 | .1125 | .1063 | .1005 | .0948 | .0894 | .0842 | .0793 | .0746 | .0701 | .0659 | 2 |
| 3 | .1954 | .1904 | .1852 | .1798 | .1743 | .1687 | .1631 | .1574 | .1517 | .1460 | .1404 | .1348 | .1293 | .1239 | .1185 | 3 |
| 4 | .1954 | .1951 | .1944 | .1933 | .1917 | .1898 | .1875 | .1849 | .1820 | .1789 | .1755 | .1719 | .1681 | .1641 | .1600 | 4 |
| 5 | .1563 | .1600 | .1633 | .1662 | .1687 | .1708 | .1725 | .1738 | .1747 | .1753 | .1755 | .1753 | .1748 | .1740 | .1728 | 5 |
| 6 | .1042 | .1093 | .1143 | .1191 | .1237 | .1281 | .1323 | .1362 | .1398 | .1432 | .1462 | .1490 | .1515 | .1537 | .1555 | 6 |
| 7 | .0595 | .0640 | .0686 | .0732 | .0778 | .0824 | .0869 | .0914 | .0959 | .1002 | .1044 | .1086 | .1125 | .1163 | .1200 | 7 |
| 8 | .0298 | .0328 | .0360 | .0393 | .0428 | .0463 | .0500 | .0537 | .0575 | .0614 | .0653 | .0692 | .0731 | .0771 | .0810 | 8 |
| 9 | .0132 | .0150 | .0168 | .0188 | .0209 | .0232 | .0255 | .0281 | .0307 | .0334 | .0363 | .0392 | .0423 | .0454 | .0486 | 9 |
| 10 | .0053 | .0061 | .0071 | .0081 | .0092 | .0104 | .0118 | .0132 | .0147 | .0164 | .0181 | .0200 | .0220 | .0241 | .0262 | 10 |
| 11 | .0019 | .0023 | .0027 | .0032 | .0037 | .0043 | .0049 | .0056 | .0064 | .0073 | .0082 | .0093 | .0104 | .0116 | .0129 | 11 |
| 12 | .0006 | .0008 | .0009 | .0011 | .0013 | .0016 | .0019 | .0022 | .0026 | .0030 | .0034 | .0039 | .0045 | .0051 | .0058 | 12 |
| 13 | .0002 | .0002 | .0003 | .0004 | .0005 | .0006 | .0007 | .0008 | .0009 | .0011 | .0013 | .0015 | .0018 | .0021 | .0024 | 13 |
| 14 | .0001 | .0001 | .0001 | .0001 | .0001 | .0002 | .0002 | .0003 | .0003 | .0004 | .0005 | .0006 | .0007 | .0008 | .0009 | 14 |
| 15 | .0000 | .0000 | .0000 | .0000 | .0000 | .0001 | .0001 | .0001 | .0001 | .0001 | .0002 | .0002 | .0002 | .0003 | .0003 | 15 |
| 16 | .0000 | .0000 | .0000 | .0000 | .0000 | .0000 | .0000 | .0000 | .0000 | .0000 | .0000 | .0001 | .0001 | .0001 | .0001 | 16 |

| μ = | 5.5 | 5.6 | 5.7 | 5.8 | 5.9 | 6.0 | 6.1 | 6.2 | 6.3 | 6.4 | 6.5 | 6.6 | 6.7 | 6.8 | 6.9 | = μ |
|---|---|---|---|---|---|---|---|---|---|---|---|---|---|---|---|---|
| Y = 0 | .0041 | .0037 | .0033 | .0030 | .0027 | .0025 | .0022 | .0020 | .0018 | .0017 | .0015 | .0014 | .0012 | .0011 | .0010 | 0 = Y |
| 1 | .0225 | .0207 | .0191 | .0176 | .0162 | .0149 | .0137 | .0126 | .0116 | .0106 | .0098 | .0090 | .0082 | .0076 | .0070 | 1 |
| 2 | .0618 | .0580 | .0544 | .0509 | .0477 | .0446 | .0417 | .0390 | .0364 | .0340 | .0318 | .0296 | .0276 | .0258 | .0240 | 2 |
| 3 | .1133 | .1082 | .1033 | .0985 | .0938 | .0892 | .0848 | .0806 | .0765 | .0726 | .0688 | .0652 | .0617 | .0584 | .0552 | 3 |
| 4 | .1558 | .1515 | .1472 | .1428 | .1383 | .1339 | .1294 | .1249 | .1205 | .1162 | .1118 | .1076 | .1034 | .0992 | .0952 | 4 |
| 5 | .1714 | .1697 | .1678 | .1656 | .1632 | .1606 | .1579 | .1549 | .1519 | .1487 | .1454 | .1420 | .1385 | .1349 | .1314 | 5 |
| 6 | .1571 | .1584 | .1594 | .1601 | .1605 | .1606 | .1605 | .1601 | .1595 | .1586 | .1575 | .1562 | .1546 | .1529 | .1511 | 6 |
| 7 | .1234 | .1267 | .1298 | .1326 | .1353 | .1377 | .1399 | .1418 | .1435 | .1450 | .1462 | .1472 | .1480 | .1486 | .1489 | 7 |
| 8 | .0849 | .0887 | .0925 | .0962 | .0998 | .1033 | .1066 | .1099 | .1130 | .1160 | .1188 | .1215 | .1240 | .1263 | .1284 | 8 |
| 9 | .0519 | .0552 | .0586 | .0620 | .0654 | .0688 | .0723 | .0757 | .0791 | .0825 | .0858 | .0891 | .0923 | .0954 | .0985 | 9 |
| 10 | .0285 | .0309 | .0334 | .0359 | .0386 | .0413 | .0441 | .0469 | .0498 | .0528 | .0558 | .0588 | .0618 | .0649 | .0679 | 10 |
| 11 | .0143 | .0157 | .0173 | .0190 | .0207 | .0225 | .0244 | .0265 | .0285 | .0307 | .0330 | .0353 | .0377 | .0401 | .0426 | 11 |
| 12 | .0065 | .0073 | .0082 | .0092 | .0102 | .0113 | .0124 | .0137 | .0150 | .0164 | .0179 | .0194 | .0210 | .0227 | .0245 | 12 |
| 13 | .0028 | .0032 | .0036 | .0041 | .0046 | .0052 | .0058 | .0065 | .0073 | .0081 | .0089 | .0099 | .0108 | .0119 | .0130 | 13 |
| 14 | .0011 | .0013 | .0015 | .0017 | .0019 | .0022 | .0025 | .0029 | .0033 | .0037 | .0041 | .0046 | .0052 | .0058 | .0064 | 14 |
| 15 | .0004 | .0005 | .0006 | .0007 | .0008 | .0009 | .0010 | .0012 | .0014 | .0016 | .0018 | .0020 | .0023 | .0026 | .0029 | 15 |
| 16 | .0001 | .0002 | .0002 | .0002 | .0003 | .0003 | .0004 | .0005 | .0005 | .0006 | .0007 | .0008 | .0010 | .0011 | .0013 | 16 |
| 17 | .0000 | .0001 | .0001 | .0001 | .0001 | .0001 | .0001 | .0002 | .0002 | .0002 | .0003 | .0003 | .0004 | .0004 | .0005 | 17 |
| 18 | .0000 | .0000 | .0000 | .0000 | .0000 | .0000 | .0000 | .0001 | .0001 | .0001 | .0001 | .0001 | .0001 | .0002 | .0002 | 18 |
| 19 | .0000 | .0000 | .0000 | .0000 | .0000 | .0000 | .0000 | .0000 | .0000 | .0000 | .0000 | .0000 | .0001 | .0001 | .0001 | 19 |

| μ = | 7.0 | 7.1 | 7.2 | 7.3 | 7.4 | 7.5 | 7.6 | 7.7 | 7.8 | 7.9 | 8.0 | 8.1 | 8.2 | 8.3 | 8.4 | = μ |
|---|---|---|---|---|---|---|---|---|---|---|---|---|---|---|---|---|
| Y = 0 | .0009 | .0008 | .0007 | .0007 | .0006 | .0006 | .0005 | .0005 | .0004 | .0004 | .0003 | .0003 | .0003 | .0002 | .0002 | 0 = Y |
| 1 | .0064 | .0059 | .0054 | .0049 | .0045 | .0041 | .0038 | .0035 | .0032 | .0029 | .0027 | .0025 | .0023 | .0021 | .0019 | 1 |
| 2 | .0223 | .0208 | .0194 | .0180 | .0167 | .0156 | .0145 | .0134 | .0125 | .0116 | .0107 | .0100 | .0092 | .0086 | .0079 | 2 |
| 3 | .0521 | .0492 | .0464 | .0438 | .0413 | .0389 | .0366 | .0345 | .0324 | .0305 | .0286 | .0269 | .0252 | .0237 | .0222 | 3 |
| 4 | .0912 | .0874 | .0836 | .0799 | .0764 | .0729 | .0696 | .0663 | .0632 | .0602 | .0573 | .0544 | .0517 | .0491 | .0466 | 4 |
| 5 | .1277 | .1241 | .1204 | .1167 | .1130 | .1094 | .1057 | .1021 | .0986 | .0951 | .0916 | .0882 | .0849 | .0816 | .0784 | 5 |
| 6 | .1490 | .1468 | .1445 | .1420 | .1394 | .1367 | .1339 | .1311 | .1282 | .1252 | .1221 | .1191 | .1160 | .1128 | .1097 | 6 |
| 7 | .1490 | .1489 | .1486 | .1481 | .1474 | .1465 | .1454 | .1442 | .1428 | .1413 | .1396 | .1378 | .1358 | .1338 | .1317 | 7 |
| 8 | .1304 | .1321 | .1337 | .1351 | .1363 | .1373 | .1381 | .1388 | .1392 | .1395 | .1396 | .1395 | .1392 | .1388 | .1382 | 8 |
| 9 | .1014 | .1042 | .1070 | .1096 | .1121 | .1144 | .1167 | .1187 | .1207 | .1224 | .1241 | .1256 | .1269 | .1280 | .1290 | 9 |
| 10 | .0710 | .0740 | .0770 | .0800 | .0829 | .0858 | .0887 | .0914 | .0941 | .0967 | .0993 | .1017 | .1040 | .1063 | .1084 | 10 |
| 11 | .0452 | .0478 | .0504 | .0531 | .0558 | .0585 | .0613 | .0640 | .0667 | .0695 | .0722 | .0749 | .0776 | .0802 | .0828 | 11 |
| 12 | .0263 | .0283 | .0303 | .0323 | .0344 | .0366 | .0388 | .0411 | .0434 | .0457 | .0481 | .0505 | .0530 | .0555 | .0579 | 12 |
| 13 | .0142 | .0154 | .0168 | .0181 | .0196 | .0211 | .0227 | .0243 | .0260 | .0278 | .0296 | .0315 | .0334 | .0354 | .0374 | 13 |
| 14 | .0071 | .0078 | .0086 | .0095 | .0104 | .0113 | .0123 | .0134 | .0145 | .0157 | .0169 | .0182 | .0196 | .0210 | .0225 | 14 |
| 15 | .0033 | .0037 | .0041 | .0046 | .0051 | .0057 | .0062 | .0069 | .0075 | .0083 | .0090 | .0098 | .0107 | .0116 | .0126 | 15 |
| 16 | .0014 | .0016 | .0019 | .0021 | .0024 | .0026 | .0030 | .0033 | .0037 | .0041 | .0045 | .0050 | .0055 | .0060 | .0066 | 16 |
| 17 | .0006 | .0007 | .0008 | .0009 | .0010 | .0012 | .0013 | .0015 | .0017 | .0019 | .0021 | .0024 | .0026 | .0029 | .0033 | 17 |
| 18 | .0002 | .0003 | .0003 | .0004 | .0004 | .0005 | .0006 | .0006 | .0007 | .0008 | .0009 | .0011 | .0012 | .0014 | .0015 | 18 |
| 19 | .0001 | .0001 | .0001 | .0001 | .0002 | .0002 | .0002 | .0003 | .0003 | .0003 | .0004 | .0005 | .0005 | .0006 | .0007 | 19 |
| 20 | .0000 | .0000 | .0000 | .0001 | .0001 | .0001 | .0001 | .0001 | .0001 | .0001 | .0002 | .0002 | .0002 | .0002 | .0003 | 20 |
| 21 | .0000 | .0000 | .0000 | .0000 | .0000 | .0000 | .0000 | .0000 | .0000 | .0001 | .0001 | .0001 | .0001 | .0001 | .0001 | 21 |

(continued on next page)

**Table A3 (continued)   Expected relative frequencies for Poisson distributions.**

| μ = | 8.5 | 8.6 | 8.7 | 8.8 | 8.9 | 9.0 | 9.1 | 9.2 | 9.3 | 9.4 | 9.5 | 9.6 | 9.7 | 9.8 | 9.9 | = μ |
|---|---|---|---|---|---|---|---|---|---|---|---|---|---|---|---|---|
| Y = 0 | .0002 | .0002 | .0002 | .0002 | .0001 | .0001 | .0001 | .0001 | .0001 | .0001 | .0001 | .0001 | .0001 | .0001 | .0001 | 0 = Y |
| 1 | .0017 | .0016 | .0014 | .0013 | .0012 | .0011 | .0010 | .0009 | .0009 | .0008 | .0007 | .0007 | .0006 | .0005 | .0005 | 1 |
| 2 | .0074 | .0068 | .0063 | .0058 | .0054 | .0050 | .0046 | .0043 | .0040 | .0037 | .0034 | .0031 | .0029 | .0027 | .0025 | 2 |
| 3 | .0208 | .0195 | .0183 | .0171 | .0160 | .0150 | .0140 | .0131 | .0123 | .0115 | .0107 | .0100 | .0093 | .0087 | .0081 | 3 |
| 4 | .0443 | .0420 | .0398 | .0377 | .0357 | .0337 | .0319 | .0302 | .0285 | .0269 | .0254 | .0240 | .0226 | .0213 | .0201 | 4 |
| 5 | .0752 | .0722 | .0692 | .0663 | .0635 | .0607 | .0581 | .0555 | .0530 | .0506 | .0483 | .0460 | .0439 | .0418 | .0398 | 5 |
| 6 | .1066 | .1034 | .1003 | .0972 | .0941 | .0911 | .0881 | .0851 | .0822 | .0793 | .0764 | .0736 | .0709 | .0682 | .0656 | 6 |
| 7 | .1294 | .1271 | .1247 | .1222 | .1197 | .1171 | .1145 | .1118 | .1091 | .1064 | .1037 | .1010 | .0982 | .0955 | .0928 | 7 |
| 8 | .1375 | .1366 | .1356 | .1344 | .1332 | .1318 | .1302 | .1286 | .1269 | .1251 | .1232 | .1212 | .1191 | .1170 | .1148 | 8 |
| 9 | .1299 | .1306 | .1311 | .1315 | .1317 | .1318 | .1317 | .1315 | .1311 | .1306 | .1300 | .1293 | .1284 | .1274 | .1263 | 9 |
| 10 | .1104 | .1123 | .1140 | .1157 | .1172 | .1186 | .1198 | .1210 | .1219 | .1228 | .1235 | .1241 | .1245 | .1249 | .1250 | 10 |
| 11 | .0853 | .0878 | .0902 | .0925 | .0948 | .0970 | .0991 | .1012 | .1031 | .1049 | .1067 | .1083 | .1098 | .1112 | .1125 | 11 |
| 12 | .0604 | .0629 | .0654 | .0679 | .0703 | .0728 | .0752 | .0776 | .0799 | .0822 | .0844 | .0866 | .0888 | .0908 | .0928 | 12 |
| 13 | .0395 | .0416 | .0438 | .0459 | .0481 | .0504 | .0526 | .0549 | .0572 | .0594 | .0617 | .0640 | .0662 | .0685 | .0707 | 13 |
| 14 | .0240 | .0256 | .0272 | .0289 | .0306 | .0324 | .0342 | .0361 | .0380 | .0399 | .0419 | .0439 | .0459 | .0479 | .0500 | 14 |
| 15 | .0136 | .0147 | .0158 | .0169 | .0182 | .0194 | .0208 | .0221 | .0235 | .0250 | .0265 | .0281 | .0297 | .0313 | .0330 | 15 |
| 16 | .0072 | .0079 | .0086 | .0093 | .0101 | .0109 | .0118 | .0127 | .0137 | .0147 | .0157 | .0168 | .0180 | .0192 | .0204 | 16 |
| 17 | .0036 | .0040 | .0044 | .0048 | .0053 | .0058 | .0063 | .0069 | .0075 | .0081 | .0088 | .0095 | .0103 | .0111 | .0119 | 17 |
| 18 | .0017 | .0019 | .0021 | .0024 | .0026 | .0029 | .0032 | .0035 | .0039 | .0042 | .0046 | .0051 | .0055 | .0060 | .0065 | 18 |
| 19 | .0008 | .0009 | .0010 | .0011 | .0012 | .0014 | .0015 | .0017 | .0019 | .0021 | .0023 | .0026 | .0028 | .0031 | .0034 | 19 |
| 20 | .0003 | .0004 | .0004 | .0005 | .0005 | .0006 | .0007 | .0008 | .0009 | .0010 | .0011 | .0012 | .0014 | .0015 | .0017 | 20 |
| 21 | .0001 | .0002 | .0002 | .0002 | .0002 | .0003 | .0003 | .0003 | .0004 | .0004 | .0005 | .0006 | .0006 | .0007 | .0008 | 21 |
| 22 | .0001 | .0001 | .0001 | .0001 | .0001 | .0001 | .0001 | .0001 | .0002 | .0002 | .0002 | .0002 | .0003 | .0003 | .0004 | 22 |
| 23 | .0000 | .0000 | .0000 | .0000 | .0000 | .0000 | .0000 | .0001 | .0001 | .0001 | .0001 | .0001 | .0001 | .0001 | .0002 | 23 |
| 24 | .0000 | .0000 | .0000 | .0000 | .0000 | .0000 | .0000 | .0000 | .0000 | .0000 | .0000 | .0000 | .0000 | .0001 | .0001 | 24 |

**Table A4 Proportions of the area under the standard normal curve.**

The table gives the proportion of the area under the curve which lies to the left of (i.e. below) a given value of z
1. Find row in left-hand margin corresponding to first decimal place of z.
2. Find column in upper margin corresponding to second decimal place of z.
3. Required proportion is in body of table where row and column intersect.
4. Proportions corresponding to values of z to three decimal places can be obtained by linear interpolation (see text).

| z | 0.00 | 0.01 | 0.02 | 0.03 | 0.04 | 0.05 | 0.06 | 0.07 | 0.08 | 0.09 |
|---|------|------|------|------|------|------|------|------|------|------|
| -3.0 | 0.0013 | 0.0010 | 0.0007 | 0.0005 | 0.0003 | 0.0002 | 0.0002 | 0.0001 | 0.0001 | 0.0000 |
| -2.9 | 0.0019 | 0.0018 | 0.0017 | 0.0017 | 0.0016 | 0.0016 | 0.0015 | 0.0015 | 0.0014 | 0.0014 |
| -2.8 | 0.0026 | 0.0025 | 0.0024 | 0.0023 | 0.0023 | 0.0022 | 0.0021 | 0.0020 | 0.0020 | 0.0019 |
| -2.7 | 0.0035 | 0.0034 | 0.0033 | 0.0032 | 0.0031 | 0.0030 | 0.0029 | 0.0028 | 0.0027 | 0.0026 |
| -2.6 | 0.0047 | 0.0045 | 0.0044 | 0.0043 | 0.0041 | 0.0040 | 0.0039 | 0.0038 | 0.0037 | 0.0036 |
| -2.5 | 0.0062 | 0.0060 | 0.0059 | 0.0057 | 0.0055 | 0.0054 | 0.0052 | 0.0051 | 0.0049 | 0.0048 |
| -2.4 | 0.0082 | 0.0080 | 0.0078 | 0.0075 | 0.0073 | 0.0071 | 0.0069 | 0.0068 | 0.0066 | 0.0064 |
| -2.3 | 0.0107 | 0.0104 | 0.0102 | 0.0099 | 0.0096 | 0.0094 | 0.0091 | 0.0089 | 0.0087 | 0.0084 |
| -2.2 | 0.0139 | 0.0136 | 0.0132 | 0.0129 | 0.0126 | 0.0122 | 0.0119 | 0.0116 | 0.0113 | 0.0110 |
| -2.1 | 0.0179 | 0.0174 | 0.0170 | 0.0166 | 0.0162 | 0.0158 | 0.0154 | 0.0150 | 0.0146 | 0.0143 |
| -2.0 | 0.0228 | 0.0222 | 0.0217 | 0.0212 | 0.0207 | 0.0202 | 0.0197 | 0.0192 | 0.0188 | 0.0183 |
| -1.9 | 0.0287 | 0.0281 | 0.0274 | 0.0268 | 0.0262 | 0.0256 | 0.0250 | 0.0244 | 0.0238 | 0.0233 |
| -1.8 | 0.0359 | 0.0352 | 0.0344 | 0.0336 | 0.0329 | 0.0322 | 0.0314 | 0.0307 | 0.0300 | 0.0294 |
| -1.7 | 0.0446 | 0.0436 | 0.0427 | 0.0418 | 0.0409 | 0.0401 | 0.0392 | 0.0384 | 0.0375 | 0.0367 |
| -1.6 | 0.0548 | 0.0537 | 0.0526 | 0.0516 | 0.0505 | 0.0495 | 0.0485 | 0.0475 | 0.0465 | 0.0455 |
| -1.5 | 0.0668 | 0.0655 | 0.0643 | 0.0630 | 0.0618 | 0.0606 | 0.0594 | 0.0582 | 0.0570 | 0.0559 |
| -1.4 | 0.0808 | 0.0793 | 0.0778 | 0.0764 | 0.0749 | 0.0735 | 0.0722 | 0.0708 | 0.0694 | 0.0681 |
| -1.3 | 0.0968 | 0.0951 | 0.0934 | 0.0918 | 0.0901 | 0.0885 | 0.0869 | 0.0853 | 0.0838 | 0.0823 |
| -1.2 | 0.1151 | 0.1131 | 0.1112 | 0.1093 | 0.1075 | 0.1056 | 0.1038 | 0.1020 | 0.1003 | 0.0985 |
| -1.1 | 0.1357 | 0.1335 | 0.1314 | 0.1292 | 0.1271 | 0.1251 | 0.1230 | 0.1210 | 0.1190 | 0.1170 |
| -1.0 | 0.1587 | 0.1562 | 0.1539 | 0.1515 | 0.1492 | 0.1469 | 0.1446 | 0.1423 | 0.1401 | 0.1379 |
| -0.9 | 0.1841 | 0.1814 | 0.1788 | 0.1762 | 0.1736 | 0.1711 | 0.1685 | 0.1660 | 0.1635 | 0.1611 |
| -0.8 | 0.2119 | 0.2090 | 0.2061 | 0.2033 | 0.2005 | 0.1977 | 0.1949 | 0.1922 | 0.1894 | 0.1867 |
| -0.7 | 0.2420 | 0.2389 | 0.2358 | 0.2327 | 0.2297 | 0.2266 | 0.2236 | 0.2206 | 0.2177 | 0.2148 |
| -0.6 | 0.2743 | 0.2709 | 0.2676 | 0.2643 | 0.2611 | 0.2578 | 0.2546 | 0.2514 | 0.2483 | 0.2451 |
| -0.5 | 0.3085 | 0.3050 | 0.3015 | 0.2981 | 0.2946 | 0.2912 | 0.2877 | 0.2843 | 0.2810 | 0.2776 |
| -0.4 | 0.3446 | 0.3409 | 0.3372 | 0.3336 | 0.3300 | 0.3264 | 0.3228 | 0.3192 | 0.3156 | 0.3121 |
| -0.3 | 0.3821 | 0.3783 | 0.3745 | 0.3707 | 0.3669 | 0.3632 | 0.3594 | 0.3557 | 0.3520 | 0.3483 |
| -0.2 | 0.4207 | 0.4168 | 0.4129 | 0.4090 | 0.4052 | 0.4013 | 0.3974 | 0.3936 | 0.3897 | 0.3859 |
| -0.1 | 0.4602 | 0.4562 | 0.4522 | 0.4483 | 0.4443 | 0.4404 | 0.4364 | 0.4325 | 0.4286 | 0.4247 |
| -0.0 | 0.5000 | 0.4960 | 0.4920 | 0.4880 | 0.4840 | 0.4801 | 0.4761 | 0.4721 | 0.4681 | 0.4641 |
| 0.0 | 0.5000 | 0.5040 | 0.5080 | 0.5120 | 0.5160 | 0.5199 | 0.5239 | 0.5279 | 0.5319 | 0.5359 |
| 0.1 | 0.5398 | 0.5438 | 0.5478 | 0.5517 | 0.5557 | 0.5596 | 0.5636 | 0.5675 | 0.5714 | 0.5753 |
| 0.2 | 0.5793 | 0.5832 | 0.5871 | 0.5910 | 0.5948 | 0.5987 | 0.6026 | 0.6064 | 0.6103 | 0.6141 |
| 0.3 | 0.6179 | 0.6217 | 0.6255 | 0.6293 | 0.6331 | 0.6368 | 0.6406 | 0.6443 | 0.6480 | 0.6517 |
| 0.4 | 0.6554 | 0.6591 | 0.6628 | 0.6664 | 0.6700 | 0.6736 | 0.6772 | 0.6808 | 0.6844 | 0.6879 |
| 0.5 | 0.6915 | 0.6950 | 0.6985 | 0.7019 | 0.7054 | 0.7088 | 0.7123 | 0.7157 | 0.7190 | 0.7224 |
| 0.6 | 0.7257 | 0.7291 | 0.7324 | 0.7357 | 0.7389 | 0.7422 | 0.7454 | 0.7486 | 0.7517 | 0.7549 |
| 0.7 | 0.7580 | 0.7611 | 0.7642 | 0.7673 | 0.7703 | 0.7734 | 0.7764 | 0.7794 | 0.7823 | 0.7852 |
| 0.8 | 0.7881 | 0.7910 | 0.7939 | 0.7967 | 0.7995 | 0.8023 | 0.8051 | 0.8078 | 0.8106 | 0.8133 |
| 0.9 | 0.8159 | 0.8186 | 0.8212 | 0.8238 | 0.8264 | 0.8289 | 0.8315 | 0.8340 | 0.8365 | 0.8389 |
| 1.0 | 0.8413 | 0.8438 | 0.8461 | 0.8485 | 0.8508 | 0.8531 | 0.8554 | 0.8577 | 0.8599 | 0.8621 |
| 1.1 | 0.8643 | 0.8665 | 0.8686 | 0.8708 | 0.8729 | 0.8749 | 0.8770 | 0.8790 | 0.8810 | 0.8830 |
| 1.2 | 0.8849 | 0.8869 | 0.8888 | 0.8907 | 0.8925 | 0.8944 | 0.8962 | 0.8980 | 0.8997 | 0.9015 |
| 1.3 | 0.9032 | 0.9049 | 0.9066 | 0.9082 | 0.9099 | 0.9115 | 0.9131 | 0.9147 | 0.9162 | 0.9177 |
| 1.4 | 0.9192 | 0.9207 | 0.9222 | 0.9236 | 0.9251 | 0.9265 | 0.9278 | 0.9292 | 0.9306 | 0.9319 |
| 1.5 | 0.9332 | 0.9345 | 0.9357 | 0.9370 | 0.9382 | 0.9394 | 0.9406 | 0.9418 | 0.9430 | 0.9441 |
| 1.6 | 0.9452 | 0.9463 | 0.9474 | 0.9484 | 0.9495 | 0.9505 | 0.9515 | 0.9525 | 0.9535 | 0.9545 |
| 1.7 | 0.9554 | 0.9564 | 0.9573 | 0.9582 | 0.9591 | 0.9599 | 0.9608 | 0.9616 | 0.9625 | 0.9633 |
| 1.8 | 0.9641 | 0.9648 | 0.9656 | 0.9664 | 0.9671 | 0.9678 | 0.9686 | 0.9693 | 0.9700 | 0.9706 |
| 1.9 | 0.9713 | 0.9719 | 0.9726 | 0.9732 | 0.9738 | 0.9744 | 0.9750 | 0.9756 | 0.9762 | 0.9767 |
| 2.0 | 0.9772 | 0.9778 | 0.9783 | 0.9788 | 0.9793 | 0.9798 | 0.9803 | 0.9808 | 0.9812 | 0.9817 |

(continued on next page)

**Table A4 (continued) Proportions of the area under the standard normal curve.**

| z | 0.00 | 0.01 | 0.02 | 0.03 | 0.04 | 0.05 | 0.06 | 0.07 | 0.08 | 0.09 |
|---|------|------|------|------|------|------|------|------|------|------|
| 2.1 | 0.9821 | 0.9826 | 0.9830 | 0.9834 | 0.9838 | 0.9842 | 0.9846 | 0.9850 | 0.9854 | 0.9857 |
| 2.2 | 0.9861 | 0.9864 | 0.9868 | 0.9871 | 0.9874 | 0.9878 | 0.9881 | 0.9884 | 0.9887 | 0.9890 |
| 2.3 | 0.9893 | 0.9896 | 0.9898 | 0.9901 | 0.9904 | 0.9906 | 0.9909 | 0.9911 | 0.9913 | 0.9916 |
| 2.4 | 0.9918 | 0.9920 | 0.9922 | 0.9925 | 0.9927 | 0.9929 | 0.9931 | 0.9932 | 0.9934 | 0.9936 |
| 2.5 | 0.9938 | 0.9940 | 0.9941 | 0.9943 | 0.9945 | 0.9946 | 0.9948 | 0.9949 | 0.9951 | 0.9952 |
| 2.6 | 0.9953 | 0.9955 | 0.9956 | 0.9957 | 0.9959 | 0.9960 | 0.9961 | 0.9962 | 0.9963 | 0.9964 |
| 2.7 | 0.9965 | 0.9966 | 0.9967 | 0.9968 | 0.9969 | 0.9970 | 0.9971 | 0.9972 | 0.9973 | 0.9974 |
| 2.8 | 0.9974 | 0.9975 | 0.9976 | 0.9977 | 0.9977 | 0.9978 | 0.9979 | 0.9979 | 0.9980 | 0.9981 |
| 2.9 | 0.9981 | 0.9982 | 0.9982 | 0.9983 | 0.9984 | 0.9984 | 0.9985 | 0.9985 | 0.9986 | 0.9986 |
| 3.0 | 0.9987 | 0.9990 | 0.9993 | 0.9995 | 0.9997 | 0.9998 | 0.9998 | 0.9999 | 0.9999 | 1.0000 |

**Table A5  Critical values of z.**

1. For confidence intervals find $z$ in column under confidence level $\gamma$.
2. For two-tailed tests, compare calculated value to critical value in column for significance level $\alpha_2$.
3. For one-tailed test, compare calculated value in column for significance level $\alpha_1$.
4. Reject the null hypothesis if the calculated value is equal to or greater than the critical value. (Only applicable if the calculation is set out as recommended in the text.)

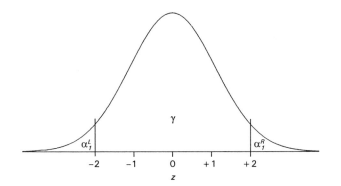

| $\alpha_1^R$ | 10% | 5% | 2.5% | 1% | 0.5% | 0.1% | 0.05% |
|---|------|------|------|------|------|------|------|
| $\alpha_2$ | 20% | 10% | 5% | 2% | 1% | 0.2% | 0.1% |
| $\gamma$ | 80% | 90% | 95% | 98% | 99% | 99.8% | 99.9% |
| z | 1.2816 | 1.6449 | 1.9600 | 2.3263 | 2.5758 | 3.0902 | 3.2905 |

327

**Table A6 Critical values of $\chi^2$.**

1. Calculate the degrees of freedom; these will depend on the data and the type of test. Check details in appropriate chapter.
2. Find row for correct degrees of freedom.
3. For **goodness-of-fit tests** and **contingency tables**: find critical value in column for **significance level $\alpha_1^R$**. Reject null hypothesis if calculated value equal to, or greater than, critical value.
4. For **test of dispersion**: find two critical values in columns for **significance level $\alpha_2$**. Reject null hypothesis if calculated value is equal to, or more extreme than, either critical value.

| $\alpha_1^R$ | | | | 10% | 5% | 2.5% | 1% | 0.5% | 0.1% |
|---|---|---|---|---|---|---|---|---|---|
| $\alpha_2$ | 1% | 5% | 10% | 20% | 10% | 5% | 2% | 1% | 0.2% |
| df = 1 | 0.000 | 0.001 | 0.004 | 2.706 | 3.841 | 5.024 | 6.635 | 7.879 | 10.828 |
| 2 | 0.010 | 0.051 | 0.103 | 4.605 | 5.991 | 7.378 | 9.210 | 10.597 | 13.816 |
| 3 | 0.072 | 0.216 | 0.352 | 6.251 | 7.815 | 9.348 | 11.345 | 12.838 | 16.266 |
| 4 | 0.207 | 0.484 | 0.711 | 7.779 | 9.488 | 11.143 | 13.277 | 14.860 | 18.467 |
| 5 | 0.412 | 0.831 | 1.145 | 9.236 | 11.070 | 12.833 | 15.086 | 16.750 | 20.515 |
| 6 | 0.676 | 1.237 | 1.635 | 10.645 | 12.592 | 14.449 | 16.812 | 18.548 | 22.458 |
| 7 | 0.989 | 1.690 | 2.167 | 12.017 | 14.067 | 16.013 | 18.475 | 20.278 | 24.322 |
| 8 | 1.344 | 2.180 | 2.733 | 13.362 | 15.507 | 17.535 | 20.090 | 21.955 | 26.124 |
| 9 | 1.735 | 2.700 | 3.325 | 14.684 | 16.919 | 19.023 | 21.666 | 23.589 | 27.877 |
| 10 | 2.156 | 3.247 | 3.940 | 15.987 | 18.307 | 20.483 | 23.209 | 25.188 | 29.588 |
| 11 | 2.603 | 3.816 | 4.575 | 17.275 | 19.675 | 21.920 | 24.725 | 26.757 | 31.264 |
| 12 | 3.074 | 4.404 | 5.226 | 18.549 | 21.026 | 23.337 | 26.217 | 28.300 | 32.909 |
| 13 | 3.565 | 5.009 | 5.892 | 19.812 | 22.362 | 24.736 | 27.688 | 29.819 | 34.528 |
| 14 | 4.075 | 5.629 | 6.571 | 21.064 | 23.685 | 26.119 | 29.141 | 31.319 | 36.123 |
| 15 | 4.601 | 6.262 | 7.261 | 22.307 | 24.996 | 27.488 | 30.578 | 32.801 | 37.697 |
| 16 | 5.142 | 6.908 | 7.962 | 23.542 | 26.296 | 28.845 | 32.000 | 34.267 | 39.252 |
| 17 | 5.697 | 7.564 | 8.672 | 24.769 | 27.587 | 30.191 | 33.409 | 35.718 | 40.790 |
| 18 | 6.265 | 8.231 | 9.390 | 25.989 | 28.869 | 31.526 | 34.805 | 37.156 | 42.312 |
| 19 | 6.844 | 8.907 | 10.117 | 27.204 | 30.144 | 32.852 | 36.191 | 38.582 | 43.820 |
| 20 | 7.434 | 9.591 | 10.851 | 28.412 | 31.410 | 34.170 | 37.566 | 39.997 | 45.315 |
| 21 | 8.034 | 10.283 | 11.591 | 29.615 | 32.671 | 35.479 | 38.932 | 41.401 | 46.797 |
| 22 | 8.643 | 10.982 | 12.338 | 30.813 | 33.924 | 36.781 | 40.289 | 42.796 | 48.268 |
| 23 | 9.260 | 11.689 | 13.091 | 32.007 | 35.172 | 38.076 | 41.638 | 44.181 | 49.728 |
| 24 | 9.886 | 12.401 | 13.848 | 33.196 | 36.415 | 39.364 | 42.980 | 45.559 | 51.179 |
| 25 | 10.520 | 13.120 | 14.611 | 34.382 | 37.652 | 40.646 | 44.314 | 46.928 | 52.620 |
| 26 | 11.160 | 13.844 | 15.379 | 35.563 | 38.885 | 41.923 | 45.642 | 48.290 | 54.052 |
| 27 | 11.808 | 14.573 | 16.151 | 36.741 | 40.113 | 43.195 | 46.963 | 49.645 | 55.476 |
| 28 | 12.461 | 15.308 | 16.928 | 37.916 | 41.337 | 44.461 | 48.278 | 50.993 | 56.892 |
| 29 | 13.121 | 16.047 | 17.708 | 39.087 | 42.557 | 45.722 | 49.588 | 52.336 | 58.301 |
| 30 | 13.787 | 16.791 | 18.493 | 40.256 | 43.773 | 46.979 | 50.892 | 53.672 | 59.703 |
| 31 | 14.458 | 17.539 | 19.281 | 41.422 | 44.985 | 48.232 | 52.191 | 55.003 | 61.098 |
| 32 | 15.134 | 18.291 | 20.072 | 42.585 | 46.194 | 49.480 | 53.486 | 56.328 | 62.487 |
| 33 | 15.815 | 19.047 | 20.867 | 43.745 | 47.400 | 50.725 | 54.776 | 57.648 | 63.870 |
| 34 | 16.501 | 19.806 | 21.664 | 44.903 | 48.602 | 51.966 | 56.061 | 58.964 | 65.247 |
| 35 | 17.192 | 20.569 | 22.465 | 46.059 | 49.802 | 53.203 | 57.342 | 60.275 | 66.619 |
| 36 | 17.887 | 21.336 | 23.269 | 47.212 | 50.998 | 54.437 | 58.619 | 61.581 | 67.985 |
| 37 | 18.586 | 22.106 | 24.075 | 48.363 | 52.192 | 55.668 | 59.893 | 62.883 | 69.346 |
| 38 | 19.289 | 22.878 | 24.884 | 49.513 | 53.384 | 56.896 | 61.162 | 64.181 | 70.703 |
| 39 | 19.996 | 23.654 | 25.695 | 50.660 | 54.572 | 58.120 | 62.428 | 65.476 | 72.055 |
| 40 | 20.707 | 24.433 | 26.509 | 51.805 | 55.758 | 59.342 | 63.691 | 66.766 | 73.402 |
| 45 | 24.311 | 28.366 | 30.612 | 57.505 | 61.656 | 65.410 | 69.957 | 73.166 | 80.077 |
| 50 | 27.991 | 32.357 | 34.764 | 63.167 | 67.505 | 71.420 | 76.154 | 79.490 | 86.661 |
| 55 | 31.735 | 36.398 | 38.958 | 68.796 | 73.311 | 77.380 | 82.292 | 85.749 | 93.168 |
| 60 | 35.535 | 40.482 | 43.188 | 74.397 | 79.082 | 83.298 | 88.379 | 91.952 | 99.607 |
| 65 | 39.383 | 44.603 | 47.450 | 79.973 | 84.821 | 89.177 | 94.422 | 98.105 | 105.988 |
| 70 | 43.275 | 48.758 | 51.739 | 85.527 | 90.531 | 95.023 | 100.425 | 104.215 | 112.317 |
| 75 | 47.206 | 52.942 | 56.054 | 91.061 | 96.217 | 100.839 | 106.393 | 110.286 | 118.599 |
| 80 | 51.172 | 57.153 | 60.391 | 96.578 | 101.879 | 106.629 | 112.329 | 116.321 | 124.839 |
| 85 | 55.170 | 61.389 | 64.749 | 102.079 | 107.522 | 112.393 | 118.236 | 122.325 | 131.041 |
| 90 | 59.196 | 65.647 | 69.126 | 107.565 | 113.145 | 118.136 | 124.116 | 128.299 | 137.208 |
| 95 | 63.250 | 69.925 | 73.520 | 113.038 | 118.752 | 123.858 | 129.973 | 134.247 | 143.344 |
| 100 | 67.328 | 74.222 | 77.929 | 118.498 | 124.342 | 129.561 | 135.807 | 140.169 | 149.449 |

**Table A7 Critical values of $t$.**

1. Calculate degrees of freedom (df). For confidence intervals and paired sample tests df $= n - 1$. For comparison of means of two samples df $= n_1 + n_2 - 2$. Find row for df.
2. For **confidence interval** find $t$ in column under **confidence level** $\gamma$.
3. For **two-tailed** tests compare calculated value to critical value in column for **significance level** $\alpha_2$.
4. For **one-tailed** tests compare calculated value to critical value in column for **significance level** $\alpha_1^R$.
5. **Reject the null hypothesis if calculated value is equal to, or greater than, critical value.** (Only applicable if the calculation set out as recommended in text.)

| $\alpha_1^R$ | 10% | 5% | 2.5% | 1% | 0.5% | 0.1% | 0.05% |
|---|---|---|---|---|---|---|---|
| $\alpha_2$ | 20% | 10% | 5% | 2% | 1% | 0.2% | 0.1% |
| $\gamma$ | 80% | 90% | 95% | 98% | 99% | 99.8% | 99.9% |
| df = 1 | 3.078 | 6.314 | 12.706 | 31.821 | 63.657 | 318.309 | 636.619 |
| 2 | 1.886 | 2.920 | 4.303 | 6.965 | 9.925 | 22.327 | 31.599 |
| 3 | 1.638 | 2.353 | 3.182 | 4.541 | 5.841 | 10.215 | 12.924 |
| 4 | 1.533 | 2.132 | 2.776 | 3.747 | 4.604 | 7.173 | 8.610 |
| 5 | 1.476 | 2.015 | 2.571 | 3.365 | 4.032 | 5.893 | 6.869 |
| 6 | 1.440 | 1.943 | 2.447 | 3.143 | 3.707 | 5.208 | 5.959 |
| 7 | 1.415 | 1.895 | 2.365 | 2.998 | 3.499 | 4.785 | 5.408 |
| 8 | 1.397 | 1.860 | 2.306 | 2.896 | 3.355 | 4.501 | 5.041 |
| 9 | 1.383 | 1.833 | 2.262 | 2.821 | 3.250 | 4.297 | 4.781 |
| 10 | 1.372 | 1.812 | 2.228 | 2.764 | 3.169 | 4.144 | 4.587 |
| 11 | 1.363 | 1.796 | 2.201 | 2.718 | 3.106 | 4.025 | 4.437 |
| 12 | 1.356 | 1.782 | 2.179 | 2.681 | 3.055 | 3.930 | 4.318 |
| 13 | 1.350 | 1.771 | 2.160 | 2.650 | 3.012 | 3.852 | 4.221 |
| 14 | 1.345 | 1.761 | 2.145 | 2.624 | 2.977 | 3.787 | 4.140 |
| 15 | 1.341 | 1.753 | 2.131 | 2.602 | 2.947 | 3.733 | 4.073 |
| 16 | 1.337 | 1.746 | 2.120 | 2.583 | 2.921 | 3.686 | 4.015 |
| 17 | 1.333 | 1.740 | 2.110 | 2.567 | 2.898 | 3.646 | 3.965 |
| 18 | 1.330 | 1.734 | 2.101 | 2.552 | 2.878 | 3.610 | 3.922 |
| 19 | 1.328 | 1.729 | 2.093 | 2.539 | 2.861 | 3.579 | 3.883 |
| 20 | 1.325 | 1.725 | 2.086 | 2.528 | 2.845 | 3.552 | 3.850 |
| 21 | 1.323 | 1.721 | 2.080 | 2.518 | 2.831 | 3.527 | 3.819 |
| 22 | 1.321 | 1.717 | 2.074 | 2.508 | 2.819 | 3.505 | 3.792 |
| 23 | 1.319 | 1.714 | 2.069 | 2.500 | 2.807 | 3.485 | 3.768 |
| 24 | 1.318 | 1.711 | 2.064 | 2.492 | 2.797 | 3.467 | 3.745 |
| 25 | 1.316 | 1.708 | 2.060 | 2.485 | 2.787 | 3.450 | 3.725 |
| 26 | 1.315 | 1.706 | 2.056 | 2.479 | 2.779 | 3.435 | 3.707 |
| 27 | 1.314 | 1.703 | 2.052 | 2.473 | 2.771 | 3.421 | 3.690 |
| 28 | 1.313 | 1.701 | 2.048 | 2.467 | 2.763 | 3.408 | 3.674 |
| 29 | 1.311 | 1.699 | 2.045 | 2.462 | 2.756 | 3.396 | 3.659 |
| 30 | 1.310 | 1.697 | 2.042 | 2.457 | 2.750 | 3.385 | 3.646 |
| 31 | 1.309 | 1.696 | 2.040 | 2.453 | 2.744 | 3.375 | 3.633 |
| 32 | 1.309 | 1.694 | 2.037 | 2.449 | 2.738 | 3.365 | 3.622 |
| 33 | 1.308 | 1.692 | 2.035 | 2.445 | 2.733 | 3.356 | 3.611 |
| 34 | 1.307 | 1.691 | 2.032 | 2.441 | 2.728 | 3.348 | 3.601 |
| 35 | 1.306 | 1.690 | 2.030 | 2.438 | 2.724 | 3.340 | 3.591 |
| 36 | 1.306 | 1.688 | 2.028 | 2.434 | 2.719 | 3.333 | 3.582 |
| 37 | 1.305 | 1.687 | 2.026 | 2.431 | 2.715 | 3.326 | 3.574 |
| 38 | 1.304 | 1.686 | 2.024 | 2.429 | 2.712 | 3.319 | 3.566 |
| 39 | 1.304 | 1.685 | 2.023 | 2.426 | 2.708 | 3.313 | 3.558 |
| 40 | 1.303 | 1.684 | 2.021 | 2.423 | 2.704 | 3.307 | 3.551 |
| 41 | 1.303 | 1.683 | 2.020 | 2.421 | 2.701 | 3.301 | 3.544 |
| 42 | 1.302 | 1.682 | 2.018 | 2.418 | 2.698 | 3.296 | 3.538 |
| 43 | 1.302 | 1.681 | 2.017 | 2.416 | 2.695 | 3.291 | 3.532 |
| 44 | 1.301 | 1.680 | 2.015 | 2.414 | 2.692 | 3.286 | 3.526 |
| 45 | 1.301 | 1.679 | 2.014 | 2.412 | 2.690 | 3.281 | 3.520 |
| 46 | 1.300 | 1.679 | 2.013 | 2.410 | 2.687 | 3.277 | 3.515 |
| 47 | 1.300 | 1.678 | 2.012 | 2.408 | 2.685 | 3.273 | 3.510 |
| 48 | 1.299 | 1.677 | 2.011 | 2.407 | 2.682 | 3.269 | 3.505 |
| 49 | 1.299 | 1.677 | 2.010 | 2.405 | 2.680 | 3.265 | 3.500 |
| 50 | 1.299 | 1.676 | 2.009 | 2.403 | 2.678 | 3.261 | 3.496 |

(continued on next page)

**Table A7 (continued)  Critical values of _t_.**

| $\alpha_1^R$ | 10% | 5% | 2.5% | 1% | 0.5% | 0.1% | 0.05% |
|---|---|---|---|---|---|---|---|
| $\alpha_2$ | 20% | 10% | 5% | 2% | 1% | 0.2% | 0.1% |
| $\gamma$ | 80% | 90% | 95% | 98% | 99% | 99.8% | 99.9% |
| df = 60 | 1.296 | 1.671 | 2.000 | 2.390 | 2.660 | 3.232 | 3.460 |
| 70 | 1.294 | 1.667 | 1.994 | 2.381 | 2.648 | 3.211 | 3.435 |
| 80 | 1.292 | 1.664 | 1.990 | 2.374 | 2.639 | 3.195 | 3.416 |
| 90 | 1.291 | 1.662 | 1.987 | 2.368 | 2.632 | 3.183 | 3.402 |
| 100 | 1.290 | 1.660 | 1.984 | 2.364 | 2.626 | 3.174 | 3.390 |
| 200 | 1.286 | 1.653 | 1.972 | 2.345 | 2.601 | 3.131 | 3.340 |
| $\infty$ | 1.2816 | 1.6449 | 1.9600 | 2.3263 | 2.5758 | 3.0902 | 3.2905 |

**Table A8  Chart giving 95% confidence limits for sample proportions.**

1. Locate sample proportion ($\hat{p}$) on horizontal axis.
2. Follow vertically upwards to two curves labelled with sample size.
3. Read off lower and upper confidence limits on vertical axis.

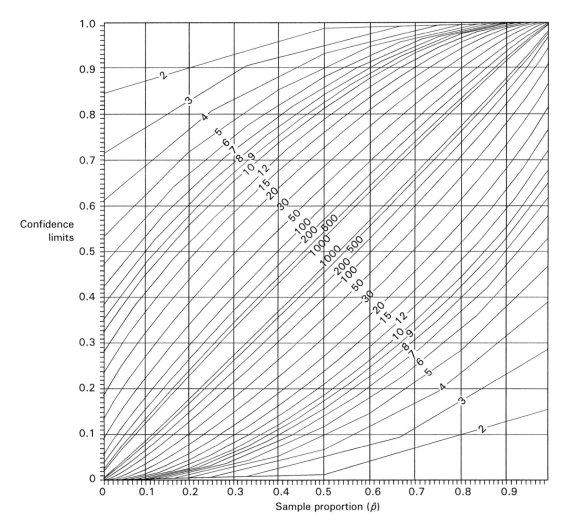

**Table A9  Confidence limits for counts from a Poisson distribution.**

This chart can be used with:

(a)  A count ($Y$) of the number of occurrences in a single sampling unit, provuiding that $Y$ is less than 30.

(b)  A mean ($\bar{Y}$) of the number of occurrences based on several sampling units, providing that $\Sigma Y$ is less than 30.

1.  Locate the value of $Y$ or $\bar{Y}$ on the horizontal axis.
2.  Follow vertically upwards to the two curves for 95% (or 99%) confidence limits.
3.  Read off lower and upper confidence limits on the vertical axis.

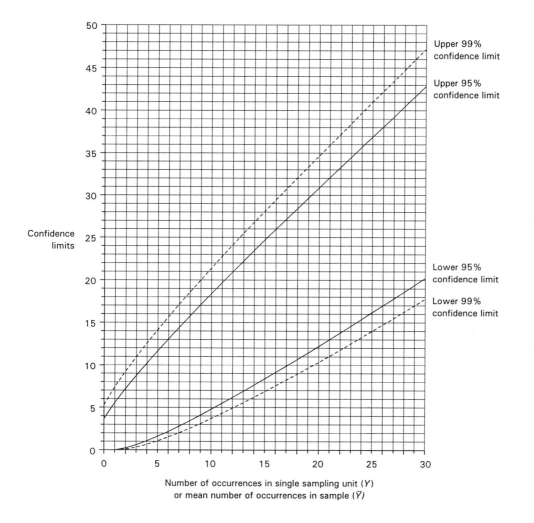

Number of occurrences in single sampling unit ($Y$)
or mean number of occurrences in sample ($\bar{Y}$)

**Table A10  Spearman's rank correlation coefficient: critical values of $r_s$.**

1. Find row corresponding to sample size ($n$)
2. For a **two-tailed** test, compare **absolute calculated value of $r_s$** (i.e. ignore a minus sign if it occurs) to critical value for **significance level $\alpha_2$**. Use decision rule 1.
3. For a **one-tailed** test in which predicted correlation is **positive**, compare calculated value of $r_s$ **with its sign** to critical value for **significance level $\alpha_1$**. Use decision rule 1.
4. For a **one-tailed** test in which predicted correlation is **negative**, compare calculated value of $r_s$ **with its sign** to critical value in column for **significance level $\alpha_1$**. Use decision rule 2.

Rule 1: **Reject null hypothesis if calculated value equal to, or greater than, calculated value.**

Rule 2: **Give critical value a minus sign.  Reject null hypothesis if calculated value is equal to, or smaller than, critical value.**

| $\alpha_1$ | 5% | 2.5% | 1% | 0.5% | 0.05% |
|---|---|---|---|---|---|
| $\alpha_2$ | 10% | 5% | 2% | 1% | 0.1% |
| n = 4 | 1.000 | | | | |
| 5 | 0.900 | 1.000 | 1.00 | | |
| 6 | 0.829 | 0.886 | 0.943 | 1.000 | |
| 7 | 0.714 | 0.786 | 0.893 | 0.929 | 1.000 |
| 8 | 0.643 | 0.738 | 0.833 | 0.881 | 0.976 |
| 9 | 0.600 | 0.700 | 0.783 | 0.833 | 0.933 |
| 10 | 0.564 | 0.648 | 0.745 | 0.794 | 0.903 |
| 11 | 0.536 | 0.618 | 0.709 | 0.755 | 0.873 |
| 12 | 0.503 | 0.587 | 0.678 | 0.727 | 0.846 |
| 13 | 0.484 | 0.560 | 0.648 | 0.703 | 0.824 |
| 14 | 0.464 | 0.538 | 0.626 | 0.679 | 0.802 |
| 15 | 0.446 | 0.521 | 0.604 | 0.654 | 0.779 |
| 16 | 0.429 | 0.503 | 0.582 | 0.635 | 0.762 |
| 17 | 0.414 | 0.485 | 0.566 | 0.615 | 0.748 |
| 18 | 0.401 | 0.472 | 0.550 | 0.600 | 0.728 |
| 19 | 0.391 | 0.460 | 0.535 | 0.584 | 0.712 |
| 20 | 0.380 | 0.447 | 0.520 | 0.570 | 0.696 |
| 21 | 0.370 | 0.435 | 0.508 | 0.556 | 0.681 |
| 22 | 0.361 | 0.425 | 0.496 | 0.544 | 0.667 |
| 23 | 0.353 | 0.415 | 0.486 | 0.532 | 0.654 |
| 24 | 0.344 | 0.406 | 0.476 | 0.521 | 0.642 |
| 25 | 0.357 | 0.398 | 0.466 | 0.511 | 0.630 |
| 26 | 0.331 | 0.390 | 0.457 | 0.501 | 0.619 |
| 27 | 0.324 | 0.382 | 0.448 | 0.491 | 0.608 |
| 28 | 0.317 | 0.375 | 0.440 | 0.483 | 0.598 |
| 29 | 0.312 | 0.368 | 0.433 | 0.475 | 0.589 |
| 30 | 0.306 | 0.362 | 0.425 | 0.467 | 0.580 |
| 31 | 0.301 | 0.356 | 0.418 | 0.459 | 0.571 |
| 32 | 0.296 | 0.350 | 0.412 | 0.452 | 0.563 |
| 33 | 0.291 | 0.345 | 0.405 | 0.446 | 0.554 |
| 34 | 0.287 | 0.340 | 0.399 | 0.439 | 0.547 |
| 35 | 0.283 | 0.335 | 0.394 | 0.433 | 0.539 |
| 36 | 0.279 | 0.330 | 0.388 | 0.427 | 0.533 |
| 37 | 0.275 | 0.325 | 0.383 | 0.421 | 0.526 |
| 38 | 0.271 | 0.321 | 0.378 | 0.415 | 0.519 |
| 39 | 0.267 | 0.317 | 0.373 | 0.410 | 0.513 |
| 40 | 0.264 | 0.313 | 0.368 | 0.405 | 0.507 |
| 41 | 0.261 | 0.309 | 0.364 | 0.400 | 0.501 |
| 42 | 0.257 | 0.305 | 0.359 | 0.395 | 0.495 |
| 43 | 0.254 | 0.301 | 0.355 | 0.391 | 0.490 |
| 44 | 0.251 | 0.298 | 0.351 | 0.386 | 0.484 |
| 45 | 0.248 | 0.294 | 0.347 | 0.382 | 0.479 |
| 46 | 0.246 | 0.291 | 0.343 | 0.378 | 0.474 |
| 47 | 0.243 | 0.288 | 0.340 | 0.374 | 0.469 |
| 48 | 0.240 | 0.285 | 0.336 | 0.370 | 0.465 |
| 49 | 0.238 | 0.282 | 0.333 | 0.366 | 0.460 |
| 50 | 0.235 | 0.279 | 0.329 | 0.363 | 0.456 |

## Table A11 Product–moment correlation coefficient: critical values of r.

1. Find row corresponding to degrees of freedom (df) $= n - 2$
2. For a **two-tailed** test, compare **absolute calculated value of r** (i.e. ignore a minus sign if it occurs) to critical value for **significance level $\alpha_2$**. Use decision rule 1.
3. For a **one-tailed** test in which predicted correlation is **positive**, compare calculated value of **r with its sign** to critical value for **significance level $\alpha_1$**. Use decision rule 1.
4. For a **one-tailed** test in which predicted correlation is **negative**, compare calculated value of **r with its sign** to critical value in column for significance level $\alpha_1$. Use decision rule 2.

Rule 1: **Reject null hypothesis if calculated value equal to, or greater than, calculated value.**

Rule 2: **Give critical value a minus sign. Reject null hypothesis if calculated value is equal to, or smaller than, critical value.**

| $\alpha_1$ | 5% | 2.5% | 1% | 0.5% | 0.05% | | $\alpha_1$ | 5% | 2.5% | 1% | 0.5% | 0.05% |
|---|---|---|---|---|---|---|---|---|---|---|---|---|
| $\alpha_2$ | 10% | 5% | 2% | 1% | 0.1% | | $\alpha_2$ | 10% | 5% | 2% | 1% | 0.1% |
| df = 1 | 0.988 | 0.997 | 1.000 | 1.000 | 1.000 | | df = 31 | 0.291 | 0.344 | 0.403 | 0.442 | 0.546 |
| 2 | 0.900 | 0.950 | 0.980 | 0.990 | 0.999 | | 32 | 0.287 | 0.339 | 0.397 | 0.436 | 0.539 |
| 3 | 0.805 | 0.878 | 0.934 | 0.959 | 0.991 | | 33 | 0.283 | 0.334 | 0.392 | 0.430 | 0.532 |
| 4 | 0.729 | 0.811 | 0.882 | 0.917 | 0.974 | | 34 | 0.279 | 0.329 | 0.386 | 0.424 | 0.525 |
| 5 | 0.669 | 0.755 | 0.833 | 0.875 | 0.951 | | 35 | 0.275 | 0.325 | 0.381 | 0.418 | 0.519 |
| 6 | 0.621 | 0.707 | 0.789 | 0.834 | 0.925 | | 36 | 0.271 | 0.320 | 0.376 | 0.413 | 0.513 |
| 7 | 0.582 | 0.666 | 0.750 | 0.798 | 0.898 | | 37 | 0.267 | 0.316 | 0.371 | 0.408 | 0.507 |
| 8 | 0.549 | 0.632 | 0.715 | 0.765 | 0.872 | | 38 | 0.264 | 0.312 | 0.367 | 0.403 | 0.501 |
| 9 | 0.521 | 0.602 | 0.685 | 0.735 | 0.847 | | 39 | 0.261 | 0.308 | 0.362 | 0.398 | 0.495 |
| 10 | 0.497 | 0.576 | 0.658 | 0.708 | 0.823 | | 40 | 0.257 | 0.304 | 0.358 | 0.393 | 0.490 |
| 11 | 0.476 | 0.553 | 0.634 | 0.684 | 0.801 | | 41 | 0.254 | 0.301 | 0.354 | 0.389 | 0.484 |
| 12 | 0.457 | 0.532 | 0.612 | 0.661 | 0.780 | | 42 | 0.251 | 0.297 | 0.350 | 0.384 | 0.479 |
| 13 | 0.441 | 0.514 | 0.592 | 0.641 | 0.760 | | 43 | 0.248 | 0.294 | 0.346 | 0.380 | 0.474 |
| 14 | 0.426 | 0.497 | 0.574 | 0.623 | 0.742 | | 44 | 0.246 | 0.291 | 0.342 | 0.376 | 0.469 |
| 15 | 0.412 | 0.482 | 0.558 | 0.606 | 0.725 | | 45 | 0.243 | 0.288 | 0.338 | 0.372 | 0.465 |
| 16 | 0.400 | 0.468 | 0.542 | 0.590 | 0.708 | | 46 | 0.240 | 0.285 | 0.335 | 0.368 | 0.460 |
| 17 | 0.389 | 0.456 | 0.529 | 0.575 | 0.693 | | 47 | 0.238 | 0.282 | 0.331 | 0.365 | 0.456 |
| 18 | 0.378 | 0.444 | 0.515 | 0.561 | 0.679 | | 48 | 0.235 | 0.279 | 0.328 | 0.361 | 0.451 |
| 19 | 0.369 | 0.433 | 0.503 | 0.549 | 0.665 | | 49 | 0.233 | 0.276 | 0.325 | 0.358 | 0.447 |
| 20 | 0.360 | 0.423 | 0.492 | 0.537 | 0.652 | | 50 | 0.231 | 0.273 | 0.322 | 0.354 | 0.443 |
| 21 | 0.352 | 0.413 | 0.482 | 0.526 | 0.640 | | 60 | 0.211 | 0.250 | 0.295 | 0.325 | 0.408 |
| 22 | 0.344 | 0.404 | 0.472 | 0.515 | 0.629 | | 70 | 0.195 | 0.232 | 0.274 | 0.302 | 0.380 |
| 23 | 0.337 | 0.396 | 0.462 | 0.505 | 0.618 | | 80 | 0.183 | 0.217 | 0.257 | 0.283 | 0.357 |
| 24 | 0.330 | 0.388 | 0.453 | 0.496 | 0.607 | | 90 | 0.173 | 0.205 | 0.242 | 0.267 | 0.338 |
| 25 | 0.323 | 0.381 | 0.445 | 0.487 | 0.597 | | 100 | 0.164 | 0.195 | 0.230 | 0.254 | 0.321 |
| 26 | 0.317 | 0.374 | 0.437 | 0.479 | 0.588 | | | | | | | |
| 27 | 0.311 | 0.367 | 0.430 | 0.471 | 0.579 | | | | | | | |
| 28 | 0.306 | 0.361 | 0.423 | 0.463 | 0.570 | | | | | | | |
| 29 | 0.301 | 0.355 | 0.416 | 0.456 | 0.562 | | | | | | | |
| 30 | 0.296 | 0.349 | 0.409 | 0.449 | 0.554 | | | | | | | |

**Table A12 The Mann–Whitney test: critical values of $U$.**

1. If two samples are same size ($n_A = n_B$), find subtable for $n_A$ or $n_B$
2. If two samples are unequal in size find subtable for larger sample size.
3. Find row corresponding to other sample size
4. For **one-tailed** test use calculated value of $U_A$ or $U_B$ which is **predicted to be the larger** and critical value in column for significance level $\alpha_1$. (N.B. Some statistical packages and tables use the smaller value of $U$.)
5. For **two-tailed** test use calculated value of $U_A$ or $U_B$ **which is the larger** of the two and critical value in column for significance level $\alpha_2$. (N.B. Some statistical packages and tables use the smaller value of $U$.)
6. **Reject the null hypothesis if calculated value is equal to, or greater than, critical value.** (N.B. Some statistical packages use the smaller value of $U$ and a different decision rule.)

| | $\alpha_1$ 10% | 5% | 2.5% | 1% | 0.5% | 0.1% | 0.05% |
|---|---|---|---|---|---|---|---|
| | $\alpha_2$ 20% | 10% | 5% | 2% | 1% | 0.2% | 0.1% |
| **$n_A$ or $n_B = 3$** | | | | | | | |
| $n_B$ or $n_A = 2$ | 6 | | | | | | |
| 3 | 8 | 9 | | | | | |
| **$n_A$ or $n_B = 4$** | | | | | | | |
| $n_B$ or $n_A = 2$ | 8 | | | | | | |
| 3 | 11 | 12 | | | | | |
| 4 | 13 | 15 | 16 | | | | |
| **$n_A$ or $n_B = 5$** | | | | | | | |
| $n_B$ or $n_A = 2$ | 9 | 10 | | | | | |
| 3 | 13 | 14 | 15 | | | | |
| 4 | 16 | 18 | 19 | 20 | | | |
| 5 | 20 | 21 | 23 | 24 | 25 | | |
| **$n_A$ or $n_B = 6$** | | | | | | | |
| $n_B$ or $n_A = 2$ | 11 | 12 | | | | | |
| 3 | 15 | 16 | 17 | | | | |
| 4 | 19 | 21 | 22 | 23 | 24 | | |
| 5 | 23 | 25 | 27 | 28 | 29 | | |
| 6 | 27 | 29 | 31 | 33 | 34 | | |
| **$n_A$ or $n_B = 7$** | | | | | | | |
| $n_B$ or $n_A = 2$ | 13 | 14 | | | | | |
| 3 | 17 | 19 | 20 | 21 | | | |
| 4 | 22 | 24 | 25 | 27 | 28 | | |
| 5 | 27 | 29 | 30 | 32 | 34 | | |
| 6 | 31 | 34 | 36 | 38 | 39 | 42 | |
| 7 | 36 | 38 | 41 | 43 | 45 | 48 | 49 |
| **$n_A$ or $n_B = 8$** | | | | | | | |
| $n_B$ or $n_A = 2$ | 14 | 15 | 16 | | | | |
| 3 | 19 | 21 | 22 | 24 | | | |
| 4 | 25 | 27 | 28 | 30 | 31 | | |
| 5 | 30 | 32 | 34 | 36 | 38 | 40 | |
| 6 | 35 | 38 | 40 | 42 | 44 | 47 | 48 |
| 7 | 40 | 43 | 46 | 49 | 50 | 54 | 55 |
| 8 | 45 | 49 | 51 | 55 | 57 | 60 | 62 |
| **$n_A$ or $n_B = 9$** | | | | | | | |
| $n_B$ or $n_A = 2$ | 9 | | | | | | |
| 2 | 16 | 17 | 18 | | | | |
| 3 | 22 | 23 | 25 | 26 | 27 | | |
| 4 | 27 | 30 | 32 | 33 | 35 | | |
| 5 | 33 | 36 | 38 | 40 | 42 | 44 | 45 |
| 6 | 39 | 42 | 44 | 47 | 49 | 52 | 53 |
| 7 | 45 | 48 | 51 | 54 | 56 | 60 | 61 |
| 8 | 50 | 54 | 57 | 61 | 63 | 67 | 68 |
| 9 | 56 | 60 | 64 | 67 | 70 | 74 | 76 |

(continued on next page)

**Table A12 (continued)  The Mann–Whitney test: critical values of *U*.**

| $\alpha_1$ | 10% | 5% | 2.5% | 1% | 0.5% | 0.1% | 0.05% |
|---|---|---|---|---|---|---|---|
| $\alpha_2$ | 20% | 10% | 5% | 2% | 1% | 0.2% | 0.1% |

| $n_A$ or $n_B$ = 10 | | | | | | | |
|---|---|---|---|---|---|---|---|
| $n_B$ or $n_A$ = 1 | 10 | | | | | | |
| 2 | 17 | 19 | 20 | | | | |
| 3 | 24 | 26 | 27 | 29 | 30 | | |
| 4 | 30 | 33 | 35 | 37 | 38 | 40 | |
| 5 | 37 | 39 | 42 | 44 | 46 | 49 | 50 |
| 6 | 43 | 46 | 49 | 52 | 54 | 57 | 58 |
| 7 | 49 | 53 | 56 | 59 | 61 | 65 | 67 |
| 8 | 56 | 60 | 63 | 67 | 69 | 74 | 75 |
| 9 | 62 | 66 | 70 | 74 | 77 | 82 | 83 |
| 10 | 68 | 73 | 77 | 81 | 84 | 90 | 92 |

| $n_A$ or $n_B$ = 11 | | | | | | | |
|---|---|---|---|---|---|---|---|
| $n_B$ or $n_A$ = 1 | 11 | | | | | | |
| 2 | 19 | 21 | 22 | | | | |
| 3 | 26 | 28 | 30 | 32 | 33 | | |
| 4 | 33 | 36 | 38 | 40 | 42 | 44 | |
| 5 | 40 | 43 | 46 | 48 | 50 | 53 | 54 |
| 6 | 47 | 50 | 53 | 57 | 59 | 62 | 64 |
| 7 | 54 | 58 | 61 | 65 | 67 | 71 | 73 |
| 8 | 61 | 65 | 69 | 73 | 75 | 80 | 82 |
| 9 | 68 | 72 | 76 | 81 | 83 | 89 | 91 |
| 10 | 74 | 79 | 84 | 88 | 92 | 98 | 100 |
| 11 | 81 | 87 | 91 | 96 | 100 | 106 | 109 |

| $n_A$ or $n_B$ = 12 | | | | | | | |
|---|---|---|---|---|---|---|---|
| $n_B$ or $n_A$ = 1 | 12 | | | | | | |
| 2 | 20 | 22 | 23 | | | | |
| 3 | 28 | 31 | 32 | 34 | 35 | | |
| 4 | 36 | 39 | 41 | 42 | 45 | 48 | |
| 5 | 43 | 47 | 49 | 52 | 54 | 58 | 59 |
| 6 | 51 | 55 | 58 | 61 | 63 | 68 | 69 |
| 7 | 58 | 63 | 66 | 70 | 72 | 77 | 79 |
| 8 | 66 | 70 | 74 | 79 | 81 | 87 | 89 |
| 9 | 73 | 78 | 82 | 87 | 90 | 96 | 98 |
| 10 | 81 | 86 | 91 | 96 | 99 | 106 | 108 |
| 11 | 88 | 94 | 99 | 104 | 108 | 115 | 117 |
| 12 | 95 | 102 | 107 | 113 | 117 | 124 | 127 |

| $n_A$ or $n_B$ = 13 | | | | | | | |
|---|---|---|---|---|---|---|---|
| $n_B$ or $n_A$ = 1 | 13 | | | | | | |
| 2 | 22 | 24 | 25 | 26 | | | |
| 3 | 30 | 33 | 35 | 37 | 38 | | |
| 4 | 39 | 42 | 44 | 47 | 49 | 51 | 52 |
| 5 | 47 | 50 | 53 | 56 | 58 | 62 | 63 |
| 6 | 55 | 59 | 62 | 66 | 68 | 73 | 74 |
| 7 | 63 | 67 | 71 | 75 | 78 | 83 | 85 |
| 8 | 71 | 76 | 80 | 84 | 87 | 93 | 95 |
| 9 | 79 | 84 | 89 | 94 | 97 | 103 | 106 |
| 10 | 87 | 93 | 97 | 103 | 106 | 113 | 116 |
| 11 | 95 | 101 | 106 | 112 | 116 | 123 | 126 |
| 12 | 103 | 109 | 115 | 121 | 125 | 133 | 136 |
| 13 | 111 | 118 | 124 | 130 | 135 | 143 | 146 |

| $n_A$ or $n_B$ = 14 | | | | | | | |
|---|---|---|---|---|---|---|---|
| $n_B$ or $n_A$ = 1 | 14 | | | | | | |
| 2 | 24 | 25 | 27 | 28 | | | |
| 3 | 32 | 35 | 37 | 40 | 41 | | |
| 4 | 41 | 45 | 47 | 50 | 52 | 55 | |
| 5 | 50 | 54 | 57 | 60 | 63 | 67 | 68 |
| 6 | 59 | 63 | 67 | 71 | 73 | 78 | 79 |
| 7 | 67 | 72 | 76 | 81 | 83 | 89 | 91 |
| 8 | 76 | 81 | 86 | 90 | 94 | 100 | 102 |
| 9 | 85 | 90 | 95 | 100 | 104 | 111 | 113 |
| 10 | 93 | 99 | 104 | 110 | 114 | 121 | 124 |
| 11 | 102 | 108 | 114 | 120 | 124 | 132 | 135 |
| 12 | 110 | 117 | 123 | 130 | 134 | 143 | 146 |
| 13 | 119 | 126 | 132 | 139 | 144 | 153 | 157 |
| 14 | 127 | 135 | 141 | 149 | 154 | 164 | 167 |

(continued on next page)

**Table A12 (continued)  The Mann–Whitney test: critical values of _U_.**

| $\alpha_1$ | 10% | 5% | 2.5% | 1% | 0.5% | 0.1% | 0.05% |
|---|---|---|---|---|---|---|---|
| $\alpha_2$ | 20% | 10% | 5% | 2% | 1% | 0.2% | 0.1% |
| $n_A$ or $n_B = 15$ | | | | | | | |
| $n_B$ or $n_A = 1$ | 15 | | | | | | |
| 2 | 25 | 27 | 29 | 30 | | | |
| 3 | 35 | 38 | 40 | 42 | 43 | | |
| 4 | 44 | 48 | 50 | 53 | 55 | 59 | |
| 5 | 53 | 57 | 61 | 64 | 67 | 71 | 72 |
| 6 | 63 | 67 | 71 | 75 | 78 | 83 | 85 |
| 7 | 72 | 77 | 81 | 86 | 89 | 95 | 97 |
| 8 | 81 | 87 | 91 | 96 | 100 | 106 | 109 |
| 9 | 90 | 96 | 101 | 107 | 111 | 118 | 120 |
| 10 | 99 | 106 | 111 | 117 | 121 | 129 | 132 |
| 11 | 108 | 115 | 121 | 128 | 132 | 141 | 144 |
| 12 | 117 | 125 | 131 | 138 | 143 | 152 | 155 |
| 13 | 127 | 134 | 141 | 148 | 153 | 163 | 167 |
| 14 | 136 | 144 | 151 | 159 | 164 | 174 | 178 |
| 15 | 145 | 153 | 161 | 169 | 174 | 185 | 189 |
| $n_A$ or $n_B = 16$ | | | | | | | |
| $n_B$ or $n_A = 1$ | 16 | | | | | | |
| 2 | 27 | 29 | 31 | 32 | | | |
| 3 | 37 | 40 | 42 | 45 | 46 | | |
| 4 | 47 | 50 | 53 | 57 | 59 | 62 | |
| 5 | 57 | 61 | 65 | 68 | 71 | 75 | 77 |
| 6 | 67 | 71 | 75 | 80 | 83 | 88 | 90 |
| 7 | 76 | 82 | 86 | 91 | 94 | 101 | 103 |
| 8 | 86 | 92 | 97 | 102 | 106 | 113 | 115 |
| 9 | 96 | 102 | 107 | 113 | 117 | 125 | 128 |
| 10 | 106 | 112 | 118 | 124 | 129 | 137 | 140 |
| 11 | 115 | 122 | 129 | 135 | 140 | 149 | 152 |
| 12 | 125 | 132 | 139 | 146 | 151 | 161 | 165 |
| 13 | 134 | 143 | 149 | 157 | 163 | 173 | 177 |
| 14 | 144 | 153 | 160 | 168 | 174 | 185 | 189 |
| 15 | 154 | 163 | 170 | 179 | 185 | 197 | 201 |
| 16 | 163 | 173 | 181 | 190 | 196 | 208 | 213 |
| $n_A$ or $n_B = 17$ | | | | | | | |
| $n_B$ or $n_A = 1$ | 17 | | | | | | |
| 2 | 28 | 31 | 32 | 34 | | | |
| 3 | 39 | 42 | 45 | 47 | 49 | 51 | |
| 4 | 50 | 53 | 57 | 60 | 62 | 66 | |
| 5 | 60 | 65 | 68 | 72 | 75 | 80 | 81 |
| 6 | 71 | 76 | 80 | 84 | 87 | 93 | 95 |
| 7 | 81 | 86 | 91 | 96 | 100 | 106 | 109 |
| 8 | 91 | 97 | 102 | 108 | 112 | 119 | 122 |
| 9 | 101 | 108 | 114 | 120 | 124 | 132 | 135 |
| 10 | 112 | 119 | 125 | 132 | 136 | 145 | 148 |
| 11 | 122 | 130 | 136 | 143 | 148 | 158 | 161 |
| 12 | 132 | 140 | 147 | 155 | 160 | 170 | 174 |
| 13 | 142 | 151 | 158 | 166 | 172 | 183 | 187 |
| 14 | 153 | 161 | 169 | 178 | 184 | 195 | 199 |
| 15 | 163 | 172 | 180 | 189 | 195 | 208 | 212 |
| 16 | 173 | 183 | 191 | 201 | 207 | 220 | 225 |
| 17 | 183 | 193 | 202 | 212 | 219 | 232 | 238 |

(continued on next page)

**Table A12 (continued)  The Mann–Whitney test: critical values of _U_.**

| $\alpha_1$ | 10% | 5% | 2.5% | 1% | 0.5% | 0.1% | 0.05% |
|---|---|---|---|---|---|---|---|
| $\alpha_2$ | 20% | 10% | 5% | 2% | 1% | 0.2% | 0.1% |

| $n_A$ or $n_B$ = 18 | | | | | | | |
|---|---|---|---|---|---|---|---|
| $n_B$ or $n_A$ = 1 | 18 | | | | | | |
| 2 | 30 | 32 | 34 | 36 | | | |
| 3 | 41 | 45 | 47 | 50 | 52 | 54 | |
| 4 | 52 | 56 | 60 | 63 | 66 | 69 | |
| 5 | 63 | 68 | 72 | 76 | 79 | 84 | 86 |
| 6 | 74 | 80 | 84 | 89 | 92 | 98 | 100 |
| 7 | 85 | 91 | 96 | 102 | 105 | 112 | 115 |
| 8 | 96 | 103 | 108 | 114 | 118 | 126 | 129 |
| 9 | 107 | 114 | 120 | 126 | 131 | 139 | 142 |
| 10 | 118 | 125 | 132 | 139 | 143 | 153 | 156 |
| 11 | 129 | 137 | 143 | 151 | 156 | 166 | 170 |
| 12 | 139 | 148 | 155 | 163 | 169 | 179 | 183 |
| 13 | 150 | 159 | 167 | 175 | 181 | 192 | 197 |
| 14 | 161 | 170 | 178 | 187 | 194 | 206 | 210 |
| 15 | 172 | 182 | 190 | 200 | 206 | 219 | 224 |
| 16 | 182 | 193 | 202 | 212 | 218 | 232 | 237 |
| 17 | 193 | 204 | 213 | 224 | 231 | 245 | 250 |
| 18 | 204 | 215 | 225 | 236 | 243 | 258 | 263 |

| $n_A$ or $n_B$ = 19 | | | | | | | |
|---|---|---|---|---|---|---|---|
| $n_B$ or $n_A$ = 1 | 18 | 19 | | | | | |
| 2 | 31 | 34 | 36 | 37 | 38 | | |
| 3 | 43 | 47 | 50 | 53 | 54 | 57 | |
| 4 | 55 | 59 | 63 | 67 | 69 | 73 | |
| 5 | 67 | 72 | 76 | 80 | 83 | 88 | 90 |
| 6 | 78 | 84 | 89 | 94 | 97 | 103 | 106 |
| 7 | 90 | 96 | 101 | 107 | 111 | 118 | 120 |
| 8 | 101 | 108 | 114 | 120 | 124 | 132 | 135 |
| 9 | 113 | 120 | 126 | 133 | 138 | 146 | 150 |
| 10 | 124 | 132 | 138 | 146 | 151 | 161 | 164 |
| 11 | 136 | 144 | 151 | 159 | 164 | 175 | 178 |
| 12 | 147 | 156 | 163 | 172 | 177 | 188 | 193 |
| 13 | 158 | 167 | 175 | 184 | 190 | 202 | 207 |
| 14 | 169 | 179 | 188 | 197 | 203 | 216 | 221 |
| 15 | 181 | 191 | 200 | 210 | 216 | 230 | 235 |
| 16 | 192 | 203 | 212 | 222 | 230 | 244 | 249 |
| 17 | 203 | 214 | 224 | 235 | 242 | 257 | 263 |
| 18 | 214 | 226 | 236 | 248 | 255 | 271 | 277 |
| 19 | 226 | 238 | 248 | 260 | 268 | 284 | 291 |

| $n_A$ or $n_B$ = 20 | | | | | | | |
|---|---|---|---|---|---|---|---|
| $n_B$ or $n_A$ = 1 | 19 | 20 | | | | | |
| 2 | 33 | 36 | 38 | 39 | 40 | | |
| 3 | 45 | 49 | 52 | 55 | 57 | 60 | |
| 4 | 58 | 62 | 66 | 70 | 72 | 77 | |
| 5 | 70 | 75 | 80 | 84 | 87 | 93 | 95 |
| 6 | 82 | 88 | 93 | 98 | 102 | 108 | 111 |
| 7 | 94 | 101 | 106 | 112 | 116 | 124 | 126 |
| 8 | 106 | 113 | 119 | 126 | 130 | 139 | 142 |
| 9 | 118 | 126 | 132 | 140 | 144 | 154 | 157 |
| 10 | 130 | 138 | 145 | 153 | 158 | 168 | 172 |
| 11 | 142 | 151 | 158 | 167 | 172 | 183 | 187 |
| 12 | 154 | 163 | 171 | 180 | 186 | 198 | 202 |
| 13 | 166 | 176 | 184 | 193 | 200 | 212 | 217 |
| 14 | 178 | 188 | 197 | 207 | 213 | 226 | 231 |
| 15 | 190 | 200 | 210 | 220 | 227 | 241 | 246 |
| 16 | 201 | 213 | 222 | 233 | 241 | 255 | 261 |
| 17 | 213 | 225 | 235 | 247 | 254 | 270 | 275 |
| 18 | 225 | 237 | 248 | 260 | 268 | 284 | 287 |
| 19 | 237 | 250 | 261 | 273 | 281 | 298 | 304 |
| 20 | 249 | 262 | 273 | 286 | 295 | 312 | 319 |

## Table A13 Critical values of $F$.

$F$ test for comparing two sample variances
1. Find subtable for degrees of freedom (df) in numerator. In **one-tailed** test these relate to size of sample **predicted** to have larger variance, in **two-tailed** test to size of sample which **has** larger variance. df = sample size − 1.
2. Find row for denominator df = (other sample size − 1)
3. For **one-tailed** test compare calculated value to critical value in column for **significance level $\alpha_1$**.
4. For **two-tailed** test compare calculated value to critical value in column for **significance level $\alpha_2$**.
5. **Reject null hypothesis if calculated value is equal to, or greater than calculated value.**

One-way analysis of variance
1. Find subtable for degrees of freedom (df) in numerator = $(k - 1)$ where $k$ = number of samples being compared.
2. Find row for denominator df = $N - k$, where $N$ = total number of observations.
3. Compare calculated value to critical value in column for **significance level $\alpha_1$**.
4. **Reject null hypothesis if calculated value is equal to, or greater than, calculated value.**

| $\alpha_1$ | 5% | 2.5% | 1% | 0.5% | 0.1% |
|---|---|---|---|---|---|
| $\alpha_2$ | 10% | 5% | 2% | 1% | 0.2% |
| | | | Numerator df = 1 | | |
| Denominator, df = 1 | 161. | 648. | 4050. | 16200. | 405000. |
| 2 | 18.5 | 38.5 | 98.5 | 199. | 999. |
| 3 | 10.1 | 17.4 | 34.1 | 55.6 | 167. |
| 4 | 7.71 | 12.2 | 21.2 | 31.3 | 74.1 |
| 5 | 6.61 | 10.0 | 16.3 | 22.8 | 47.2 |
| 6 | 5.99 | 8.81 | 13.7 | 18.6 | 35.5 |
| 7 | 5.59 | 8.07 | 12.2 | 16.2 | 29.2 |
| 8 | 5.32 | 7.57 | 11.3 | 14.7 | 25.4 |
| 9 | 5.12 | 7.21 | 10.6 | 13.6 | 22.9 |
| 10 | 4.96 | 6.94 | 10.0 | 12.8 | 21.0 |
| 11 | 4.84 | 6.72 | 9.65 | 12.2 | 19.7 |
| 12 | 4.75 | 6.55 | 9.33 | 11.8 | 18.6 |
| 13 | 4.67 | 6.41 | 9.07 | 11.4 | 17.8 |
| 14 | 4.60 | 6.30 | 8.86 | 11.1 | 17.1 |
| 15 | 4.54 | 6.20 | 8.68 | 10.8 | 16.6 |
| 16 | 4.49 | 6.12 | 8.53 | 10.6 | 16.1 |
| 17 | 4.45 | 6.04 | 8.40 | 10.4 | 15.7 |
| 18 | 4.41 | 5.98 | 8.29 | 10.2 | 15.4 |
| 19 | 4.38 | 5.92 | 8.18 | 10.1 | 15.1 |
| 20 | 4.35 | 5.87 | 8.10 | 9.94 | 14.8 |
| 21 | 4.32 | 5.83 | 8.02 | 9.83 | 14.6 |
| 22 | 4.30 | 5.79 | 7.95 | 9.73 | 14.4 |
| 23 | 4.28 | 5.75 | 7.88 | 9.63 | 14.2 |
| 24 | 4.26 | 5.72 | 7.82 | 9.55 | 14.0 |
| 25 | 4.24 | 5.69 | 7.77 | 9.48 | 13.9 |
| 40 | 4.08 | 5.42 | 7.31 | 8.83 | 12.6 |
| 60 | 4.00 | 5.29 | 7.08 | 8.49 | 12.0 |
| 120 | 3.92 | 5.15 | 6.85 | 8.18 | 11.4 |
| ∞ | 3.84 | 5.02 | 6.64 | 7.88 | 10.8 |

(continued on next page)

## Table A13 (continued)  Critical values of *F*.

| $\alpha_1$ | 5% | 2.5% | 1% | 0.5% | 0.1% |
|---|---|---|---|---|---|
| $\alpha_2$ | 10% | 5% | 2% | 1% | 0.2% |

| | Numerator df = 2 | | | | |
|---|---|---|---|---|---|
| Denominator, df = 1 | 200. | 800. | 5000. | 20000. | 500000 |
| 2 | 19.0 | 39.0 | 99.0 | 199. | 999. |
| 3 | 9.55 | 16.0 | 30.8 | 49.8 | 149. |
| 4 | 6.94 | 10.6 | 18.0 | 26.3 | 61.2 |
| 5 | 5.79 | 8.43 | 13.3 | 18.3 | 37.1 |
| 6 | 5.14 | 7.26 | 10.9 | 14.5 | 27.0 |
| 7 | 4.74 | 6.54 | 9.55 | 12.4 | 21.7 |
| 8 | 4.46 | 6.06 | 8.65 | 11.0 | 18.5 |
| 9 | 4.26 | 5.71 | 8.02 | 10.1 | 16.4 |
| 10 | 4.10 | 5.46 | 7.56 | 9.43 | 14.9 |
| 11 | 3.98 | 5.26 | 7.21 | 8.91 | 13.8 |
| 12 | 3.89 | 5.10 | 6.93 | 8.51 | 13.0 |
| 13 | 3.81 | 4.97 | 6.70 | 8.19 | 12.3 |
| 14 | 3.74 | 4.86 | 6.51 | 7.92 | 11.8 |
| 15 | 3.68 | 4.77 | 6.36 | 7.70 | 11.3 |
| 16 | 3.63 | 4.69 | 6.23 | 7.51 | 11.0 |
| 17 | 3.59 | 4.62 | 6.11 | 7.35 | 10.7 |
| 18 | 3.55 | 4.56 | 6.01 | 7.21 | 10.4 |
| 19 | 3.52 | 4.51 | 5.93 | 7.09 | 10.2 |
| 20 | 3.49 | 4.46 | 5.85 | 6.99 | 9.95 |
| 21 | 3.47 | 4.42 | 5.78 | 6.89 | 9.77 |
| 22 | 3.44 | 4.38 | 5.72 | 6.81 | 9.61 |
| 23 | 3.42 | 4.35 | 5.66 | 6.73 | 9.47 |
| 24 | 3.40 | 4.32 | 5.61 | 6.66 | 9.34 |
| 25 | 3.39 | 4.29 | 5.57 | 6.60 | 9.22 |
| 40 | 3.23 | 4.05 | 5.18 | 6.07 | 8.25 |
| 60 | 3.15 | 3.93 | 4.98 | 5.79 | 7.77 |
| 120 | 3.07 | 3.80 | 4.79 | 5.54 | 7.32 |
| ∞ | 3.00 | 3.69 | 4.61 | 5.30 | 6.91 |

| | Numerator df = 3 | | | | |
|---|---|---|---|---|---|
| Denominator, df = 1 | 216. | 864. | 5400. | 21600. | 540000. |
| 2 | 19.2 | 39.2 | 99.2 | 199. | 999. |
| 3 | 9.28 | 15.4 | 29.5 | 47.5 | 141. |
| 4 | 6.59 | 9.98 | 16.7 | 24.3 | 56.2 |
| 5 | 5.41 | 7.76 | 12.1 | 16.5 | 33.2 |
| 6 | 4.76 | 6.60 | 9.78 | 12.9 | 23.7 |
| 7 | 4.35 | 5.89 | 8.45 | 10.9 | 18.8 |
| 8 | 4.07 | 5.42 | 7.59 | 9.60 | 15.8 |
| 9 | 3.86 | 5.08 | 6.99 | 8.72 | 13.9 |
| 10 | 3.71 | 4.83 | 6.55 | 8.08 | 12.6 |
| 11 | 3.59 | 4.63 | 6.22 | 7.60 | 11.6 |
| 12 | 3.49 | 4.47 | 5.95 | 7.23 | 10.8 |
| 13 | 3.41 | 4.35 | 5.74 | 6.93 | 10.2 |
| 14 | 3.34 | 4.24 | 5.56 | 6.68 | 9.73 |
| 15 | 3.29 | 4.15 | 5.42 | 6.48 | 9.34 |
| 16 | 3.24 | 4.08 | 5.29 | 6.30 | 9.01 |
| 17 | 3.20 | 4.01 | 5.19 | 6.16 | 8.73 |
| 18 | 3.16 | 3.95 | 5.09 | 6.03 | 8.49 |
| 19 | 3.13 | 3.90 | 5.01 | 5.92 | 8.28 |
| 20 | 3.10 | 3.86 | 4.94 | 5.82 | 8.10 |
| 21 | 3.07 | 3.82 | 4.87 | 5.73 | 7.94 |
| 22 | 3.05 | 3.78 | 4.82 | 5.65 | 7.80 |
| 23 | 3.03 | 3.75 | 4.76 | 5.58 | 7.67 |
| 24 | 3.01 | 3.72 | 4.72 | 5.52 | 7.55 |
| 25 | 2.99 | 3.69 | 4.68 | 5.46 | 7.45 |
| 40 | 2.84 | 3.46 | 4.31 | 4.98 | 6.59 |
| 60 | 2.76 | 3.34 | 4.13 | 4.73 | 6.17 |
| 120 | 2.68 | 3.23 | 3.95 | 4.50 | 5.78 |
| ∞ | 2.61 | 3.12 | 3.78 | 4.28 | 5.42 |

(continued on next page)

**Table A13 (continued)  Critical values of *F*.**

| $\alpha_1$ | 5% | 2.5% | 1% | 0.5% | 0.1% |
|---|---|---|---|---|---|
| $\alpha_2$ | 10% | 5% | 2% | 1% | 0.2% |

| | | Numerator df = 4 | | | |
|---|---|---|---|---|---|
| Denominator, df = 1 | 225. | 900. | 5620. | 22500. | 562000. |
| 2 | 19.2 | 39.2 | 99.2 | 199. | 999. |
| 3 | 9.12 | 15.1 | 28.7 | 46.2 | 137. |
| 4 | 6.39 | 9.60 | 16.0 | 23.2 | 53.4 |
| 5 | 5.19 | 7.39 | 11.4 | 15.6 | 31.1 |
| 6 | 4.53 | 6.23 | 9.15 | 12.0 | 21.9 |
| 7 | 4.12 | 5.52 | 7.85 | 10.1 | 17.2 |
| 8 | 3.84 | 5.05 | 7.01 | 8.81 | 14.4 |
| 9 | 3.63 | 4.72 | 6.42 | 7.96 | 12.6 |
| 10 | 3.48 | 4.47 | 5.99 | 7.34 | 11.3 |
| 11 | 3.36 | 4.28 | 5.67 | 6.88 | 10.3 |
| 12 | 3.26 | 4.12 | 5.41 | 6.52 | 9.63 |
| 13 | 3.18 | 4.00 | 5.21 | 6.23 | 9.07 |
| 14 | 3.11 | 3.89 | 5.04 | 6.00 | 8.62 |
| 15 | 3.06 | 3.80 | 4.89 | 5.80 | 8.25 |
| 16 | 3.01 | 3.73 | 4.77 | 5.64 | 7.94 |
| 17 | 2.96 | 3.66 | 4.67 | 5.50 | 7.68 |
| 18 | 2.93 | 3.61 | 4.58 | 5.37 | 7.46 |
| 19 | 2.90 | 3.56 | 4.50 | 5.27 | 7.27 |
| 20 | 2.87 | 3.51 | 4.43 | 5.17 | 7.10 |
| 21 | 2.84 | 3.48 | 4.37 | 5.09 | 6.95 |
| 22 | 2.82 | 3.44 | 4.31 | 5.02 | 6.81 |
| 23 | 2.80 | 3.41 | 4.26 | 4.95 | 6.70 |
| 24 | 2.78 | 3.38 | 4.22 | 4.89 | 6.59 |
| 25 | 2.76 | 3.35 | 4.18 | 4.84 | 6.49 |
| 40 | 2.61 | 3.13 | 3.83 | 4.37 | 5.70 |
| 60 | 2.53 | 3.01 | 3.65 | 4.14 | 5.31 |
| 120 | 2.45 | 2.89 | 3.48 | 3.92 | 4.95 |
| ∞ | 2.37 | 2.79 | 3.32 | 3.72 | 4.62 |

| | | Numerator df = 5 | | | |
|---|---|---|---|---|---|
| Denominator, df = 1 | 230. | 922. | 5760. | 23100. | 576000. |
| 2 | 19.3 | 39.3 | 99.3 | 199. | 999. |
| 3 | 9.01 | 14.9 | 28.2 | 45.4 | 135. |
| 4 | 6.26 | 9.36 | 15.5 | 22.5 | 51.7 |
| 5 | 5.05 | 7.15 | 11.0 | 14.9 | 29.8 |
| 6 | 4.39 | 5.99 | 8.75 | 11.5 | 20.8 |
| 7 | 3.97 | 5.29 | 7.46 | 9.52 | 16.2 |
| 8 | 3.69 | 4.82 | 6.63 | 8.30 | 13.5 |
| 9 | 3.48 | 4.48 | 6.06 | 7.47 | 11.7 |
| 10 | 3.33 | 4.24 | 5.64 | 6.87 | 10.5 |
| 11 | 3.20 | 4.04 | 5.32 | 6.42 | 9.58 |
| 12 | 3.11 | 3.89 | 5.06 | 6.07 | 8.89 |
| 13 | 3.03 | 3.77 | 4.86 | 5.79 | 8.35 |
| 14 | 2.96 | 3.66 | 4.69 | 5.56 | 7.92 |
| 15 | 2.90 | 3.58 | 4.56 | 5.37 | 7.57 |
| 16 | 2.85 | 3.50 | 4.44 | 5.21 | 7.27 |
| 17 | 2.81 | 3.44 | 4.34 | 5.07 | 7.02 |
| 18 | 2.77 | 3.38 | 4.25 | 4.96 | 6.81 |
| 19 | 2.74 | 3.33 | 4.17 | 4.85 | 6.62 |
| 20 | 2.71 | 3.29 | 4.10 | 4.76 | 6.46 |
| 21 | 2.68 | 3.25 | 4.04 | 4.68 | 6.32 |
| 22 | 2.66 | 3.22 | 3.99 | 4.61 | 6.19 |
| 23 | 2.64 | 3.18 | 3.94 | 4.54 | 6.08 |
| 24 | 2.62 | 3.15 | 3.90 | 4.49 | 5.98 |
| 25 | 2.60 | 3.13 | 3.85 | 4.43 | 5.89 |
| 40 | 2.45 | 2.90 | 3.51 | 3.99 | 5.13 |
| 60 | 2.37 | 2.79 | 3.34 | 3.76 | 4.76 |
| 120 | 2.29 | 2.67 | 3.17 | 3.55 | 4.42 |
| ∞ | 2.21 | 2.57 | 3.02 | 3.35 | 4.10 |

(continued on next page)

**Table A13 (continued)  Critical values of _F_.**

| $\alpha_1$ | 5% | 2.5% | 1% | 0.5% | 0.1% |
|---|---|---|---|---|---|
| $\alpha_2$ | 10% | 5% | 2% | 1% | 0.2% |
| Numerator df = 6 | | | | | |
| Denominator, df = 1 | 234. | 937. | 5860. | 23400. | 586000. |
| 2 | 19.3 | 39.3 | 99.3 | 199. | 999. |
| 3 | 8.94 | 14.7 | 27.9 | 44.8 | 133. |
| 4 | 6.16 | 9.20 | 15.2 | 22.0 | 50.5 |
| 5 | 4.95 | 6.98 | 10.7 | 14.5 | 28.8 |
| 6 | 4.28 | 5.82 | 8.47 | 11.1 | 20.0 |
| 7 | 3.87 | 5.12 | 7.19 | 9.16 | 15.5 |
| 8 | 3.58 | 4.65 | 6.37 | 7.95 | 12.9 |
| 9 | 3.37 | 4.32 | 5.80 | 7.13 | 11.1 |
| 10 | 3.22 | 4.07 | 5.39 | 6.54 | 9.93 |
| 11 | 3.09 | 3.88 | 5.07 | 6.10 | 9.05 |
| 12 | 3.00 | 3.73 | 4.82 | 5.76 | 8.38 |
| 13 | 2.92 | 3.60 | 4.62 | 5.48 | 7.86 |
| 14 | 2.85 | 3.50 | 4.46 | 5.26 | 7.44 |
| 15 | 2.79 | 3.41 | 4.32 | 5.07 | 7.09 |
| 16 | 2.74 | 3.34 | 4.20 | 4.91 | 6.80 |
| 17 | 2.70 | 3.28 | 4.10 | 4.78 | 6.56 |
| 18 | 2.66 | 3.22 | 4.01 | 4.66 | 6.35 |
| 19 | 2.63 | 3.17 | 3.94 | 4.56 | 6.18 |
| 20 | 2.60 | 3.13 | 3.87 | 4.47 | 6.02 |
| 21 | 2.57 | 3.09 | 3.81 | 4.39 | 5.88 |
| 22 | 2.55 | 3.05 | 3.76 | 4.32 | 5.76 |
| 23 | 2.53 | 3.02 | 3.71 | 4.26 | 5.65 |
| 24 | 2.51 | 2.99 | 3.67 | 4.20 | 5.55 |
| 25 | 2.49 | 2.97 | 3.63 | 4.15 | 5.46 |
| 40 | 2.34 | 2.74 | 3.29 | 3.71 | 4.73 |
| 60 | 2.25 | 2.63 | 3.12 | 3.49 | 4.37 |
| 120 | 2.18 | 2.52 | 2.96 | 3.28 | 4.04 |
| ∞ | 2.10 | 2.41 | 2.80 | 3.09 | 3.74 |
| Numerator df = 7 | | | | | |
| Denominator, df = 1 | 237. | 948. | 5930. | 23700. | 593000. |
| 2 | 19.4 | 39.4 | 99.4 | 199. | 999. |
| 3 | 8.89 | 14.6 | 27.7 | 44.4 | 132. |
| 4 | 6.09 | 9.07 | 15.0 | 21.6 | 49.7 |
| 5 | 4.88 | 6.85 | 10.5 | 14.2 | 28.2 |
| 6 | 4.21 | 5.70 | 8.26 | 10.8 | 19.5 |
| 7 | 3.79 | 4.99 | 6.99 | 8.89 | 15.0 |
| 8 | 3.50 | 4.53 | 6.18 | 7.69 | 12.4 |
| 9 | 3.29 | 4.20 | 5.61 | 6.88 | 10.7 |
| 10 | 3.14 | 3.95 | 5.20 | 6.30 | 9.52 |
| 11 | 3.01 | 3.76 | 4.89 | 5.86 | 8.66 |
| 12 | 2.91 | 3.61 | 4.64 | 5.52 | 8.00 |
| 13 | 2.83 | 3.48 | 4.44 | 5.25 | 7.49 |
| 14 | 2.76 | 3.38 | 4.28 | 5.03 | 7.08 |
| 15 | 2.71 | 3.29 | 4.14 | 4.85 | 6.74 |
| 16 | 2.66 | 3.22 | 4.03 | 4.69 | 6.46 |
| 17 | 2.61 | 3.16 | 3.93 | 4.56 | 6.22 |
| 18 | 2.58 | 3.10 | 3.84 | 4.44 | 6.02 |
| 19 | 2.54 | 3.05 | 3.77 | 4.34 | 5.85 |
| 20 | 2.51 | 3.01 | 3.70 | 4.26 | 5.69 |
| 21 | 2.49 | 2.97 | 3.64 | 4.18 | 5.56 |
| 22 | 2.46 | 2.93 | 3.59 | 4.11 | 5.44 |
| 23 | 2.44 | 2.90 | 3.54 | 4.05 | 5.33 |
| 24 | 2.42 | 2.87 | 3.50 | 3.99 | 5.23 |
| 25 | 2.40 | 2.85 | 3.46 | 3.94 | 5.15 |
| 40 | 2.25 | 2.62 | 3.12 | 3.51 | 4.44 |
| 60 | 2.17 | 2.51 | 2.95 | 3.29 | 4.09 |
| 120 | 2.09 | 2.39 | 2.79 | 3.09 | 3.77 |
| ∞ | 2.01 | 2.29 | 2.64 | 2.90 | 3.47 |

(continued on next page)

**Table A13 (continued)  Critical values of _F_.**

| $\alpha_1$ | 5% | 2.5% | 1% | 0.5% | 0.1% |
|---|---|---|---|---|---|
| $\alpha_2$ | 10% | 5% | 2% | 1% | 0.2% |

| | | Numerator df = 8 | | | |
|---|---|---|---|---|---|
| Denominator, df = 1 | 239. | 957. | 5980. | 23900. | 598000. |
| 2 | 19.4 | 39.4 | 99.4 | 199. | 999. |
| 3 | 8.85 | 14.5 | 27.5 | 44.1 | 131. |
| 4 | 6.04 | 8.98 | 14.8 | 21.4 | 49.0 |
| 5 | 4.82 | 6.76 | 10.3 | 14.0 | 27.6 |
| 6 | 4.15 | 5.60 | 8.10 | 10.6 | 19.0 |
| 7 | 3.73 | 4.90 | 6.84 | 8.68 | 14.6 |
| 8 | 3.44 | 4.43 | 6.03 | 7.50 | 12.0 |
| 9 | 3.23 | 4.10 | 5.47 | 6.69 | 10.4 |
| 10 | 3.07 | 3.85 | 5.06 | 6.12 | 9.20 |
| 11 | 2.95 | 3.66 | 4.74 | 5.68 | 8.35 |
| 12 | 2.85 | 3.51 | 4.50 | 5.35 | 7.71 |
| 13 | 2.77 | 3.39 | 4.30 | 5.08 | 7.21 |
| 14 | 2.70 | 3.29 | 4.14 | 4.86 | 6.80 |
| 15 | 2.64 | 3.20 | 4.00 | 4.67 | 6.47 |
| 16 | 2.59 | 3.12 | 3.89 | 4.52 | 6.19 |
| 17 | 2.55 | 3.06 | 3.79 | 4.39 | 5.96 |
| 18 | 2.51 | 3.01 | 3.71 | 4.28 | 5.76 |
| 19 | 2.48 | 2.96 | 3.63 | 4.18 | 5.59 |
| 20 | 2.45 | 2.91 | 3.56 | 4.09 | 5.44 |
| 21 | 2.42 | 2.87 | 3.51 | 4.01 | 5.31 |
| 22 | 2.40 | 2.84 | 3.45 | 3.94 | 5.19 |
| 23 | 2.37 | 2.81 | 3.41 | 3.88 | 5.09 |
| 24 | 2.36 | 2.78 | 3.36 | 3.83 | 4.99 |
| 25 | 2.34 | 2.75 | 3.32 | 3.78 | 4.91 |
| 40 | 2.18 | 2.53 | 2.99 | 3.35 | 4.21 |
| 60 | 2.10 | 2.41 | 2.82 | 3.13 | 3.86 |
| 120 | 2.02 | 2.30 | 2.66 | 2.93 | 3.55 |
| ∞ | 1.94 | 2.19 | 2.51 | 2.74 | 3.27 |

| | | Numerator df = 9 | | | |
|---|---|---|---|---|---|
| Denominator, df = 1 | 241. | 963. | 6020. | 24100. | 602000. |
| 2 | 19.4 | 39.4 | 99.4 | 199. | 999. |
| 3 | 8.81 | 14.5 | 27.3 | 43.9 | 130. |
| 4 | 6.00 | 8.90 | 14.7 | 21.1 | 48.5 |
| 5 | 4.77 | 6.68 | 10.2 | 13.8 | 27.2 |
| 6 | 4.10 | 5.52 | 7.98 | 10.4 | 18.7 |
| 7 | 3.68 | 4.82 | 6.72 | 8.51 | 14.3 |
| 8 | 3.39 | 4.36 | 5.91 | 7.34 | 11.8 |
| 9 | 3.18 | 4.03 | 5.35 | 6.54 | 10.1 |
| 10 | 3.02 | 3.78 | 4.94 | 5.97 | 8.96 |
| 11 | 2.90 | 3.59 | 4.63 | 5.54 | 8.12 |
| 12 | 2.80 | 3.44 | 4.39 | 5.20 | 7.48 |
| 13 | 2.71 | 3.31 | 4.19 | 4.94 | 6.98 |
| 14 | 2.65 | 3.21 | 4.03 | 4.72 | 6.58 |
| 15 | 2.59 | 3.12 | 3.89 | 4.54 | 6.26 |
| 16 | 2.54 | 3.05 | 3.78 | 4.38 | 5.98 |
| 17 | 2.49 | 2.98 | 3.68 | 4.25 | 5.75 |
| 18 | 2.46 | 2.93 | 3.60 | 4.14 | 5.56 |
| 19 | 2.42 | 2.88 | 3.52 | 4.04 | 5.39 |
| 20 | 2.39 | 2.84 | 3.46 | 3.96 | 5.24 |
| 21 | 2.37 | 2.80 | 3.40 | 3.88 | 5.11 |
| 22 | 2.34 | 2.76 | 3.35 | 3.81 | 4.99 |
| 23 | 2.32 | 2.73 | 3.30 | 3.75 | 4.89 |
| 24 | 2.30 | 2.70 | 3.26 | 3.69 | 4.80 |
| 25 | 2.28 | 2.68 | 3.22 | 3.64 | 4.71 |
| 40 | 2.12 | 2.45 | 2.89 | 3.22 | 4.02 |
| 60 | 2.04 | 2.33 | 2.72 | 3.01 | 3.69 |
| 120 | 1.96 | 2.22 | 2.56 | 2.81 | 3.38 |
| ∞ | 1.88 | 2.11 | 2.41 | 2.62 | 3.10 |

(continued on next page)

## Table A13 (continued)  Critical values of *F*.

| $\alpha_1$ | 5% | 2.5% | 1% | 0.5% | 0.1% |
|---|---|---|---|---|---|
| $\alpha_2$ | 10% | 5% | 2% | 1% | 0.2% |

| | Numerator df = 10 | | | | |
|---|---|---|---|---|---|
| Denominator, df = 1 | 242. | 969. | 6060. | 24200. | 606000. |
| 2 | 19.4 | 39.4 | 99.4 | 199. | 999. |
| 3 | 8.79 | 14.4 | 27.2 | 43.7 | 129. |
| 4 | 5.96 | 8.84 | 14.5 | 21.0 | 48.1 |
| 5 | 4.74 | 6.62 | 10.1 | 13.6 | 26.9 |
| 6 | 4.06 | 5.46 | 7.87 | 10.3 | 18.4 |
| 7 | 3.64 | 4.76 | 6.62 | 8.38 | 14.1 |
| 8 | 3.35 | 4.30 | 5.81 | 7.21 | 11.5 |
| 9 | 3.14 | 3.96 | 5.26 | 6.42 | 9.89 |
| 10 | 2.98 | 3.72 | 4.85 | 5.85 | 8.75 |
| 11 | 2.85 | 3.53 | 4.54 | 5.42 | 7.92 |
| 12 | 2.75 | 3.37 | 4.30 | 5.09 | 7.29 |
| 13 | 2.67 | 3.25 | 4.10 | 4.82 | 6.80 |
| 14 | 2.60 | 3.15 | 3.94 | 4.60 | 6.40 |
| 15 | 2.54 | 3.06 | 3.80 | 4.42 | 6.08 |
| 16 | 2.49 | 2.99 | 3.69 | 4.27 | 5.81 |
| 17 | 2.45 | 2.92 | 3.59 | 4.14 | 5.58 |
| 18 | 2.41 | 2.87 | 3.51 | 4.03 | 5.39 |
| 19 | 2.38 | 2.82 | 3.43 | 3.93 | 5.22 |
| 20 | 2.35 | 2.77 | 3.37 | 3.85 | 5.08 |
| 21 | 2.32 | 2.73 | 3.31 | 3.77 | 4.95 |
| 22 | 2.30 | 2.70 | 3.26 | 3.70 | 4.83 |
| 23 | 2.27 | 2.67 | 3.21 | 3.64 | 4.73 |
| 24 | 2.25 | 2.64 | 3.17 | 3.59 | 4.64 |
| 25 | 2.24 | 2.61 | 3.13 | 3.54 | 4.56 |
| 40 | 2.08 | 2.39 | 2.80 | 3.12 | 3.87 |
| 60 | 1.99 | 2.27 | 2.63 | 2.90 | 3.54 |
| 120 | 1.91 | 2.16 | 2.47 | 2.71 | 3.24 |
| $\infty$ | 1.83 | 2.05 | 2.32 | 2.52 | 2.96 |

| | Numerator df = 11 | | | | |
|---|---|---|---|---|---|
| Denominator, df = 1 | 243. | 973. | 6080 | 24300. | 608000. |
| 2 | 19.4 | 39.4 | 99.4 | 199. | 999. |
| 3 | 8.76 | 14.4 | 27.1 | 43.5 | 129. |
| 4 | 5.94 | 8.79 | 14.5 | 20.8 | 47.7 |
| 5 | 4.70 | 6.57 | 9.96 | 13.5 | 26.6 |
| 6 | 4.03 | 5.41 | 7.79 | 10.1 | 18.2 |
| 7 | 3.60 | 4.71 | 6.54 | 8.27 | 13.9 |
| 8 | 3.31 | 4.24 | 5.73 | 7.10 | 11.4 |
| 9 | 3.10 | 3.91 | 5.18 | 6.31 | 9.72 |
| 10 | 2.94 | 3.66 | 4.77 | 5.75 | 8.59 |
| 11 | 2.82 | 3.47 | 4.46 | 5.32 | 7.76 |
| 12 | 2.72 | 3.32 | 4.22 | 4.99 | 7.14 |
| 13 | 2.63 | 3.20 | 4.02 | 4.72 | 6.65 |
| 14 | 2.57 | 3.09 | 3.86 | 4.51 | 6.26 |
| 15 | 2.51 | 3.01 | 3.73 | 4.33 | 5.94 |
| 16 | 2.46 | 2.93 | 3.62 | 4.18 | 5.67 |
| 17 | 2.41 | 2.87 | 3.52 | 4.05 | 5.44 |
| 18 | 2.37 | 2.81 | 3.43 | 3.94 | 5.25 |
| 19 | 2.34 | 2.76 | 3.36 | 3.84 | 5.08 |
| 20 | 2.31 | 2.72 | 3.29 | 3.76 | 4.94 |
| 21 | 2.28 | 2.68 | 3.24 | 3.68 | 4.81 |
| 22 | 2.26 | 2.65 | 3.18 | 3.61 | 4.70 |
| 23 | 2.24 | 2.62 | 3.14 | 3.55 | 4.60 |
| 24 | 2.22 | 2.59 | 3.09 | 3.50 | 4.51 |
| 25 | 2.20 | 2.56 | 3.06 | 3.45 | 4.42 |
| 40 | 2.04 | 2.33 | 2.73 | 3.03 | 3.75 |
| 60 | 1.95 | 2.22 | 2.56 | 2.82 | 3.42 |
| 120 | 1.87 | 2.10 | 2.40 | 2.62 | 3.12 |
| $\infty$ | 1.79 | 1.99 | 2.25 | 2.43 | 2.84 |

(continued on next page)

| $\alpha_1$ | 5% | 2.5% | 1% | 0.5% | 0.1% |
|---|---|---|---|---|---|
| $\alpha_2$ | 10% | 5% | 2% | 1% | 0.2% |
| **Numerator df = 12** | | | | | |
| Denominator, df = 1 | 244. | 977. | 6110. | 24400. | 611000. |
| 2 | 19.4 | 39.4 | 99.4 | 199. | 999. |
| 3 | 8.74 | 14.3 | 27.1 | 43.4 | 128. |
| 4 | 5.91 | 8.75 | 14.4 | 20.7 | 47.4 |
| 5 | 4.68 | 6.52 | 9.89 | 13.4 | 26.4 |
| 6 | 4.00 | 5.37 | 7.72 | 10.0 | 18.0 |
| 7 | 3.57 | 4.67 | 6.47 | 8.18 | 13.7 |
| 8 | 3.28 | 4.20 | 5.67 | 7.01 | 11.2 |
| 9 | 3.07 | 3.87 | 5.11 | 6.23 | 9.57 |
| 10 | 2.91 | 3.62 | 4.71 | 5.66 | 8.45 |
| 11 | 2.79 | 3.43 | 4.40 | 5.24 | 7.63 |
| 12 | 2.69 | 3.28 | 4.16 | 4.91 | 7.00 |
| 13 | 2.60 | 3.15 | 3.96 | 4.64 | 6.52 |
| 14 | 2.53 | 3.05 | 3.80 | 4.43 | 6.13 |
| 15 | 2.48 | 2.96 | 3.67 | 4.25 | 5.81 |
| 16 | 2.42 | 2.89 | 3.55 | 4.10 | 5.55 |
| 17 | 2.38 | 2.82 | 3.46 | 3.97 | 5.32 |
| 18 | 2.34 | 2.77 | 3.37 | 3.86 | 5.13 |
| 19 | 2.31 | 2.72 | 3.30 | 3.76 | 4.97 |
| 20 | 2.28 | 2.68 | 3.23 | 3.68 | 4.82 |
| 21 | 2.25 | 2.64 | 3.17 | 3.60 | 4.70 |
| 22 | 2.23 | 2.60 | 3.12 | 3.54 | 4.58 |
| 23 | 2.20 | 2.57 | 3.07 | 3.47 | 4.48 |
| 24 | 2.18 | 2.54 | 3.03 | 3.42 | 4.39 |
| 25 | 2.16 | 2.51 | 2.99 | 3.37 | 4.31 |
| 40 | 2.00 | 2.29 | 2.66 | 2.95 | 3.64 |
| 60 | 1.92 | 2.17 | 2.50 | 2.74 | 3.32 |
| 120 | 1.83 | 2.05 | 2.34 | 2.54 | 3.02 |
| ∞ | 1.75 | 1.94 | 2.18 | 2.36 | 2.74 |
| **Numerator df = 15** | | | | | |
| Denominator, df = 1 | 246. | 985. | 6160. | 24600. | 616000. |
| 2 | 19.4 | 39.4 | 99.4 | 199. | 999. |
| 3 | 8.70 | 14.3 | 26.9 | 43.1 | 127. |
| 4 | 5.86 | 8.66 | 14.2 | 20.4 | 46.8 |
| 5 | 4.62 | 6.43 | 9.72 | 13.1 | 25.9 |
| 6 | 3.94 | 5.27 | 7.56 | 9.81 | 17.6 |
| 7 | 3.51 | 4.57 | 6.31 | 7.97 | 13.3 |
| 8 | 3.22 | 4.10 | 5.52 | 6.81 | 10.8 |
| 9 | 3.01 | 3.77 | 4.96 | 6.03 | 9.24 |
| 10 | 2.85 | 3.52 | 4.56 | 5.47 | 8.13 |
| 11 | 2.72 | 3.33 | 4.25 | 5.05 | 7.32 |
| 12 | 2.62 | 3.18 | 4.01 | 4.72 | 6.71 |
| 13 | 2.53 | 3.05 | 3.82 | 4.46 | 6.23 |
| 14 | 2.46 | 2.95 | 3.66 | 4.25 | 5.85 |
| 15 | 2.40 | 2.86 | 3.52 | 4.07 | 5.54 |
| 16 | 2.35 | 2.79 | 3.41 | 3.92 | 5.27 |
| 17 | 2.31 | 2.72 | 3.31 | 3.79 | 5.05 |
| 18 | 2.27 | 2.67 | 3.23 | 3.68 | 4.87 |
| 19 | 2.23 | 2.62 | 3.15 | 3.59 | 4.70 |
| 20 | 2.20 | 2.57 | 3.09 | 3.50 | 4.56 |
| 21 | 2.18 | 2.53 | 3.03 | 3.43 | 4.44 |
| 22 | 2.15 | 2.50 | 2.98 | 3.36 | 4.33 |
| 23 | 2.13 | 2.47 | 2.93 | 3.30 | 4.23 |
| 24 | 2.11 | 2.44 | 2.89 | 3.25 | 4.14 |
| 25 | 2.09 | 2.41 | 2.85 | 3.20 | 4.06 |
| 40 | 1.92 | 2.18 | 2.52 | 2.78 | 3.40 |
| 60 | 1.84 | 2.06 | 2.35 | 2.57 | 3.08 |
| 120 | 1.75 | 1.94 | 2.19 | 2.37 | 2.78 |
| ∞ | 1.67 | 1.83 | 2.04 | 2.19 | 2.51 |

(continued on next page)

# Table A13 (continued)  Critical values of *F*.

| $\alpha_1$ | 5% | 2.5% | 1% | 0.5% | 0.1% |
|---|---|---|---|---|---|
| $\alpha_2$ | 10% | 5% | 2% | 1% | 0.2% |
| | | | Numerator df = 20 | | |
| Denominator, df = 1 | 248. | 993. | 6210. | 24800. | 621000. |
| 2 | 19.4 | 39.4 | 99.4 | 199. | 999. |
| 3 | 8.66 | 14.2 | 26.7 | 42.8 | 126. |
| 4 | 5.80 | 8.56 | 14.0 | 20.2 | 46.1 |
| 5 | 4.56 | 6.33 | 9.55 | 12.9 | 25.4 |
| 6 | 3.87 | 5.17 | 7.40 | 9.59 | 17.1 |
| 7 | 3.44 | 4.47 | 6.16 | 7.75 | 12.9 |
| 8 | 3.15 | 4.00 | 5.36 | 6.61 | 10.5 |
| 9 | 2.94 | 3.67 | 4.81 | 5.83 | 8.90 |
| 10 | 2.77 | 3.42 | 4.41 | 5.27 | 7.80 |
| 11 | 2.65 | 3.23 | 4.10 | 4.86 | 7.01 |
| 12 | 2.54 | 3.07 | 3.86 | 4.53 | 6.40 |
| 13 | 2.46 | 2.95 | 3.66 | 4.27 | 5.93 |
| 14 | 2.39 | 2.84 | 3.51 | 4.06 | 5.56 |
| 15 | 2.33 | 2.76 | 3.37 | 3.88 | 5.25 |
| 16 | 2.28 | 2.68 | 3.26 | 3.73 | 4.99 |
| 17 | 2.23 | 2.62 | 3.16 | 3.61 | 4.78 |
| 18 | 2.19 | 2.56 | 3.08 | 3.50 | 4.59 |
| 19 | 2.16 | 2.51 | 3.00 | 3.40 | 4.43 |
| 20 | 2.12 | 2.46 | 2.94 | 3.32 | 4.29 |
| 21 | 2.10 | 2.42 | 2.88 | 3.24 | 4.17 |
| 22 | 2.07 | 2.39 | 2.83 | 3.18 | 4.06 |
| 23 | 2.05 | 2.36 | 2.78 | 3.12 | 3.96 |
| 24 | 2.03 | 2.33 | 2.74 | 3.06 | 3.87 |
| 25 | 2.01 | 2.30 | 2.70 | 3.01 | 3.79 |
| 40 | 1.84 | 2.07 | 2.37 | 2.60 | 3.14 |
| 60 | 1.75 | 1.94 | 2.20 | 2.39 | 2.83 |
| 120 | 1.66 | 1.82 | 2.03 | 2.19 | 2.53 |
| ∞ | 1.57 | 1.71 | 1.88 | 2.00 | 2.27 |
| | | | Numerator df = 24 | | |
| Denominator, df = 1 | 249. | 997. | 6230 | 24900. | 623000. |
| 2 | 19.5 | 39.5 | 99.5 | 199. | 999. |
| 3 | 8.64 | 14.1 | 26.6 | 42.6 | 126. |
| 4 | 5.77 | 8.51 | 13.9 | 20.0 | 45.8 |
| 5 | 4.53 | 6.28 | 9.47 | 12.8 | 25.1 |
| 6 | 3.84 | 5.12 | 7.31 | 9.47 | 16.9 |
| 7 | 3.41 | 4.41 | 6.07 | 7.64 | 12.7 |
| 8 | 3.12 | 3.95 | 5.28 | 6.50 | 10.3 |
| 9 | 2.90 | 3.61 | 4.73 | 5.73 | 8.72 |
| 10 | 2.74 | 3.37 | 4.33 | 5.17 | 7.64 |
| 11 | 2.61 | 3.17 | 4.02 | 4.76 | 6.85 |
| 12 | 2.51 | 3.02 | 3.78 | 4.43 | 6.25 |
| 13 | 2.42 | 2.89 | 3.59 | 4.17 | 5.78 |
| 14 | 2.35 | 2.79 | 3.43 | 3.96 | 5.41 |
| 15 | 2.29 | 2.70 | 3.29 | 3.79 | 5.10 |
| 16 | 2.24 | 2.63 | 3.18 | 3.64 | 4.85 |
| 17 | 2.19 | 2.56 | 3.08 | 3.51 | 4.63 |
| 18 | 2.15 | 2.50 | 3.00 | 3.40 | 4.45 |
| 19 | 2.11 | 2.45 | 2.92 | 3.31 | 4.29 |
| 20 | 2.08 | 2.41 | 2.86 | 3.22 | 4.15 |
| 21 | 2.05 | 2.37 | 2.80 | 3.15 | 4.03 |
| 22 | 2.03 | 2.33 | 2.75 | 3.08 | 3.92 |
| 23 | 2.01 | 2.30 | 2.70 | 3.02 | 3.82 |
| 24 | 1.98 | 2.27 | 2.66 | 2.97 | 3.74 |
| 25 | 1.96 | 2.24 | 2.62 | 2.92 | 3.66 |
| 40 | 1.79 | 2.01 | 2.29 | 2.50 | 3.01 |
| 60 | 1.70 | 1.88 | 2.12 | 2.29 | 2.69 |
| 120 | 1.61 | 1.76 | 1.95 | 2.09 | 2.40 |
| ∞ | 1.52 | 1.64 | 1.79 | 1.90 | 2.13 |

(continued on next page)

# Table A13 (continued)  Critical values of *F*.

| $\alpha_1$ | 5% | 2.5% | 1% | 0.5% | 0.1% |
|---|---|---|---|---|---|
| $\alpha_2$ | 10% | 5% | 2% | 1% | 0.2% |

| Numerator df = 30 | | | | | |
|---|---|---|---|---|---|
| Denominator, df = 1 | 250. | 1000. | 6260. | 25000. | 626000. |
| 2 | 19.5 | 39.5 | 99.5 | 199. | 999. |
| 3 | 8.62 | 14.1 | 26.5 | 42.5 | 125. |
| 4 | 5.75 | 8.46 | 13.8 | 19.9 | 45.4 |
| 5 | 4.50 | 6.23 | 9.38 | 12.7 | 24.9 |
| 6 | 3.81 | 5.07 | 7.23 | 9.36 | 16.7 |
| 7 | 3.38 | 4.36 | 5.99 | 7.53 | 12.5 |
| 8 | 3.08 | 3.89 | 5.20 | 6.40 | 10.1 |
| 9 | 2.86 | 3.56 | 4.65 | 5.62 | 8.55 |
| 10 | 2.70 | 3.31 | 4.25 | 5.07 | 7.47 |
| 11 | 2.57 | 3.12 | 3.94 | 4.65 | 6.68 |
| 12 | 2.47 | 2.96 | 3.70 | 4.33 | 6.09 |
| 13 | 2.38 | 2.84 | 3.51 | 4.07 | 5.63 |
| 14 | 2.31 | 2.73 | 3.35 | 3.86 | 5.25 |
| 15 | 2.25 | 2.64 | 3.21 | 3.69 | 4.95 |
| 16 | 2.19 | 2.57 | 3.10 | 3.54 | 4.70 |
| 17 | 2.15 | 2.50 | 3.00 | 3.41 | 4.48 |
| 18 | 2.11 | 2.44 | 2.92 | 3.30 | 4.30 |
| 19 | 2.07 | 2.39 | 2.84 | 3.21 | 4.14 |
| 20 | 2.04 | 2.35 | 2.78 | 3.12 | 4.00 |
| 21 | 2.01 | 2.31 | 2.72 | 3.05 | 3.88 |
| 22 | 1.98 | 2.27 | 2.67 | 2.98 | 3.78 |
| 23 | 1.96 | 2.24 | 2.62 | 2.92 | 3.68 |
| 24 | 1.94 | 2.21 | 2.58 | 2.87 | 3.59 |
| 25 | 1.92 | 2.18 | 2.54 | 2.82 | 3.52 |
| 40 | 1.74 | 1.94 | 2.20 | 2.40 | 2.87 |
| 60 | 1.65 | 1.82 | 2.03 | 2.19 | 2.55 |
| 120 | 1.55 | 1.69 | 1.86 | 1.98 | 2.26 |
| ∞ | 1.46 | 1.57 | 1.70 | 1.79 | 1.99 |

| Numerator df = 40 | | | | | |
|---|---|---|---|---|---|
| Denominator, df = 1 | 251. | 1010. | 6290. | 25100. | 629000. |
| 2 | 19.5 | 39.5 | 99.5 | 199. | 999. |
| 3 | 8.59 | 14.0 | 26.4 | 42.3 | 125. |
| 4 | 5.72 | 8.41 | 13.7 | 19.8 | 45.1 |
| 5 | 4.46 | 6.18 | 9.29 | 12.5 | 24.6 |
| 6 | 3.77 | 5.01 | 7.14 | 9.24 | 16.4 |
| 7 | 3.34 | 4.31 | 5.91 | 7.42 | 12.3 |
| 8 | 3.04 | 3.84 | 5.12 | 6.29 | 9.92 |
| 9 | 2.83 | 3.51 | 4.57 | 5.52 | 8.37 |
| 10 | 2.66 | 3.26 | 4.17 | 4.97 | 7.30 |
| 11 | 2.53 | 3.06 | 3.86 | 4.55 | 6.52 |
| 12 | 2.43 | 2.91 | 3.62 | 4.23 | 5.93 |
| 13 | 2.34 | 2.78 | 3.43 | 3.97 | 5.47 |
| 14 | 2.27 | 2.67 | 3.27 | 3.76 | 5.10 |
| 15 | 2.20 | 2.59 | 3.13 | 3.58 | 4.80 |
| 16 | 2.15 | 2.51 | 3.02 | 3.44 | 4.54 |
| 17 | 2.10 | 2.44 | 2.92 | 3.31 | 4.33 |
| 18 | 2.06 | 2.38 | 2.84 | 3.20 | 4.15 |
| 19 | 2.03 | 2.33 | 2.76 | 3.11 | 3.99 |
| 20 | 1.99 | 2.29 | 2.69 | 3.02 | 3.86 |
| 21 | 1.96 | 2.25 | 2.64 | 2.95 | 3.74 |
| 22 | 1.94 | 2.21 | 2.58 | 2.88 | 3.63 |
| 23 | 1.91 | 2.18 | 2.54 | 2.82 | 3.53 |
| 24 | 1.89 | 2.15 | 2.49 | 2.77 | 3.45 |
| 25 | 1.87 | 2.12 | 2.45 | 2.72 | 3.37 |
| 40 | 1.69 | 1.88 | 2.11 | 2.30 | 2.73 |
| 60 | 1.59 | 1.74 | 1.94 | 2.08 | 2.41 |
| 120 | 1.50 | 1.61 | 1.76 | 1.87 | 2.11 |
| ∞ | 1.39 | 1.48 | 1.59 | 1.67 | 1.84 |

(continued on next page)

**Table A13 (continued)  Critical values of _F_.**

| $\alpha_1$ | 5% | 2.5% | 1% | 0.5% | 0.1% |
|---|---|---|---|---|---|
| $\alpha_2$ | 10% | 5% | 2% | 1% | 0.2% |

| Numerator df = 60 | | | | | |
|---|---|---|---|---|---|
| Denominator, df = 1 | 252. | 1010. | 6310. | 25300. | 631000. |
| 2 | 19.5 | 39.5 | 99.5 | 199. | 999. |
| 3 | 8.57 | 14.0 | 26.3 | 42.1 | 124. |
| 4 | 5.69 | 8.36 | 13.7 | 19.6 | 44.7 |
| 5 | 4.43 | 6.12 | 9.20 | 12.4 | 24.3 |
| 6 | 3.74 | 4.96 | 7.06 | 9.12 | 16.2 |
| 7 | 3.30 | 4.25 | 5.82 | 7.31 | 12.1 |
| 8 | 3.01 | 3.78 | 5.03 | 6.18 | 9.73 |
| 9 | 2.79 | 3.45 | 4.48 | 5.41 | 8.19 |
| 10 | 2.62 | 3.20 | 4.08 | 4.86 | 7.12 |
| 11 | 2.49 | 3.00 | 3.78 | 4.45 | 6.35 |
| 12 | 2.38 | 2.85 | 3.54 | 4.12 | 5.76 |
| 13 | 2.30 | 2.72 | 3.34 | 3.87 | 5.30 |
| 14 | 2.22 | 2.61 | 3.18 | 3.66 | 4.94 |
| 15 | 2.16 | 2.52 | 3.05 | 3.48 | 4.64 |
| 16 | 2.11 | 2.45 | 2.93 | 3.33 | 4.39 |
| 17 | 2.06 | 2.38 | 2.83 | 3.21 | 4.18 |
| 18 | 2.02 | 2.32 | 2.75 | 3.10 | 4.00 |
| 19 | 1.98 | 2.27 | 2.67 | 3.00 | 3.84 |
| 20 | 1.95 | 2.22 | 2.61 | 2.92 | 3.70 |
| 21 | 1.92 | 2.18 | 2.55 | 2.84 | 3.58 |
| 22 | 1.89 | 2.14 | 2.50 | 2.77 | 3.48 |
| 23 | 1.86 | 2.11 | 2.45 | 2.71 | 3.38 |
| 24 | 1.84 | 2.08 | 2.40 | 2.66 | 3.29 |
| 25 | 1.82 | 2.05 | 2.36 | 2.61 | 3.22 |
| 40 | 1.64 | 1.80 | 2.02 | 2.18 | 2.57 |
| 60 | 1.53 | 1.67 | 1.84 | 1.96 | 2.25 |
| 120 | 1.43 | 1.53 | 1.66 | 1.75 | 1.95 |
| $\infty$ | 1.32 | 1.39 | 1.47 | 1.53 | 1.66 |

| Numerator df = 120 | | | | | |
|---|---|---|---|---|---|
| Denominator, df = 1 | 253. | 1010. | 6340. | 25400. | 634000. |
| 2 | 19.5 | 39.5 | 99.5 | 199. | 999. |
| 3 | 8.55 | 13.9 | 26.2 | 42.0 | 124. |
| 4 | 5.66 | 8.31 | 13.6 | 19.5 | 44.4 |
| 5 | 4.40 | 6.07 | 9.11 | 12.3 | 24.1 |
| 6 | 3.70 | 4.90 | 6.97 | 9.00 | 16.0 |
| 7 | 3.27 | 4.20 | 5.74 | 7.19 | 11.9 |
| 8 | 2.97 | 3.73 | 4.95 | 6.06 | 9.53 |
| 9 | 2.75 | 3.39 | 4.40 | 5.30 | 8.00 |
| 10 | 2.58 | 3.14 | 4.00 | 4.75 | 6.94 |
| 11 | 2.45 | 2.94 | 3.69 | 4.34 | 6.18 |
| 12 | 2.34 | 2.79 | 3.45 | 4.01 | 5.59 |
| 13 | 2.25 | 2.66 | 3.25 | 3.76 | 5.14 |
| 14 | 2.18 | 2.55 | 3.09 | 3.55 | 4.77 |
| 15 | 2.11 | 2.46 | 2.96 | 3.37 | 4.47 |
| 16 | 2.06 | 2.38 | 2.84 | 3.22 | 4.23 |
| 17 | 2.01 | 2.32 | 2.75 | 3.10 | 4.02 |
| 18 | 1.97 | 2.26 | 2.66 | 2.99 | 3.84 |
| 19 | 1.93 | 2.20 | 2.58 | 2.89 | 3.68 |
| 20 | 1.90 | 2.16 | 2.52 | 2.81 | 3.54 |
| 21 | 1.87 | 2.11 | 2.46 | 2.73 | 3.42 |
| 22 | 1.84 | 2.08 | 2.40 | 2.66 | 3.32 |
| 23 | 1.81 | 2.04 | 2.35 | 2.60 | 3.22 |
| 24 | 1.79 | 2.01 | 2.31 | 2.55 | 3.14 |
| 25 | 1.77 | 1.98 | 2.27 | 2.50 | 3.06 |
| 40 | 1.58 | 1.72 | 1.92 | 2.06 | 2.41 |
| 60 | 1.47 | 1.58 | 1.73 | 1.83 | 2.08 |
| 120 | 1.35 | 1.43 | 1.53 | 1.61 | 1.77 |
| $\infty$ | 1.22 | 1.27 | 1.32 | 1.36 | 1.45 |

(continued on next page)

| $\alpha_1$ | 5% | 2.5% | 1% | 0.5% | 0.1% |
|---|---|---|---|---|---|
| $\alpha_2$ | 10% | 5% | 2% | 1% | 0.2% |
| | | | Numerator df $= \infty$ | | |
| Denominator, df = 1 | 254. | 1020. | 6370. | 25500. | 637000. |
| 2 | 19.5 | 39.5 | 99.5 | 199. | 999. |
| 3 | 8.53 | 13.9 | 26.1 | 41.8 | 123. |
| 4 | 5.63 | 8.26 | 13.5 | 19.3 | 44.0 |
| 5 | 4.37 | 6.02 | 9.02 | 12.1 | 23.8 |
| 6 | 3.67 | 4.85 | 6.88 | 8.88 | 15.7 |
| 7 | 3.23 | 4.14 | 5.65 | 7.08 | 11.7 |
| 8 | 2.93 | 3.67 | 4.86 | 5.95 | 9.33 |
| 9 | 2.71 | 3.33 | 4.31 | 5.19 | 7.81 |
| 10 | 2.54 | 3.08 | 3.91 | 4.64 | 6.76 |
| 11 | 2.40 | 2.88 | 3.60 | 4.23 | 6.00 |
| 12 | 2.30 | 2.72 | 3.36 | 3.90 | 5.42 |
| 13 | 2.21 | 2.60 | 3.17 | 3.65 | 4.97 |
| 14 | 2.13 | 2.49 | 3.00 | 3.44 | 4.60 |
| 15 | 2.07 | 2.40 | 2.87 | 3.26 | 4.31 |
| 16 | 2.01 | 2.32 | 2.75 | 3.11 | 4.06 |
| 17 | 1.96 | 2.25 | 2.65 | 2.98 | 3.85 |
| 18 | 1.92 | 2.19 | 2.57 | 2.87 | 3.67 |
| 19 | 1.88 | 2.13 | 2.49 | 2.78 | 3.51 |
| 20 | 1.84 | 2.09 | 2.42 | 2.69 | 3.38 |
| 21 | 1.81 | 2.04 | 2.36 | 2.61 | 3.26 |
| 22 | 1.78 | 2.00 | 2.31 | 2.55 | 3.15 |
| 23 | 1.76 | 1.97 | 2.26 | 2.48 | 3.05 |
| 24 | 1.73 | 1.94 | 2.21 | 2.43 | 2.97 |
| 25 | 1.71 | 1.91 | 2.17 | 2.38 | 2.89 |
| 40 | 1.51 | 1.64 | 1.80 | 1.93 | 2.23 |
| 60 | 1.39 | 1.48 | 1.60 | 1.69 | 1.89 |
| 120 | 1.25 | 1.31 | 1.38 | 1.43 | 1.54 |
| $\infty$ | 1.00 | 1.00 | 1.00 | 1.00 | 1.00 |

**Table A14** The sign test for paired samples: critical values of *S*.

1. Find row corresponding to number of pairs (*n*).
2. For **one-tailed** test use calculated value of *S* which is **predicted to be the larger** and critical value in column for **significance level** $\alpha_1$. (N.B. Some statistical packages and tables use the smaller value of *S*.)
3. For **two-tailed** test use calculated value of S which **is the larger** of the two and critical value in column for **significance level** $\alpha_2$. (N.B. Some statistical packages and tables use the smaller value of *S*.)
4. **Reject the null hypothesis if calculated value is equal to, or greater than, critical value.** (N.B. Some statistical packages and tables use the smaller value of *S* and a different decision rule.)

| $\alpha_1$ | 5% | 2.5% | 1% | 0.5% | 0.05% |
|---|---|---|---|---|---|
| $\alpha_2$ | 10% | 5% | 2% | 1% | 0.1% |
| n = 5 | 5 | – | – | – | – |
| 6 | 6 | 6 | – | – | – |
| 7 | 7 | 7 | 7 | – | – |
| 8 | 7 | 8 | 8 | 8 | – |
| 9 | 8 | 9 | 9 | 9 | – |
| 10 | 9 | 9 | 10 | 10 | – |
| 11 | 9 | 10 | 10 | 11 | 11 |
| 12 | 10 | 10 | 11 | 11 | 12 |
| 13 | 10 | 11 | 12 | 12 | 13 |
| 14 | 11 | 12 | 12 | 13 | 14 |
| 15 | 12 | 12 | 13 | 13 | 14 |
| 16 | 12 | 13 | 14 | 14 | 15 |
| 17 | 13 | 13 | 14 | 15 | 16 |
| 18 | 13 | 14 | 15 | 15 | 17 |
| 19 | 14 | 15 | 15 | 16 | 17 |
| 20 | 15 | 15 | 16 | 17 | 18 |
| 21 | 15 | 16 | 17 | 17 | 19 |
| 22 | 16 | 17 | 17 | 18 | 19 |
| 23 | 16 | 17 | 18 | 19 | 20 |
| 24 | 17 | 18 | 19 | 19 | 21 |
| 25 | 18 | 18 | 19 | 20 | 21 |
| 26 | 18 | 19 | 20 | 20 | 22 |
| 27 | 19 | 20 | 20 | 21 | 23 |
| 28 | 19 | 20 | 21 | 22 | 23 |
| 29 | 20 | 21 | 22 | 22 | 24 |
| 30 | 20 | 21 | 22 | 23 | 25 |
| 31 | 21 | 22 | 23 | 24 | 25 |
| 32 | 22 | 23 | 24 | 24 | 26 |
| 33 | 22 | 23 | 24 | 25 | 27 |
| 34 | 23 | 24 | 25 | 25 | 27 |
| 35 | 23 | 24 | 25 | 26 | 28 |
| 36 | 24 | 25 | 26 | 27 | 29 |
| 37 | 24 | 25 | 27 | 27 | 29 |
| 38 | 25 | 26 | 27 | 28 | 30 |
| 39 | 26 | 27 | 28 | 28 | 31 |
| 40 | 26 | 27 | 28 | 29 | 31 |
| 41 | 27 | 28 | 29 | 30 | 32 |
| 42 | 27 | 28 | 29 | 30 | 32 |
| 43 | 28 | 29 | 30 | 31 | 33 |
| 44 | 28 | 29 | 31 | 31 | 34 |
| 45 | 29 | 30 | 31 | 32 | 34 |
| 46 | 30 | 31 | 32 | 33 | 35 |
| 47 | 30 | 31 | 32 | 33 | 36 |
| 48 | 31 | 32 | 33 | 34 | 36 |
| 49 | 31 | 32 | 34 | 34 | 37 |
| 50 | 32 | 33 | 34 | 35 | 37 |

## Table A15 Wilcoxon's signed rank test: critical values of *T*.

1. Find row corresponding to number of pairs (*n*).
2. For **one-tailed** test use calculated value of *T* which is **predicted to be the larger** and critical value in column for **significance level** $\alpha_1$. (N.B. Some statistical packages and tables use the smaller value of *T*.)
3. For **two-tailed** test use calculated value of *T* which **is the larger** of the two and critical value in column for **significance level** $\alpha_2$. (N.B. Some statistical packages and tables use the smaller value of *T*.)
4. **Reject null hypothesis if calculated value is equal to, or greater than, critical value.** (N.B. Some statistical packages and tables use the smaller value of *T* and a different decision rule.)

| $\alpha_1$ | 5% | 2.5% | 1% | 0.5% | 0.05% |
|---|---|---|---|---|---|
| $\alpha_2$ | 10% | 5% | 2% | 1% | 0.1% |
| n = 4 | – | – | – | – | – |
| 5 | 15 | – | – | – | – |
| 6 | 19 | 21 | – | – | – |
| 7 | 25 | 26 | 28 | | – |
| 8 | 31 | 33 | 35 | 36 | – |
| 9 | 37 | 40 | 42 | 44 | – |
| 10 | 45 | 47 | 50 | 52 | – |
| 11 | 53 | 56 | 59 | 61 | 66 |
| 12 | 61 | 65 | 69 | 71 | 77 |
| 13 | 70 | 74 | 79 | 82 | 89 |
| 14 | 80 | 84 | 90 | 93 | 101 |
| 15 | 90 | 95 | 101 | 105 | 114 |
| 16 | 101 | 107 | 113 | 117 | 128 |
| 17 | 112 | 119 | 126 | 130 | 142 |
| 18 | 124 | 131 | 139 | 144 | 157 |
| 19 | 137 | 144 | 153 | 158 | 172 |
| 20 | 150 | 158 | 167 | 173 | 189 |
| 21 | 164 | 173 | 182 | 189 | 206 |
| 22 | 178 | 188 | 198 | 205 | 223 |
| 23 | 193 | 203 | 214 | 222 | 241 |
| 24 | 209 | 219 | 231 | 239 | 260 |
| 25 | 225 | 236 | 249 | 257 | 280 |
| 26 | 241 | 253 | 267 | 276 | 300 |
| 27 | 259 | 271 | 286 | 295 | 321 |
| 28 | 276 | 290 | 305 | 315 | 342 |
| 29 | 295 | 309 | 325 | 335 | 364 |
| 30 | 314 | 328 | 345 | 356 | 387 |
| 31 | 333 | 349 | 366 | 378 | 410 |
| 32 | 353 | 369 | 388 | 400 | 434 |
| 33 | 374 | 391 | 410 | 423 | 459 |
| 34 | 395 | 413 | 433 | 447 | 484 |
| 35 | 417 | 435 | 457 | 471 | 510 |
| 36 | 439 | 458 | 481 | 495 | 536 |
| 37 | 462 | 482 | 505 | 521 | 563 |
| 38 | 485 | 506 | 530 | 547 | 591 |
| 39 | 509 | 531 | 556 | 573 | 619 |
| 40 | 534 | 556 | 582 | 600 | 648 |
| 41 | 559 | 582 | 609 | 628 | 678 |
| 42 | 584 | 609 | 637 | 656 | 708 |
| 43 | 610 | 636 | 665 | 685 | 739 |
| 44 | 637 | 663 | 694 | 714 | 770 |
| 45 | 664 | 692 | 723 | 744 | 802 |
| 46 | 692 | 720 | 753 | 774 | 835 |
| 47 | 721 | 750 | 783 | 806 | 868 |
| 48 | 750 | 780 | 814 | 837 | 902 |
| 49 | 779 | 810 | 846 | 870 | 936 |
| 50 | 809 | 841 | 878 | 902 | 971 |

**Table A16  Kruskal–Wallis test: critical values of $H$ (for samples of equal sizes).**

1. Find subtable for value of $k$ ( = total number of samples being compared) (see p. 277 if $k > 6$).
2. Find row for sample size $n$ (see p. 278 if sample sizes are unequal).
3. Compare calculated value to critical value for **significance level** $\alpha$.
4. **Reject null hypothesis if calculated value equal to, or greater than, critical value.**

| | $k = 3$ | | | $k = 4$ | | | $k = 5$ | | | $k = 6$ | | |
|---|---|---|---|---|---|---|---|---|---|---|---|---|
| $\alpha$ | 10% | 5% | 1% | 10% | 5% | 1% | 10% | 5% | 1% | 10% | 5% | 1% |
| n = 2 | 4.571 | – | – | 5.667 | 6.167 | 6.667 | 6.982 | 7.418 | 8.291 | 8.154 | 8.846 | 9.846 |
| 3 | 4.622 | 5.600 | 7.200 | 6.026 | 7.000 | 8.538 | 7.333 | 8.333 | 10.20 | 8.620 | 9.789 | 11.82 |
| 4 | 4.654 | 5.692 | 7.654 | 6.088 | 7.235 | 9.287 | 7.457 | 8.685 | 11.07 | 8.800 | 10.14 | 12.72 |
| 5 | 4.560 | 5.780 | 8.000 | 6.120 | 7.377 | 9.789 | 7.532 | 8.876 | 11.57 | 8.902 | 10.36 | 13.26 |
| 6 | 4.643 | 5.801 | 8.222 | 6.127 | 7.453 | 10.09 | 7.557 | 9.002 | 11.91 | 8.958 | 10.50 | 13.60 |
| 7 | 4.594 | 5.819 | 8.378 | 6.141 | 7.501 | 10.25 | 7.600 | 9.080 | 12.14 | 8.992 | 10.59 | 13.84 |
| 8 | 4.595 | 5.805 | 8.465 | 6.148 | 7.534 | 10.42 | 7.624 | 9.126 | 12.29 | 9.037 | 10.66 | 13.99 |
| 9 | 4.586 | 5.831 | 8.529 | 6.161 | 7.557 | 10.53 | 7.637 | 9.166 | 12.41 | 9.057 | 10.71 | 14.13 |
| 10 | 4.581 | 5.853 | 8.607 | 6.167 | 7.586 | 10.62 | 7.650 | 9.200 | 12.50 | 9.078 | 10.75 | 14.24 |
| 11 | 4.587 | 5.885 | 8.648 | 6.163 | 7.623 | 10.69 | 7.660 | 9.242 | 12.58 | 9.093 | 10.76 | 14.32 |
| 12 | 4.578 | 5.872 | 8.712 | 6.185 | 7.629 | 10.75 | 7.675 | 9.274 | 12.63 | 9.105 | 10.79 | 14.38 |
| 13 | 4.601 | 5.901 | 8.735 | 6.191 | 7.645 | 10.80 | 7.685 | 9.303 | 12.69 | 9.115 | 10.83 | 14.44 |
| 14 | 4.592 | 5.896 | 8.754 | 6.198 | 7.658 | 10.84 | 7.695 | 9.307 | 12.74 | 9.125 | 10.84 | 14.49 |
| 15 | 4.591 | 5.902 | 8.821 | 6.201 | 7.676 | 10.87 | 7.701 | 9.302 | 12.77 | 9.133 | 10.86 | 14.53 |
| 16 | 4.595 | 5.909 | 8.822 | 6.205 | 7.678 | 10.90 | 7.705 | 9.313 | 12.79 | 9.140 | 10.88 | 14.56 |
| 17 | 4.593 | 5.915 | 8.856 | 6.206 | 7.682 | 10.92 | 7.709 | 9.325 | 12.83 | 9.144 | 10.88 | 14.60 |
| 18 | 4.596 | 5.932 | 8.865 | 6.212 | 7.698 | 10.95 | 7.714 | 9.334 | 12.85 | 9.149 | 10.89 | 14.63 |
| 19 | 4.598 | 5.923 | 8.887 | 6.212 | 7.701 | 10.98 | 7.717 | 9.342 | 12.87 | 9.156 | 10.90 | 14.64 |
| 20 | 4.594 | 5.926 | 8.905 | 6.216 | 7.703 | 10.98 | 7.719 | 9.353 | 12.91 | 9.159 | 10.92 | 14.67 |
| 21 | 4.597 | 5.930 | 8.918 | 6.218 | 7.709 | 11.01 | 7.723 | 9.356 | 12.92 | 9.164 | 10.93 | 14.70 |
| 22 | 4.597 | 5.932 | 8.928 | 6.215 | 7.714 | 11.03 | 7.724 | 9.362 | 12.92 | 9.168 | 10.94 | 14.72 |
| 23 | 4.598 | 5.937 | 8.947 | 6.220 | 7.719 | 11.03 | 7.727 | 9.368 | 12.94 | 9.171 | 10.93 | 14.74 |
| 24 | 4.598 | 5.936 | 8.964 | 6.221 | 7.724 | 11.06 | 7.729 | 9.375 | 12.96 | 9.170 | 10.93 | 14.74 |
| 25 | 4.599 | 5.942 | 8.975 | 6.222 | 7.727 | 11.07 | 7.730 | 9.377 | 12.96 | 9.177 | 10.94 | 14.77 |
| ∞ | 4.605 | 5.991 | 9.210 | 6.251 | 7.815 | 11.34 | 7.779 | 9.488 | 13.28 | 9.236 | 11.07 | 15.09 |

**Table A17 Critical values of $Q$ for multiple comparisons following the Kruskal–Wallis test.**

1. Find row for correct value of $k$ (the total number of samples being compared)
2. Compare calculated value to critical value for **significance level $\alpha$.**
3. **Reject null hypothesis if calculated value equal to, or greater than, critical value.**

| $\alpha$ | 10% | 5% | 1% | 0.01% |
|---|---|---|---|---|
| $k = 2$ | 1.645 | 1.960 | 2.576 | 3.291 |
| 3 | 2.128 | 2.394 | 2.936 | 3.588 |
| 4 | 2.394 | 2.639 | 3.144 | 3.765 |
| 5 | 2.576 | 2.807 | 3.291 | 3.891 |
| 6 | 2.713 | 2.936 | 3.403 | 3.988 |
| 7 | 2.823 | 3.038 | 3.494 | 4.067 |
| 8 | 2.914 | 3.124 | 3.570 | 4.134 |
| 9 | 2.992 | 3.197 | 3.635 | 4.191 |
| 10 | 3.059 | 3.261 | 3.692 | 4.241 |
| 11 | 3.119 | 3.317 | 3.743 | 4.286 |
| 12 | 3.172 | 3.368 | 3.789 | 4.326 |
| 13 | 3.220 | 3.414 | 3.830 | 4.363 |
| 14 | 3.264 | 3.456 | 3.868 | 4.397 |
| 15 | 3.304 | 3.494 | 3.902 | 4.428 |
| 16 | 3.342 | 3.529 | 3.935 | 4.456 |
| 17 | 3.376 | 3.562 | 3.965 | 4.483 |
| 18 | 3.409 | 3.593 | 3.993 | 4.508 |
| 19 | 3.439 | 3.662 | 4.019 | 4.532 |
| 20 | 3.467 | 3.649 | 4.044 | 4.554 |

**Table A18 Critical values of $F_{max}$.**

1. Find subtable for value of $k$ (the total number of samples).
2. Find row for correct degrees of freedom (df), which will be one less than sample size ($n - 1$). If two samples have different sizes use degrees of freedom for smaller sample.
3. Compare calculated value to critical value in column for **significance level α**.
4. **Reject null hypothesis if calculated value equal to, or greater than, critical value.**

| | $k = 2$ | | $k = 3$ | | $k = 4$ | | $k = 5$ | | $k = 6$ | | $k = 7$ | |
|---|---|---|---|---|---|---|---|---|---|---|---|---|
| α | 5% | 1% | 5% | 1% | 5% | 1% | 5% | 1% | 5% | 1% | 5% | 1% |
| df = 2 | 39.0 | 199.0 | 87.5 | 448.0 | 142.0 | 729.0 | 202.0 | 1036.0 | 266.0 | 1362.0 | 333.0 | 1705.0 |
| 3 | 15.4 | 47.5 | 27.8 | 85.0 | 39.2 | 120.0 | 50.7 | 151.0 | 62.0 | 184.0 | 72.9 | 21(6) |
| 4 | 9.60 | 23.2 | 15.5 | 37.0 | 20.6 | 49.0 | 25.2 | 59.0 | 29.5 | 69.0 | 33.6 | 79.0 |
| 5 | 7.15 | 14.9 | 10.8 | 22.0 | 13.7 | 28.0 | 16.3 | 33.0 | 18.7 | 38.0 | 20.8 | 42.0 |
| 6 | 5.82 | 11.1 | 8.38 | 15.5 | 10.4 | 19.1 | 12.1 | 22.0 | 13.7 | 25.0 | 15.0 | 27.0 |
| 7 | 4.99 | 8.89 | 6.94 | 12.1 | 8.44 | 14.5 | 9.70 | 16.5 | 10.8 | 18.4 | 11.8 | 20.0 |
| 8 | 4.43 | 7.50 | 6.00 | 9.9 | 7.18 | 11.7 | 8.12 | 13.2 | 9.03 | 14.5 | 9.78 | 15.8 |
| 9 | 4.03 | 6.54 | 5.34 | 8.5 | 6.31 | 9.9 | 7.11 | 11.1 | 7.80 | 12.1 | 8.41 | 13.1 |
| 10 | 3.72 | 5.85 | 4.85 | 7.4 | 5.67 | 8.6 | 6.34 | 9.6 | 6.92 | 10.4 | 7.42 | 11.1 |
| 12 | 3.28 | 4.91 | 4.16 | 6.1 | 4.79 | 6.9 | 5.30 | 7.6 | 5.72 | 8.2 | 6.09 | 8.7 |
| 15 | 2.86 | 4.07 | 3.54 | 4.9 | 4.01 | 5.5 | 4.37 | 6.1 | 4.68 | 6.4 | 4.95 | 6.7 |
| 20 | 2.46 | 3.32 | 2.95 | 3.8 | 3.29 | 4.3 | 3.54 | 4.6 | 3.76 | 4.9 | 3.94 | 5.1 |
| 30 | 2.07 | 2.63 | 2.40 | 3.0 | 2.61 | 3.3 | 2.78 | 3.4 | 2.91 | 3.6 | 3.02 | 3.7 |
| 60 | 1.67 | 1.96 | 1.85 | 2.2 | 1.96 | 2.3 | 2.04 | 2.4 | 2.11 | 2.4 | 2.17 | 2.5 |
| ∞ | 1.00 | 1.00 | 1.00 | 1.00 | 1.00 | 1.00 | 1.00 | 1.00 | 1.00 | 1.00 | 1.00 | 1.00 |

| | $k = 8$ | | $k = 9$ | | $k = 10$ | | $k = 11$ | | $k = 12$ | |
|---|---|---|---|---|---|---|---|---|---|---|
| α | 5% | 1% | 5% | 1% | 5% | 1% | 5% | 1% | 5% | 1% |
| df = 2 | 403. | 2063. | 475. | 2432. | 550. | 2813. | 626. | 3204. | 704. | 3605. |
| 3 | 83.5 | 24(9) | 93.9 | 28(1) | 104. | 31(0) | 114. | 33(7) | 124. | 36(1) |
| 4 | 37.5 | 89. | 41.1 | 97. | 44.6 | 106. | 48.0 | 113. | 51.4 | 120. |
| 5 | 22.9 | 46. | 24.7 | 50. | 26.5 | 54. | 28.2 | 57. | 29.9 | 60. |
| 6 | 16.3 | 30. | 17.5 | 32. | 18.6 | 34. | 19.7 | 36. | 20.7 | 37. |
| 7 | 12.7 | 22. | 13.5 | 23. | 14.3 | 24. | 15.1 | 26. | 15.8 | 27. |
| 8 | 10.5 | 16.9 | 11.1 | 17.9 | 11.7 | 18.9 | 12.2 | 19.8 | 12.7 | 21. |
| 9 | 8.95 | 13.9 | 9.45 | 14.7 | 9.91 | 15.3 | 10.3 | 16.0 | 10.7 | 16.6 |
| 10 | 7.87 | 11.8 | 8.28 | 12.4 | 8.66 | 12.9 | 9.01 | 13.4 | 9.34 | 13.9 |
| 12 | 6.42 | 9.1 | 6.72 | 9.5 | 7.00 | 9.9 | 7.25 | 10.2 | 7.48 | 10.6 |
| 15 | 5.19 | 7.1 | 5.40 | 7.3 | 5.59 | 7.5 | 5.77 | 7.8 | 5.93 | 8.0 |
| 20 | 4.10 | 5.3 | 4.24 | 5.5 | 4.37 | 5.6 | 4.49 | 5.8 | 4.59 | 5.9 |
| 30 | 3.12 | 3.8 | 3.21 | 3.9 | 3.29 | 4.0 | 3.36 | 4.1 | 3.39 | 4.2 |
| 60 | 2.22 | 2.5 | 2.26 | 2.6 | 2.30 | 2.6 | 2.33 | 2.7 | 2.36 | 2.7 |
| ∞ | 1.00 | 1.00 | 1.00 | 1.00 | 1.00 | 1.00 | 1.00 | 1.00 | 1.00 | 1.00 |

**Table A19 Critical values of q for multiple comparisons following analysis of variance.**

1. Find subtable for value of k (the total number of means being compared)
2. Find row for correct degrees of freedom (df). These will be the same as those used in the denominator when calculating F.
3. Compare calculated value to critical value in column for **significance level α**.
4. **Reject null hypothesis if calculated value equal to, or greater than, critical value.**

| | k = 2 | | k = 3 | | k = 4 | | k = 5 | | k = 6 | | k = 7 | |
|---|---|---|---|---|---|---|---|---|---|---|---|---|
| α | 5% | 1% | 5% | 1% | 5% | 1% | 5% | 1% | 5% | 1% | 5% | 1% |
| df = 1 | 17.97 | 90.03 | 26.98 | 135.0 | 32.82 | 164.3 | 37.08 | 185.6 | 40.41 | 202.2 | 43.12 | 215.8 |
| 2 | 6.085 | 14.04 | 8.331 | 19.02 | 9.798 | 22.29 | 10.88 | 24.72 | 11.75 | 26.63 | 12.44 | 28.20 |
| 3 | 4.501 | 8.261 | 5.910 | 10.62 | 6.825 | 12.17 | 7.502 | 13.33 | 8.037 | 14.24 | 8.478 | 15.00 |
| 4 | 3.927 | 6.512 | 5.040 | 8.120 | 5.757 | 9.173 | 6.287 | 9.958 | 6.707 | 10.58 | 7.053 | 11.10 |
| 5 | 3.635 | 5.702 | 4.602 | 6.976 | 5.218 | 7.804 | 5.673 | 8.421 | 6.033 | 8.913 | 6.330 | 9.321 |
| 6 | 3.461 | 5.243 | 4.339 | 6.331 | 4.896 | 7.033 | 5.305 | 7.556 | 5.628 | 7.973 | 5.895 | 8.318 |
| 7 | 3.344 | 4.949 | 4.165 | 5.919 | 4.681 | 6.543 | 5.060 | 7.005 | 5.359 | 7.373 | 5.606 | 7.679 |
| 8 | 3.261 | 4.746 | 4.041 | 5.635 | 4.529 | 6.204 | 4.886 | 6.625 | 5.167 | 6.960 | 5.399 | 7.237 |
| 9 | 3.199 | 4.596 | 3.949 | 5.428 | 4.415 | 5.957 | 4.756 | 6.348 | 5.024 | 6.658 | 5.244 | 6.915 |
| 10 | 3.151 | 4.482 | 3.877 | 5.270 | 4.327 | 5.769 | 4.654 | 6.136 | 4.912 | 6.428 | 5.124 | 6.669 |
| 11 | 3.113 | 4.392 | 3.820 | 5.146 | 4.256 | 5.621 | 4.574 | 5.970 | 4.823 | 6.247 | 5.028 | 6.476 |
| 12 | 3.082 | 4.320 | 3.773 | 5.046 | 4.199 | 5.502 | 4.508 | 5.836 | 4.751 | 6.101 | 4.950 | 6.321 |
| 13 | 3.055 | 4.260 | 3.735 | 4.964 | 4.151 | 5.404 | 4.453 | 5.727 | 4.690 | 5.981 | 4.885 | 6.192 |
| 14 | 3.033 | 4.210 | 3.702 | 4.895 | 4.111 | 5.322 | 4.407 | 5.634 | 4.639 | 5.881 | 4.829 | 6.085 |
| 15 | 3.014 | 4.168 | 3.674 | 4.836 | 4.076 | 5.252 | 4.367 | 5.556 | 4.595 | 5.796 | 4.782 | 5.994 |
| 16 | 2.998 | 4.131 | 3.649 | 4.786 | 4.046 | 5.192 | 4.333 | 5.489 | 4.557 | 5.722 | 4.741 | 5.915 |
| 17 | 2.984 | 4.099 | 3.628 | 4.742 | 4.020 | 5.140 | 4.303 | 5.430 | 4.524 | 5.659 | 4.705 | 5.847 |
| 18 | 2.971 | 4.071 | 3.609 | 4.703 | 3.997 | 5.094 | 4.277 | 5.379 | 4.495 | 5.603 | 4.673 | 5.788 |
| 19 | 2.960 | 4.046 | 3.593 | 4.670 | 3.977 | 5.054 | 4.253 | 5.334 | 4.469 | 5.554 | 4.645 | 5.735 |
| 20 | 2.950 | 4.024 | 3.578 | 4.639 | 3.958 | 5.018 | 4.232 | 5.294 | 4.445 | 5.510 | 4.620 | 5.688 |
| 24 | 2.919 | 3.956 | 3.532 | 4.546 | 3.901 | 4.907 | 4.166 | 5.168 | 4.373 | 5.374 | 4.541 | 5.542 |
| 30 | 2.888 | 3.889 | 3.486 | 4.455 | 3.845 | 4.799 | 4.102 | 5.048 | 4.302 | 5.242 | 4.464 | 5.401 |
| 40 | 2.858 | 3.825 | 3.442 | 4.367 | 3.791 | 4.696 | 4.039 | 4.931 | 4.232 | 5.114 | 4.389 | 5.265 |
| 60 | 2.829 | 3.762 | 3.399 | 4.282 | 3.737 | 4.595 | 3.977 | 4.818 | 4.163 | 4.991 | 4.314 | 5.133 |
| 120 | 2.800 | 3.702 | 3.356 | 4.200 | 3.685 | 4.497 | 3.917 | 4.709 | 4.096 | 4.872 | 4.241 | 5.005 |
| ∞ | 2.772 | 3.643 | 3.314 | 4.120 | 3.633 | 4.403 | 3.858 | 4.603 | 4.030 | 4.757 | 4.170 | 4.882 |

| | k = 8 | | k = 9 | | k = 10 | | k = 11 | | k = 12 | | k = 13 | |
|---|---|---|---|---|---|---|---|---|---|---|---|---|
| α | 5% | 1% | 5% | 1% | 5% | 1% | 5% | 1% | 5% | 1% | 5% | 1% |
| df = 1 | 45.40 | 227.2 | 47.36 | 237.0 | 49.07 | 245.6 | 50.59 | 253.2 | 51.96 | 260.0 | 53.20 | 266.2 |
| 2 | 13.03 | 29.53 | 13.54 | 30.68 | 13.99 | 31.69 | 14.39 | 32.59 | 14.75 | 33.40 | 15.08 | 34.13 |
| 3 | 8.853 | 15.64 | 9.177 | 16.20 | 9.462 | 16.69 | 9.717 | 17.13 | 9.946 | 17.53 | 10.15 | 17.89 |
| 4 | 7.347 | 11.55 | 7.602 | 11.93 | 7.826 | 12.27 | 8.027 | 12.57 | 8.208 | 12.84 | 8.373 | 13.09 |
| 5 | 6.582 | 9.669 | 6.802 | 9.972 | 6.995 | 10.24 | 7.168 | 10.48 | 7.324 | 10.70 | 7.466 | 10.89 |
| 6 | 6.122 | 8.613 | 6.319 | 8.869 | 6.493 | 9.097 | 6.649 | 9.301 | 6.789 | 9.485 | 6.917 | 9.653 |
| 7 | 5.815 | 7.939 | 5.998 | 8.166 | 6.158 | 8.368 | 6.302 | 8.548 | 6.431 | 8.711 | 6.550 | 8.860 |
| 8 | 5.597 | 7.474 | 5.767 | 7.681 | 5.918 | 7.863 | 6.054 | 8.027 | 6.175 | 8.176 | 6.287 | 8.312 |
| 9 | 5.432 | 7.134 | 5.595 | 7.325 | 5.739 | 7.495 | 5.867 | 7.647 | 5.983 | 7.784 | 6.089 | 7.910 |
| 10 | 5.305 | 6.875 | 5.461 | 7.055 | 5.599 | 7.213 | 5.722 | 7.356 | 5.833 | 7.485 | 5.935 | 7.603 |
| 11 | 5.202 | 6.672 | 5.353 | 6.842 | 5.487 | 6.992 | 5.605 | 7.128 | 5.713 | 7.250 | 5.811 | 7.362 |
| 12 | 5.119 | 6.507 | 5.265 | 6.670 | 5.395 | 6.814 | 5.511 | 6.943 | 5.615 | 7.060 | 5.710 | 7.167 |
| 13 | 5.049 | 6.372 | 5.192 | 6.528 | 5.318 | 6.667 | 5.431 | 6.791 | 5.533 | 6.903 | 5.625 | 7.006 |
| 14 | 4.990 | 6.258 | 5.131 | 6.409 | 5.254 | 6.543 | 5.364 | 6.664 | 5.463 | 6.772 | 5.554 | 6.871 |
| 15 | 4.940 | 6.162 | 5.077 | 6.309 | 5.198 | 6.439 | 5.306 | 6.555 | 5.404 | 6.660 | 5.493 | 6.757 |
| 16 | 4.897 | 6.079 | 5.031 | 6.222 | 5.150 | 6.349 | 5.256 | 6.462 | 5.352 | 6.564 | 5.439 | 6.658 |
| 17 | 4.858 | 6.007 | 4.991 | 6.147 | 5.108 | 6.270 | 5.212 | 6.381 | 5.307 | 6.480 | 5.392 | 6.572 |
| 18 | 4.824 | 5.944 | 4.956 | 6.081 | 5.071 | 6.201 | 5.174 | 6.310 | 5.267 | 6.407 | 5.352 | 6.497 |
| 19 | 4.794 | 5.889 | 4.924 | 6.022 | 5.038 | 6.141 | 5.140 | 6.247 | 5.231 | 6.342 | 5.315 | 6.430 |
| 20 | 4.768 | 5.839 | 4.896 | 5.970 | 5.008 | 6.087 | 5.108 | 6.191 | 5.199 | 6.285 | 5.282 | 6.371 |
| 24 | 4.684 | 5.685 | 4.807 | 5.809 | 4.915 | 5.919 | 5.012 | 6.017 | 5.099 | 6.106 | 5.179 | 6.186 |
| 30 | 4.602 | 5.536 | 4.720 | 5.653 | 4.824 | 5.756 | 4.917 | 5.849 | 5.001 | 5.932 | 5.077 | 6.008 |
| 40 | 4.521 | 5.392 | 4.635 | 5.502 | 4.735 | 5.599 | 4.824 | 5.686 | 4.904 | 5.764 | 4.977 | 5.835 |
| 60 | 4.441 | 5.253 | 4.550 | 5.356 | 4.646 | 5.447 | 4.732 | 5.528 | 4.808 | 5.601 | 4.878 | 5.667 |
| 120 | 4.363 | 5.118 | 4.468 | 5.214 | 4.560 | 5.299 | 4.641 | 5.375 | 4.714 | 5.443 | 4.781 | 5.505 |
| ∞ | 4.286 | 4.987 | 4.387 | 5.078 | 4.474 | 5.157 | 4.552 | 5.227 | 4.622 | 5.290 | 4.685 | 5.348 |

(continued on next page)

**Table A19 (continued)  Critical values of _q_ for multiple comparisons following analysis of variance.**

| | k = 14 | | k = 15 | | k = 16 | | k = 17 | | k = 18 | | k = 19 | | k = 20 | |
|---|---|---|---|---|---|---|---|---|---|---|---|---|---|---|
| α | 5% | 1% | 5% | 1% | 5% | 1% | 5% | 1% | 5% | 1% | 5% | 1% | 5% | 1% |
| df = 1 | 54.33 | 271.8 | 55.36 | 277.0 | 56.32 | 281.8 | 57.22 | 286.3 | 58.04 | 290.4 | 58.83 | 294.3 | 59.56 | 298.0 |
| 2 | 15.38 | 34.81 | 15.65 | 35.43 | 15.91 | 36.00 | 16.14 | 36.53 | 16.37 | 37.03 | 16.57 | 37.50 | 16.77 | 37.95 |
| 3 | 10.35 | 18.22 | 10.53 | 18.52 | 10.69 | 18.81 | 10.84 | 19.07 | 10.98 | 19.32 | 11.11 | 19.55 | 11.24 | 19.77 |
| 4 | 8.525 | 13.32 | 8.664 | 13.53 | 8.794 | 13.73 | 8.914 | 13.91 | 9.028 | 14.08 | 9.134 | 14.24 | 9.233 | 14.40 |
| 5 | 7.596 | 11.08 | 7.717 | 11.24 | 7.828 | 11.40 | 7.932 | 11.55 | 8.030 | 11.68 | 8.122 | 11.81 | 8.208 | 11.93 |
| 6 | 7.034 | 9.808 | 7.143 | 9.951 | 7.244 | 10.08 | 7.338 | 10.21 | 7.426 | 10.32 | 7.508 | 10.43 | 7.587 | 10.54 |
| 7 | 6.658 | 8.997 | 6.759 | 9.124 | 6.852 | 9.242 | 6.939 | 9.353 | 7.020 | 9.456 | 7.097 | 9.554 | 7.170 | 9.646 |
| 8 | 6.389 | 8.436 | 6.483 | 8.552 | 6.571 | 8.659 | 6.653 | 8.760 | 6.729 | 8.854 | 6.802 | 8.943 | 6.870 | 9.027 |
| 9 | 6.186 | 8.025 | 6.276 | 8.132 | 6.359 | 8.232 | 6.437 | 8.325 | 6.510 | 8.412 | 6.579 | 8.495 | 6.644 | 8.573 |
| 10 | 6.028 | 7.712 | 6.114 | 7.812 | 6.194 | 7.906 | 6.269 | 7.993 | 6.339 | 8.076 | 6.405 | 8.153 | 6.467 | 8.226 |
| 11 | 5.901 | 7.465 | 5.984 | 7.560 | 6.062 | 7.649 | 6.134 | 7.732 | 6.202 | 7.809 | 6.265 | 7.883 | 6.326 | 7.952 |
| 12 | 5.798 | 7.265 | 5.878 | 7.356 | 5.953 | 7.441 | 6.023 | 7.520 | 6.089 | 7.594 | 6.151 | 7.665 | 6.209 | 7.731 |
| 13 | 5.711 | 7.101 | 5.789 | 7.188 | 5.862 | 7.269 | 5.931 | 7.345 | 5.995 | 7.417 | 6.055 | 7.485 | 6.112 | 7.548 |
| 14 | 5.637 | 6.962 | 5.714 | 7.047 | 5.786 | 7.126 | 5.852 | 7.199 | 5.915 | 7.268 | 5.974 | 7.333 | 6.029 | 7.395 |
| 15 | 5.574 | 6.845 | 5.649 | 6.927 | 5.720 | 7.003 | 5.785 | 7.074 | 5.846 | 7.142 | 5.904 | 7.204 | 5.958 | 7.264 |
| 16 | 5.520 | 6.744 | 5.593 | 6.823 | 5.662 | 6.898 | 5.727 | 6.967 | 5.786 | 7.032 | 5.843 | 7.093 | 5.897 | 7.152 |
| 17 | 5.471 | 6.656 | 5.544 | 6.734 | 5.612 | 6.806 | 5.675 | 6.873 | 5.734 | 6.937 | 5.790 | 6.997 | 5.842 | 7.053 |
| 18 | 5.429 | 6.579 | 5.501 | 6.655 | 5.568 | 6.725 | 5.630 | 6.792 | 5.688 | 6.854 | 5.743 | 6.912 | 5.794 | 6.968 |
| 19 | 5.391 | 6.510 | 5.462 | 6.585 | 5.528 | 6.654 | 5.589 | 6.719 | 5.647 | 6.780 | 5.701 | 6.837 | 5.752 | 6.891 |
| 20 | 5.357 | 6.450 | 5.427 | 6.523 | 5.493 | 6.591 | 5.553 | 6.654 | 5.610 | 6.714 | 5.663 | 6.771 | 5.714 | 6.823 |
| 24 | 5.251 | 6.261 | 5.319 | 6.330 | 5.381 | 6.394 | 5.439 | 6.453 | 5.494 | 6.510 | 5.545 | 6.563 | 5.594 | 6.612 |
| 30 | 5.147 | 6.078 | 5.211 | 6.143 | 5.271 | 6.203 | 5.327 | 6.259 | 5.379 | 6.311 | 5.429 | 6.361 | 5.475 | 6.407 |
| 40 | 5.044 | 5.900 | 5.106 | 5.961 | 5.163 | 6.017 | 5.216 | 6.069 | 5.266 | 6.119 | 5.313 | 6.165 | 5.358 | 6.209 |
| 60 | 4.942 | 5.728 | 5.001 | 5.785 | 5.056 | 5.837 | 5.107 | 5.886 | 5.154 | 5.931 | 5.199 | 5.974 | 5.241 | 6.015 |
| 120 | 4.842 | 5.562 | 4.898 | 5.614 | 4.950 | 5.662 | 4.998 | 5.708 | 5.044 | 5.750 | 5.086 | 5.790 | 5.126 | 5.827 |
| ∞ | 4.743 | 5.400 | 4.796 | 5.448 | 4.845 | 5.493 | 4.891 | 5.535 | 4.934 | 5.750 | 4.974 | 5.611 | 5.012 | 5.645 |

# Further reading

Elliot, J. M. 1977. *Some methods for the statistical analysis of samples of benthic inverte-brates*, 2nd edn. Ambleside: Freshwater Biological Association, Publication No. 25. Very good on methods for dealing with data on patterns of spatial dispersion and testing for differences between means of non-normally distributed variables using transformed data.

Fowler, J. & Cohen, L. 1990. *Practical statistics for field biologists*. Milton Keynes: Open University Press. Good on the analysis of data on patterns of spatial dispersion and on extensions to basic analysis of variance and regression. Clearly worked examples but restricted to ecological topics.

Neave, H. R. 1985. *Elementary statistics tables*. London: George Allen & Unwin. Good set of statistical tables, including values of $H$ for Kruskal–Wallis test with unequal sample sizes.

Neave, H. R. & Worthington, P. L. 1988. *Distribution-free tests*. London: Unwin Hyman. Very readable and clearly explained account of a wide range of statistical tests which do not rely on the variables following any specified distribution. Examples drawn from wide range of disciplines including biology.

Sokal, R. R & Rohlf, F. J. 1981. *Biometry*. San Francisco: W. H. Freeman. Excellent advanced book which has good explanations and clearly worked examples of tests. Particularly good on advanced topics in analysis of variance and regression.

Zar, J. H. 1984. *Biostatistical analysis*. New York: Prentice-Hall. Advanced text at the same sort of level as that by Sokal & Rohlf (1981), with a wider variety of tests described although not as well explained. Good worked examples and very comprehensive set of tables of critical values.

# Glossary

**acceptance region**   The proportion of the sampling distribution which contains values of the test statistic which are judged likely to occur if the null hypothesis is true; obtaining a value in this region gives no evidence for the alternative hypothesis

**accuracy**   The closeness of a measurement to the true value

**alternative hypothesis ($H_1$)**   The hypothesis that the population(s) from which the sample or samples have been drawn has a distribution or parameters different from those specified in the null hypothesis

**analysis of variance**   A parametric statistical test of the null hypothesis that three or more populations have the same mean

**anova**   Abbreviation for analysis of variance

**arithmetic mean**   The most widely used measure of the value at which a sample or population of variates is centred or located; the sum of all the variates divided by the total number of variates; sample mean ($\bar{Y}$), population mean ($\mu$)

**association**   The tendency for the frequencies in the categories of two nominal variables to vary together in a systematic (i.e. non-random) manner

**associated probability ($P$)**   The probability, under the null hypothesis, of obtaining the calculated value of the test statistic and all more extreme values

**back-transformation**   The process of converting a transformed variable back into its original units

**bar chart**   A form of graphical presentation used to display the frequency distribution of a nominal, ordinal or discontinuous variable; frequencies are represented by vertical bars which do not touch one another

**between-sample sum of squares**   Sum of squared deviations of sample means from grand mean in analysis of variance; used to calculate between-sample variance

**between-sample variance**   The measure of the amount of variation between sample means in an analysis of variance

**bias**   A systematic error in measuring or counting, such that values tend to be either all too high or all too low

**bimodal**   Describes a frequency distribution with two peaks, which strictly should be of the same height

**binomial distribution**   A frequency distribution of the number of times an outcome occurs in an experiment of a number of trials, when each trial has two mutually exclusive, independent outcomes

**binomial test**   A statistical test of the null hypothesis that a sample of a nominal variable has been taken from a binomial population with some specified value of $p$

**bivariate**   The data which results when each sampling unit is described with respect to two variables

**blind experiment**   An experiment conducted in such a way that either the subject or the experimenter has no indication of what the treatments are and so cannot affect or prejudge the results

**calculated value (of test statistic)**   The value obtained from the data and which is compared with the critical value

**cells**   The intersection of a column and a row in a contingency table which contains an observed frequency

**chi-squared distribution ($\chi^2$)**   A sampling distribution of chi-squared

**chi-squared test**   Any statistical test which uses critical values of chi-squared; most commonly employed with nominal variables

**class interval**   The distance between the implied limits of a class

**class mark**   A value of a continuous variable which lies half way between the implied limits of the class and which is used to label/identify all the values in that class

**clumped**   A pattern of spatial dispersion in which events or organisms occur in groups or clusters which are more pronounced than would be the case for a random pattern

**coefficient of determination ($r^2$)**   The proportion of the variation in one variable which is accounted for by variation in another in correlation; given by squaring the correlation coefficient

**column total**   The sum of the observed frequencies in a column of a contingency table

**completely randomized design**   An experiment in which sampling units are allocated completely at random and not organized into blocks

**confidence interval (CI)**   A range of values calculated from sample data and used to quantify the reliability of an estimate; has the property that, in repeated sampling, a specified percentage of the confidence intervals calculated will contain the population parameter.

**confidence level ($\gamma$)**   The percentage specified when calculating a confidence interval

**confidence limit (CL)**   The upper and lower limits of a confidence interval

**confounded**   Two or more treatments whose effects cannot be separated, because of an inadequate experimental design

**conservative**   A decision made when exact critical values are not known and which uses critical values which are more extreme; it errs on the side of caution and results in a significance level which is smaller than that specified by the nominal significance level

**contagious**   Another word for clumped dispersion

**contingency table**   The two-dimensional arrangement of the frequencies with which the categories of two nominal variables occur

**continuous variable**   A variable which can take any non-integer value, between certain limits

**control**   One of the treatments in an experiment in which the factor of interest is either omitted or is at a naturally occurring level; provides a contrast to treatments in which the factor of interest is present or present at unusual levels

**correction term (CT)**   A quantity used in calculating two of the sums of squares in analysis of variance

**correction for continuity**   A correction used in the calculation of the test statistic in a $\chi^2$ test to correct for the fact that $\chi^2$ is based on a continuous variable; only used when there is one degree of freedom

**correlation**   The tendency of two ordinal, discontinuous or continuous variables to vary together in a systematic (i.e. non-random) manner

**critical value**   The value(s) of the test statistic which mark the border between the acceptance and rejection region; the least extreme values which are significant at a given significance level

**critical region**   Another name for the rejection region

**cross-classified**   Data classified on the basis of two (or more) variables, e.g. bivariate data

**cumulative frequency**   The frequency of occurrence of the specified value and all more extreme values

**cumulative probability**   The probability of obtaining the specified value and all more extreme values

**data**   The observations made, the results (i.e. the numerical values which are obtained); variates

**datum**   A single observation or result; a variate

**decision rule**   A statement which describes what the relationship between the calculated and critical value of a test statistic must be in order to reject the null hypothesis

**degrees of freedom (df)**   A number used in calculating a variance or in finding a critical value. It is related to the sample size and takes into account the number of pieces of information from the data which are used as estimates in the calculations

**denominator**   The expression or quantity on the bottom line of an equation

**dependent variable**   The variable which is described in an experiment under fixed conditions of an independent variable

**derived variable**   A variable produced by combining information from two or more variables, e.g. a ratio

**deviate**   *See Deviation from the mean*

**deviation from the mean**   The difference between a variate and the mean

**discontinuous variable**   A variable which can only take integer (i.e. whole number) values

**distribution free**   A statistical test which makes few or no assumptions about the distribution of the variable

**estimate**   Any measure calculated from a sample and used to make inferences about the equivalent measure in the population

**expected frequency**   The frequency with which a result should occur in the long run

**expected relative frequency**   An expected frequency expressed as a fraction of the total sample size

**experiment**   *See* manipulative experiment

**experimental design**   The procedure(s) adopted to ensure that an experiment yields data which (a) meet the assumptions of the statistical test to be used, and (b) can unambiguously show that any statistically significant effects are due to differences between the experimental treatments

**experimental error**   The differences between replicates in a treatment which arise because of unavoidable differences in (a) their origin and preparation, and (b) the experimental conditions and measurement

**experimental variation**   The variation between the values recorded for each replicate which arises because of experimental errors

**extrinsic hypothesis**   A null hypothesis in a goodness-of-fit test in which the expected ratios are derived from an external theory, e.g. Mendelian genetics, and not from the data

*F* **test**   A statistical test of the null hypothesis that two populations have the same variance

$F_{max}$ **test**   A statistical test of the null hypothesis that three or more populations have the same variance; uses the ratio of the largest: smallest sample variance

**factor**   Something which it is suspected may cause change in the variable of interest; an independent variable

**field experiment**   A manipulative experiment carried out in the natural environment

**frequency (f)**   The number of times a variate or value occurs

**frequency distribution**   A tabular or graphical presentation which shows the frequency with which each value of a variate occurs

**frequency polygon**   A form of graphical presentation used to display the frequency distribution of a continuous variable; frequencies are represented by points above the classmarks, joined by straight lines

**goodness-of-fit test**   A statistical test based on $\chi^2$ which compares the observed frequencies in each category to those expected under the null hypothesis

**grand mean**   The mean of all the observations in two or more samples

**grand total**   The total number of observations in a contingency table (*N*); the total of all the variates in an analysis of variance (*GT*)

**histogram**   A form of graphical presentation used to display the frequency distribution of a continuous variable; frequencies are represented by vertical bars which touch adjacent bars

**homogeneity**   Two or more populations show homogeneity if they are identical with respect to a property such as a variance or a proportion

**hypothesis test**   Any statistical test which examines whether the null hypothesis is a reasonable explanation for the result(s) obtained from an experiment or sampling programme

**implied limits**   The limits between which the true but unknowable value of a continuous variable lies

**independence**   The situation in which the probability of one event is not influenced by whether some other event has or has not occurred

**independent events**   Events for which the probability of occurrence is not influenced by whether other events have or have not occurred

**independent samples**   Samples in which the units or observations in one sample have been selected without reference to those in another sample; c.f. matched samples

**independent variable**   A variable whose values are chosen, changed or manipulated by the experimenter and which it is thought may cause change in another (dependent) variable

**index**   A descriptive measure which combines information from several variables in a single numerical value

**index of dispersion** ($I$)   The ratio of the sample variance:sample mean; used to describe patterns of spatial dispersion

**inference**   A statement made about some aspect of a population on the basis of a sample

**intercept**   The point where the regression line crosses the $y$ axis; the value of $Y$ when $X = 0$

**interference**   Any unwanted interaction between experimental units during the course of an experiment which alters their response to the treatments

**interval estimate**   A range of values, such as a standard error or a confidence interval, calculated from a sample which quantifies the reliability of an estimate such as a mean

**intrinsic hypothesis**   A null hypothesis in a goodness-of-fit test in which the expected frequencies are related to and calculated from information in the sample

**Kruskal–Wallis test**   A statistical test of the null hypothesis that three or more populations have the same median; non-parametric and based on ranked variates

**level**   A particular, defined condition of an independent variable (e.g. light, pH, fertilizer) to which experimental material is subjected

**line of best fit**   A line fitted mathematically through a set of points in such a way as to minimize the total squared deviations of the points from the line

**linear interpolation**   A procedure for finding critical values in tables, when the value required is not given but lies between two other values

**manipulative experiment**   An investigation in which the experimenter manipulates the factor(s) of interest, while keeping other factors constant, and observes the effect on the variable of interest

**Mann–Whitney $U$ test**   A statistical test of the null hypothesis that two populations have the same median; non-parametric and uses ranked variates

**marginal total**   Total of the observed frequencies in a row or a column of a contingency table

**matched samples**   Two (or more) samples in which each sampling unit in one sample is matched with one in the other sample; c.f. independent samples

**mean ($\bar{Y}$, $\mu$)**   See Arithmetic mean

**mean square (MS)**   Another term for variance

**median**   The value of the variable which is in the middle of the array when all the values are put in order

**mode**   The most commonly occurring value of a variable in a sample

**monotonic**   A relationship between two variables which can be described by a curved or straight line which only slopes in one direction

**multiple comparisons**   A statistical test which is used when an analysis of variance or a Kruskal–Wallis test has shown that not all of the samples have been drawn from the same population. Pairwise combinations of samples are used to detect which populations are different from which

**multistage random sampling**   Procedure in which groups of sampling units are chosen at random first, then subgroups within groups and finally sampling units within subgroups; used when there are too many sampling units to identify/number each one

**negative association/correlation**   The tendency of high frequencies/values of one variable to occur with low frequencies/values of another variable more often than expected on the basis of chance

**negative skew**   An asymmetrical frequency distribution with a left-hand, lower tail which is more pronounced or elongated than the right-hand tail and with the mode displaced towards large values of $Y$

**non-parametric**   Statistical tests which do not involve the estimation of parameters and the testing of hypotheses about the parameters

**non-significant (n.s.)**   Describes a result which is likely to occur if the null hypothesis is true; a result which gives no evidence for the alternative hypothesis

**nominal significance level**   Nearest, larger conventional significance level (e.g. 5%) to the actual significance level being used (e.g. 3.42%); referred to when sampling distribution is discontinuous and significance level cannot be made exactly the desired conventional size, but is made smaller for a conservative test

**nominal variable**   A variable where the difference between sampling units cannot be quantified but is described by putting them into named categories, e.g. male and female

**normal distribution**   A smooth, symmetrical bell-shaped frequency/probability distribution

**null hypothesis ($H_0$)**   The hypothesis that the population from which the sample has been drawn has some specified parameters or shape of distribution or both; often formulated so as to be rejected

**numerator**   The expression or quantity on the top line of an equation, which is divided by the denominator on the bottom line

**observation**   A single value of a variate; a datum

**observational study**   A study which makes use of naturally occurring variation in the factor of interest and examines whether this is associated with change in the variable of interest

**one-tailed (one-sided) test**   A statistical test in which the alternative hypothesis predicts the direction of any difference or relationship

**ordinal variable**   A variable in which the difference between sampling units is described by putting them in order and giving each a number or rank which specifies its position in the order

**outcome** The result of a single trial in which there are two possible results

**paired samples/design**   A sampling programme or experimental design in which a each sampling unit in one sample is paired or matched with a sampling unit in the other sample so that members of each pair are similar

**parameter**   A characteristic (e.g. the mean) of a population or the numerical value of the characteristic; denoted by a Greek letter

**parametric**   A statistical test which involves the estimation of parameters and the testing of hypotheses about them

**percentage**   A proportion expressed as a number out of 100

**point estimate**   A single value, calculated from a sample, which is used as an estimate of the corresponding population parameter

**Poisson distribution**   A frequency or probability distribution of the number of times that a rare independent event will occur in a sample

**population**   The very large, often infinite, set of sampling units which could in theory be observed or the set of variates, i.e. observations which would result

**positive association/correlation**   The tendency of high frequencies/values of one variable to occur with high frequencies/values of another variable more often than expected on the basis of chance

**positive skew**   An asymmetrical frequency distribution with a right-hand upper tail which is more pronounced or elongated than the left-hand lower tail and with the mode displaced towards lower values of $Y$

**power**   The capacity of a statistical test to reject the null hypothesis when it is not true

**precision**   The limits within which the true value of a measurement variable is known to lie; the nearness of repeated measures of the same quantity

**probability ($p$)**   The chance or likelihood that a particular event will occur or that a particular result will be obtained; measured on scale of 0–1

**proportion ($p$)**   One quantity expressed as a fraction of another on scale of 0–1

**pseudoreplication**   Replicates which are not independent of one another

**random numbers**   A set of digits (0, 1, …, 9) arranged in such a way that any particular digit is equally likely to be adjacent to any other digit including itself. Any sequence of these digits is a random sequence

**random sampling**   A method of choosing sampling units and/or allocating them to treatments, so that every unit has the same chance of being chosen and of being allocated to each treatment

**random spatial dispersion**   The spatial pattern which results when every point has an equal chance of being occupied

**randomized blocks**   A way of putting together similar sampling units into groups or blocks and then within each block assigning units at random to different treatments

**randomized design**   A method for allocating sampling units at random to experimental treatments

**range**   The difference between the largest and the smallest variate (observation) in a sample

**rank**   A number which specifies the position of an observation in the order, when all the observations are placed in order of decreasing or increasing magnitude

**ratio**   The simple relationship between two numbers, e.g. 3:1

**raw data**   Data, that is individual values of variates, in the form and order in which they were obtained from the experiment or sampling; not processed in any way

**rectilinear**   A relationship between two variables which is described by a straight line

**regression**   The statistical procedure for producing the line of best fit, which describes the relationship between a dependent and independent variable

**regular**   A pattern of spatial dispersion in which objects or events are more regularly spaced than in a random pattern

**rejection region**   That proportion of the sampling distribution which includes values of the test statistic which are judged to be unlikely to be found if the null hypothesis is correct; obtaining a value in this region is evidence for the alternative hypothesis

**related samples**   Samples in which each sampling unit in one sample has been paired or matched with a similar unit in the other sample

**reliability**   Closeness of the sample statistic to the population parameter

**replicate**   A sampling unit

**research hypothesis**   The question of biological interest that an experiment or sampling programme is designed to answer

**row total ($r$)**   The sum of the observed frequencies in a row of a contingency table

**sample**   A collection of sampling units made in a specified way and under specified conditions; the data which result

**sample size ($n$)**   The number of sampling units or replicates in a sample

**sampling distribution under null hypothesis**   The frequency or probability distribution of all possible values of a test statistic under the null hypothesis

**sample standard deviation ($s$)**   Square root of the sample variance; used to quantify the variation of the observations in a sample

**sample variance ($s^2$)**   The value of the variance calculated from a sample of variates and used as an estimate of the population variance; the sum of squares divided by ($n - 1$)

**sampling error**   Another term for sampling variation

**sampling programme or strategy**   A plan for collecting samples in an observational study so as to ensure that it yields data which (a) meet the assumptions of the statistical test to be used, and (b) can unambiguously show that any statistically significant effects are related to differences in the factor of interest

**sampling unit**   The unit on which an observation is made and the description is based

**sampling variation**   Chance variation in the make-up of a sample, such that two samples drawn from the same population may yield different values

**scatter diagram**   A graphical presentation for bivariate data in which one variable is plotted on each axis; the value of the two variables for each sampling unit is represented by a single point

**sign test**   A statistical test for paired or related samples which only requires that the direction of the differences between pairs is known

**significance level ($\alpha$)**   The proportion (%) of the sampling distribution which is in the rejection region(s)

**significance test**   Any statistical test which examines whether the null hypothesis is a reasonable explanation for the result(s) obtained from an experiment or sampling programme

**significant result**   Describes a result which is judged unlikely to occur if the null hypothesis is true; a result which does give evidence for the alternative hypothesis

**simple random sampling**   A method for choosing sampling units which does not involve strata

**skewed distribution**   An asymmetrical frequency distribution with one tail longer than the other and a mode which is not at the centre

**slope**   A number which indicates the extent to which a regression line departs from the horizontal and the direction (positive or negative) of that departure

**Spearman's rank correlation**   A distribution-free procedure for quantifying the strength and the direction of the correlation between two ordinal variables in a sample and for testing whether this is significantly different from zero

**squared deviation from mean**   The squared difference between the value of an observation and the mean of the sample

**standard deviation**   A measure of the variability of the observations around the mean in a sample ($s$) or a population ($\sigma$) obtained by taking the square root of the variance; has the same units as the variable and the mean and used in reporting variability

**standard error (s.e.)**   The standard deviation of a distribution of a statistic such as a mean; quantifies the extent to which values of the statistic are spread out around the value of the parameter and so gives a measure of the reliability

**standard error of the difference between two means**   The standard deviation of a sampling distribution of all possible differences between a pair of sample means

**standardization**   The conversion of a value of $Y$ from a particular, specified normal distribution into a value of $z$ which relates to the standard normal distribution

**statistic**   A number, calculated from a sample, which describes some characteristic of that sample, e.g. the mean

**stratified random sampling**   A form of sampling in which the sampling units available are organized so that units which are similar are grouped together in a stratum; a random sample is then taken from each stratum

**stratum**   A group of sampling units which are similar in some way

**sum of products (SP or $\Sigma XY$)**   Used when each sampling unit has been described with respect to two variables $Y_1$ and $Y_2$ (or $X$ and $Y$) and obtained by calculating $Y_1 \times Y_2$ (or $X \times Y$) for each sampling unit and adding results together

**sum of squares (SS)** The sum of the squared deviations from the sample mean of all the observations in a sample

**sum of squared deviations from the mean** *See Sum of squares*

**sum of squared values of $Y$ ($\Sigma Y^2$)** The quantity obtained by squaring each value of $Y$ and then adding the results together

**sum of the values of $Y$ squared ($\Sigma Y)^2$** The quantity obtained by adding together all the values of $Y$ and then squaring the answer

*t* **test** A parametric statistical test of the null hypothesis that two populations have the same mean

**test statistic** A number calculated from a sample which has a known sampling distribution under the null hypothesis and which can therefore be used to test whether the null hypothesis is a reasonable explanation for the sample obtained

**tied** Two or more observations of an ordinal variable which have the same rank

**transformation** A mathematical procedure for converting variates which do not follow a normal distribution into a form which does (more nearly) follow a normal distribution

**treatment** A level of one independent variable or a particular combination of levels of two or more independent variables to which experimental material is exposed to in an experiment

**type I error** An error in decision making which occurs when the null hypothesis is rejected when it is true; caused by chance occurrence of very atypical value of sample statistic which is not close to value predicted under null hypothesis

**type II error** An error in decision making which occurs when the alternative hypothesis is true but this is not detected; caused by chance occurrence of very atypical value of sample statistic which is close to value predicted under null hypothesis

**two-tailed (two-sided) test** A statistical test in which the alternative hypothesis does not predict the direction of any difference or correlation

**unimodal** A frequency distribution with a single peak or mode

**univariate** The data produced when each sampling unit is described with respect to a single variable

**variable** Any property with respect to which sampling units vary

**variance** A measure of the variability of a set of values around the mean of a sample ($s^2$) or a population ($\sigma^2$)

**variance ratio ($F$)** The ratio of the variances of two samples

**variate** The numerical value of an individual observation or datum

**within-sample sum of squares** Sum of the squared deviations of the variates in several samples around their own sample means; used in analysis of variance

**within-sample variance** A measure of the variability of the items around the mean of a sample; used in analysis of variance

**Wilcoxon's signed rank test** A non-parametric test for paired samples

**Yates' correction** Another name for the continuity correction

*z* **test** A parametric test of the null hypothesis that two populations have the same mean; strictly requires that the variances are known but used as an approximation when samples are large

# Glossary of symbols

$a$   Value of intercept of regression line calculated from sample

$a, b, c, d$   Observed frequencies in $2 \times 2$ contingency table

$b$   Value of slope of regression line calculated from sample

$C$   Number of categories in goodness-of-fit test or contingency table

**CI**   Confidence interval

**CL**   Confidence limits

**CT**   Correction term in analysis of variance

$d$   Difference in ranks in Spearman's rank correlation and Wilcoxon's signed rank test

**df**   Degrees of freedom

$e$   A constant = 2.171828

$f$   Frequency

$F$   A value of the variance ratio

**GT**   Grand total (in analysis of variance)

$H$   Kruskal–Wallis test statistic

$H_0$   Null hypothesis

$H_1$   Alternative hypothesis

$i$   Number of pieces of information needed to calculate expected values in goodness-of-fit tests

$I$   Index of dispersion

$k$   Number of samples

**MS**   Mean square

$n$   Sample size

$n_A, n_B$   Sample sizes in Mann–Whitney $U$ test

$n_1, n_2$   Sample sizes in two sample tests

**n.s.**   Not significant

$N$   Grand total, total number of observations in all samples

$p$   Probability of one outcome in binomial distribution, population proportion

$\hat{p}$ ($p$ **hat**)   sample proportion

$P$   Associated probability

$q$   Value of test statistic in multiple comparisons in analysis of variance

$Q$   Value of test statistic in multiple comparisons in Kruskal–Wallis test

$r$   Row total in contingency table

$r$   Value of product–moment correlation coefficient

$r_s$   Value of Spearman's rank correlation coefficient

$r^2$   Coefficient of determination

$R$   Sum of ranks in a sample in Kruskal–Wallis test

$s$   Sample standard deviation

$s_1, s_2$    Standard deviation of sample 1, sample 2, etc.

$s^2$    Sample variance

$S$    Value of test statistic for sign test

**s.e.**    Standard error

**SS**    Sum of squares

$t$    Value of test statistic for $t$ test; value of $t$ for calculating confidence limits

$T$    Value of test statistic for Wilcoxon's signed rank test

$U$    Value of test statistic for Mann–Whitney $U$ test

$Y$    A dependent variable; value of an individual variate

$\bar{Y}$    Mean value of $Y$

$Y_1, Y_2$, **etc.**    Univariate data: values of observation 1, observation 2, etc. Bivariate data: values of the two variables for a single sampling unit

$X$    Value of independent variable

$\bar{X}$    mean value of $X$

$X^2$    Value of test statistic in $\chi^2$ tests

$z$    Value of test statistic in $z$ test; value of $z$ for calculating confidence limits

$\alpha$ **(alpha)**    Significance level, size of rejection region

$\alpha_1$    Significance level, size of rejection region in one-tailed test

$\alpha_2$    Significance level, size of rejection region in two-tailed test

$\alpha_1^R$    Significance level, size of rejection region in right-hand tail

$\alpha_1^L$    Significance level, size of rejection region in left-hand tail

$\beta$ **(beta)**    Value of slope of regression line in population

$\sigma$ **(sigma)**    Population standard deviation

$\sigma_1, \sigma_2$    Standard deviations of populations 1 and 2

$\sigma^2$    Population variance

$\Sigma$ **(sigma)**    Sum of

$\gamma$ **(gamma)**    Confidence level

$\mu$ **(mu)**    (pronounced mew) Population mean

$\pi$    Constant = 3.14159

$\rho$ **(rho)**    (pronounced roe) Value of product-moment correlation coefficient in population

$\chi^2$    chi-squared statistic, test or distribution

$\pm$    Plus or minus

$/$    Divide by

$\approx$    Approximately equal to

$\neq$    Not equal to

$<$    The quantity before the sign is less than the quantity after it

$>$    The quantity before the sign is greater than the quantity after it

$\leq$    The quantity before the sign is less than or equal to the quantity after it

$\geq$    The quantity before the sign is equal to or greater than the quantity after it

To remember these, note that the narrow end of the symbol always points towards the smaller value

$|\ |$    Take the absolute value of what is between the lines, i.e. ignore a minus sign if it occurs

# Index